T0296443

LONDON MATHEMATICAL SOCIETY LECTURE NOTE SERIES

Managing Editor: Professor J.W.S. Cassels, Department of Pure Mathematics and Mathematical Statistics,
University of Cambridge, 16 Mill Lane, Cambridge CB2 1SB, England

The titles below are available from booksellers, or, in case of difficulty, from Cambridge University Press.

London Mathematical Society Lecture Note Series. 250

Characters and Blocks of Finite Groups

G. Navarro
Universidad de Valencia

CAMBRIDGE UNIVERSITY PRESS
Cambridge, New York, Melbourne, Madrid, Cape Town, Singapore, São Paulo

Cambridge University Press
The Edinburgh Building, Cambridge CB2 2RU, UK

Published in the United States of America by Cambridge University Press, New York

www.cambridge.org
Information on this title: www.cambridge.org/9780521595131

First published 1998

A catalogue record for this publication is available from the British Library

ISBN-13 978-0-521-59513-1 paperback
ISBN-10 0-521-59513-4 paperback

Transferred to digital printing 2005

Para Javier, Gabo, Nacho e Isabel

Contents

viii   *Contents*

Preface

This set of notes grew out of a course that I gave at Ohio University in the spring of 1996. My aim was to give graduate students who were familiar with ordinary character theory an introduction to Brauer characters and blocks of finite groups.

To do that I chose an objective: the Glauberman Z^*-theorem. This theorem gives an excellent excuse for introducing modular representation theory to students interested in groups. Glauberman's outstanding result is one of the major applications of the theory to finite groups. However, to be able to prove it, one needs to proceed from the very basic facts to the three main theorems of R. Brauer.

In Chapter 1, I prove what is absolutely necessary to get started. Assuming that the students have already had a course on ordinary characters, I use this chapter to remind them of some familiar ideas while introducing some new ones.

In Chapter 2, I introduce Brauer characters (in the same spirit as in the book of M. Isaacs) and develop their basic properties.

In Chapter 3, I introduce blocks and, in Chapter 4, Brauer's first main theorem is given. The second main theorem is proven in Chapter 5 and its proof is a new "elementary" proof by Isaacs based on work by A. Juhász and Y. Tsushima. The third main theorem is given in a very general form and its proof is due to T. Okuyama. Once the third main theorem has been proven, we are ready for the Z^*-theorem.

After Glauberman's theorem is completed, I include Chapter 8 on the basic behaviour of Brauer characters. Blocks and Brauer characters of p-solvable groups are studied in Chapter 10.

The relationship between blocks and normal subgroups (which is needed in Chapter 10, but not for the Z^*-theorem) is covered in Chapter 9.

Finally, in Chapter 11, I develop one of the highlights of the whole theory: the description of the p-blocks of the groups with a Sylow p-subgroup of order p.

When writing down this set of notes, I could not resist introducing several topics which were not necessary to accomplish my objective, but which have interest on their own. Perhaps some of them may be taught if time allows it.

For many years, modular representation theory of finite groups was developed only through the incredible talent of Richard Brauer. I take this opportunity to express my deepest admiration for his work.

These notes would not have been possible without the help of Martin Isaacs, to whom I am very much indebted. Thomas Keller read the complete set of notes which have thus benefited from his comments. Chris Puin helped me with the English.

I also extend my thanks to G. Glauberman, M. Lewis, J. Muñoz, F. Pérez Monasor, L. Sanus, W. Willems and T. Wolf.

Valencia
December, 1997

1. Algebras

We will assume that all of our rings have an identity. If R is a ring, an abelian group M is a **left R-module** if for every $r \in R$ and $m \in M$, there is a unique element $rm \in M$ such that

$$r(a + b) = ra + rb\,,$$

$$(r + s)a = ra + sa\,,$$

$$(rs)a = r(sa)\,,$$

$$1_R a = a$$

for all $r, s \in R$ and $a, b \in M$. In the same way, but multiplying on the right, we define a **right R-module**.

If M and N are left R-modules, a map $f : M \to N$ is **R-linear** if f is additive and $f(rm) = rf(m)$ for all $m \in M$ and $r \in R$.

(1.1) DEFINITION. Suppose that R is a commutative ring and suppose that A is a left R-module. If A also is a ring such that

$$(ra)b = r(ab) = a(rb)$$

for all $r \in R$ and all $a, b \in A$, we say that A is an **R-algebra**.

When we think of R-algebras, we have two important examples in mind: $\mathrm{Mat}(n, R)$, the R-algebra of $n \times n$ matrices with entries in R and, for every finite group G, the **group algebra**

$$RG = \{\sum_{g \in G} a_g g \,|\, a_g \in R\}$$

with the multiplication of G extended linearly to RG. (In fact, representation theory studies the homomorphisms between RG and $\mathrm{Mat}(n, R)$.)

1

Also, if A is any ring and R is a subring of $\mathbf{Z}(A) = \{a \in A \mid ax = xa$ for all $x \in A\}$ (with the identity of A inside R), then A is an R-algebra.

If A and B are R-algebras, an **algebra homomorphism** is an R-linear, multiplicative map $f : A \to B$ such that $f(1_A) = 1_B$.

For the rest of this chapter, A is an R-algebra.

(1.2) DEFINITION. A left R-module V is said to be an A-**module** if V is a right A-module (A considered as a ring) such that for all $v \in V$, $r \in R$ and $a \in A$, we have that

$$(rv)a = r(va) = v(ra) \, .$$

One of the most important examples of an A-module is A itself with right multiplication. This is usually called the **regular** A-module.

If V is an A-module, a subgroup W of V is an A-**submodule** if $wa \in W$ for all $w \in W$ and $a \in A$. Notice that A-submodules are necessarily R-submodules since $rv = v(r1_A)$ for $r \in R$ and $v \in V$. Observe that the A-submodules of the regular A-module are the right ideals of A.

If W is an A-submodule of V, then V/W is an A-module via

$$(v + W)a = va + W$$

for $v \in V$ and $a \in A$.

(1.3) DEFINITION. We say that a nonzero A-module V is **simple** if its only A-submodules are 0 and V. (It is also common to say, in this case, that V is **irreducible**.)

If V and W are A-modules, an additive map $f : V \to W$ such that

$$f(va) = f(v)a$$

for all $v \in V$ and $a \in A$ is an A-**homomorphism** of modules. A bijective A-homomorphism is an **isomorphism** and we write $V \cong W$ in this case.

Notice that A-homomorphisms are necessarily R-linear since $f(rv) = f(v(r1_A)) = f(v)(r1_A) = rf(v)$ for $r \in R$ and $v \in V$.

If $f : V \to W$ is an A-homomorphism, then $\ker(f) = \{v \in V \mid f(v) = 0\}$ and $\mathrm{Im}(f)$ are A-submodules of V and W, respectively. Also, the map $v + \ker(f) \mapsto f(v)$ defines an isomorphism $V/\ker(f) \cong \mathrm{Im}(f)$.

If V and W are A-modules, we write $\mathrm{Hom}_A(V, W)$ for the abelian group of all A-homomorphisms $V \to W$. If $r \in R$ and $f \in \mathrm{Hom}_A(V, W)$, then $\mathrm{Hom}_A(V, W)$ is a left R-module via $(rf)(v) = rf(v)$ for $v \in V$. The set of all A-homomorphisms $V \to V$ is denoted by $\mathrm{End}_A(V)$. It is easy to check that $\mathrm{End}_A(V)$ is an R-algebra. Furthermore, R and $R1_V = \{r1_V \mid r \in R\}$ may be identified whenever R is a field.

(1.4) LEMMA (Schur). *Suppose that V and W are simple A-modules. Then every nonzero A-homomorphism $f : V \to W$ is invertible. As a consequence, if R is an algebraically closed field and $\dim_F(V)$ is finite, then $\mathrm{End}_A(V) = R$.*

Proof. Suppose that $f : V \to W$ is nonzero. Since $\ker(f)$ and $\mathrm{Im}(f)$ are A-submodules of V and W, respectively, it follows that $\ker(f) = 0$ and $\mathrm{Im}(f) = W$. Then f is bijective. To prove the latter assertion, we choose $0 \neq \lambda \in R$, an eigenvalue of f. Then $f - \lambda 1_V \in \mathrm{End}_A(V)$ is not invertible and therefore $f = \lambda 1_V$, by applying the first part. ∎

Sometimes, we use the fact that if R is an algebraically closed field and $f : V \to W$ is a nonzero A-homomorphism between two simple finite dimensional A-modules, then $\mathrm{Hom}_A(V, W) = \{rf \,|\, r \in R\}$. This easily follows from the second part in Schur's lemma.

If I is an ideal (we always mean double sided) of A, it is straightforward to check that A/I is an R-algebra.

The **annihilator** of an A-module V is $\mathrm{ann}(V) = \{a \in A \,|\, va = 0$ for all $v \in V\}$. This is an ideal of A, and notice that we may view V as an $(A/\mathrm{ann}(V))$-module.

We define the **Jacobson radical** of an R-algebra A to be the intersection of all $\mathrm{ann}(V)$ where V runs over all the simple A-modules. It is denoted by $\mathbf{J}(A)$, and certainly it is an ideal of A.

The next result tells us where to find the simple A-modules.

(1.5) THEOREM. *If A is an R-algebra and V is a simple A-module, then there exists a maximal right ideal I of A such that V and A/I are isomorphic. In fact, $\mathbf{J}(A)$ is the intersection of all maximal right ideals of A.*

Proof. If $0 \neq v \in V$, then $v \in vA = \{va \,|\, a \in A\}$. Thus, $vA = V$ since vA is a nonzero A-submodule of V. Now, the map $a \mapsto va$ from A onto V is an A-homomorphism of A-modules. Since $I = \mathrm{ann}(v) = \{a \in A \,|\, va = 0\}$ is the kernel of the map, A/I is isomorphic to V. The fact that V is simple makes I a maximal right ideal of A. Now, if J is the intersection of all maximal right ideals of A, we have that

$$J \subseteq \bigcap_{v \in V} \mathrm{ann}(v) = \mathrm{ann}(V)$$

and thus $\mathbf{J}(A)$ contains J. Now, if L is any maximal right ideal, then A/L is a simple A-module. Also, $\mathrm{ann}(A/L) \subseteq \mathrm{ann}(1 + L) = L$. Hence, $\mathbf{J}(A) \subseteq L$ for all such L. Thus $\mathbf{J}(A) \subseteq J$ and the proof of the theorem is completed. ∎

If $\mathbf{J}(A)$ is the unique maximal ideal of A, we say that A is **local**.

There is a useful fact about the elements of the Jacobson radical which will be used later on.

(1.6) THEOREM. *If A is an R-algebra and $a \in \mathbf{J}(A)$, then $1 - a$ is invertible.*

Proof. If $(1 - a)A < A$, then the right ideal $(1 - a)A$ is contained in some maximal right ideal M of A. In this case, since $\mathbf{J}(A) \subseteq M$, we have that $a \in M$ and we conclude that $1 \in M$, a contradiction. Therefore, we see that $(1 - a)A = A$. Thus, we may find $1 - b \in A$ such that $(1 - a)(1 - b) = 1$. We just need to prove that $(1 - b)(1 - a) = 1$. Since $(1 - a)(1 - b) = 1$, we see that $b = a(b - 1) \in \mathbf{J}(A)$. Hence, by the same reasoning as before, $1 - b$ has a right inverse, say c. Now, $1 - a = (1 - a)((1 - b)c) = ((1 - a)(1 - b))c = c$ and therefore $1 - b$ is a left and right inverse of $1 - a$, as required. ∎

If V is an A-module, $W \subseteq V$ and I is a right ideal of A, then WI denotes the additive subgroup of V generated by all the products wx with $w \in W$ and $x \in I$. Notice that WI is an A-submodule of V. By repeated application of this definition (with $V = A$), we can define I^n for every positive integer n. The right ideal I is **nilpotent** if there is an n with $I^n = 0$. (Note that $I^n = 0$ if and only if every product of n elements of I is zero.)

An A-module V is **finitely generated** if there exist $v_1, \ldots, v_n \in V$ such that

$$V = v_1 A + \ldots + v_n A.$$

(1.7) LEMMA (Nakayama). *Suppose that W is an A-submodule of V such that V/W is finitely generated over A. If $V = W + V\mathbf{J}(A)$, then $V = W$.*

Proof. It suffices to show the lemma for the case $W = 0$ and afterwards to apply it to V/W. So we have that V is a finitely generated A-module such that $V\mathbf{J}(A) = V$ and we wish to prove that $V = 0$. If $V \neq 0$, let $X \neq \emptyset$ be a minimal A-generating subset of V. Now,

$$V = V\mathbf{J}(A) = (\sum_{x \in X} xA)\mathbf{J}(A) = \sum_{x \in X} x\mathbf{J}(A).$$

If $y \in X$, then we may write

$$y = \sum_{x \in X} xa_x,$$

where $a_x \in \mathbf{J}(A)$. Now,

$$y(1 - a_y) = \sum_{x \in X - \{y\}} xa_x$$

and thus, by applying Theorem (1.6), we have that

$$y = \sum_{x \in X - \{y\}} x a_x (1 - a_y)^{-1}.$$

Therefore, $X - \{y\}$ generates V over A which contradicts the minimality of X. ∎

From now until the end of this chapter, we assume that $R = F$ is a field. Hence, from now on, every F-algebra is a vector space over F. We will assume that A and, in general, every A-module have finite dimension over F. Notice that, in this case, we may also assume that $F \subseteq \mathbf{Z}(A)$ since the map $f \mapsto f 1_A$ is an injective ring homomorphism from F into $\mathbf{Z}(A)$.

(1.8) THEOREM. *Suppose that A is an F-algebra. Then $\mathbf{J}(A)$ is the unique maximal nilpotent right ideal of A. Moreover,*

$$\mathbf{J}(\mathbf{Z}(A)) = \mathbf{J}(A) \cap \mathbf{Z}(A).$$

Proof. We have that $\mathbf{J}(A)^n$ is an F-subspace of A, and thus it is finitely generated over A (since A contains F). By Nakayama's lemma (1.7), we have that $\mathbf{J}(A)^{n+1}$ is smaller than $\mathbf{J}(A)^n$, if this is nonzero. Hence, since the dimension of A is finite, we see that $\mathbf{J}(A)$ is necessarily nilpotent. Now, if I is a nilpotent right ideal of A and V is a simple A-module, then $VI = 0$ or $VI = V$ since VI is an A-submodule of V. If $VI = V$, then $VI^2 = (VI)I = VI = V$ and, in general, $VI^n = V$. But this is impossible because there is an integer m with $I^m = 0$. Thus, $VI = 0$ for all simple A-modules V and hence, $I \subseteq \mathbf{J}(A)$, as desired.

Finally, $\mathbf{J}(A) \cap \mathbf{Z}(A)$ is a nilpotent ideal of $\mathbf{Z}(A)$ and therefore, by the first part, it is contained in $\mathbf{J}(\mathbf{Z}(A))$. Now, let $z \in \mathbf{J}(\mathbf{Z}(A))$. Since $\mathbf{J}(\mathbf{Z}(A))$ is nilpotent and commutes with the elements of A, we have that zA is a nilpotent right ideal of A. Then $z \in \mathbf{J}(A)$. This proves that $\mathbf{J}(\mathbf{Z}(A)) \subseteq \mathbf{J}(A) \cap \mathbf{Z}(A)$. ∎

(1.9) DEFINITION. If A is an F-algebra, we say that A is **semisimple** if $\mathbf{J}(A) = 0$. Also, we say that A is **simple** if it has no proper (two sided) ideals.

Since A and $A/\mathbf{J}(A)$ have the same set of simple A-modules, it follows that $A/\mathbf{J}(A)$ is semisimple.

An A-module V is said to be **completely reducible** if it is the direct sum of simple A-submodules. (It is also common in this case to say that V is **semisimple**.) In fact, there is no difference between completely reducible modules and modules which may be written as a sum (not necessarily direct) of simple submodules.

(1.10) LEMMA. *Let V be an A-module and suppose $V = \sum V_i$, where the V_i's are simple submodules. Then V is the direct sum of some of the V_i's.*

Proof. Since V has finite dimension, we let W be an A-submodule of V maximal with respect to the property that W is the direct sum of some of the V_i's. If W is proper, then there exists a V_j not contained in W. But then, since V_j is simple, we have that $V_j \cap W = 0$. Then $W + V_j$ is a direct sum, which contradicts the maximality of W. ∎

More interesting is the next result.

(1.11) THEOREM. *If A is an F-algebra and V is an A-module, then the following conditions are equivalent.*

 (a) V is completely reducible.

 (b) If U is an A-submodule of V, then there is an A-submodule W such that $V = U \oplus W$.

Proof. Write $V = \sum V_i$, where the V_i's are simple submodules, and suppose that U is an A-submodule of V. Since V has finite dimension, let W be an A-submodule of V maximal such that $U + W = U \oplus W$. If $U + W$ is proper, then there is some V_j not contained in $U + W$. Since V_j is simple, $V_j \cap (U + W) = 0$. Therefore, $U + (V_j + W) = U \oplus (V_j + W)$, which contradicts the maximality of W. This proves that (a) implies (b).

Assume (b) and, since V is finite dimensional, let U be an A-submodule of V maximal such that U is a sum of simple A-submodules. By hypothesis, there is an A-submodule W of V such that $V = U \oplus W$. If $W \neq 0$, since V is finite dimensional, we may find W_0, a simple submodule of V inside W. Then $U + W_0 > U$, which contradicts the maximality of U. Hence, $U = V$ is completely reducible. ∎

(1.12) COROLLARY. *Suppose that V is a completely reducible A-module. If U is an A-submodule of V, then U and V/U are completely reducible.*

Proof. By Theorem (1.11), we have that V/U is isomorphic to a submodule of V. Hence, it suffices to show the first part. If W is an A-submodule of U, again by Theorem (1.11) we know that there exists an A-submodule W_0 of V such that $V = W \oplus W_0$. Then $U = W \oplus (U \cap W_0)$. This proves that U is completely reducible. ∎

If $V_1, \ldots, V_n$ are A-modules, we may form the **external direct sum** of $V_1, \ldots, V_n$, which is denoted by $V_1 \oplus \ldots \oplus V_n$, by setting $V_1 \oplus \ldots \oplus V_n = V_1 \times \ldots \times V_n$ with the action

$$(v_1, \ldots, v_n)a = (v_1 a, \ldots, v_n a)$$

for $v_i \in V_i$ and $a \in A$. It is clear that if V_i is a simple A-module for all i, then $V_1 \oplus \ldots \oplus V_n$ is a completely reducible A-module.

(1.13) THEOREM. *Suppose that A is an F-algebra. Then A is semisimple if and only if every A-module is completely reducible.*

Proof. Assume first that every A-module is completely reducible. If we consider A, the regular A-module, by hypothesis we have that $A = \sum_i I_i$ is a sum of minimal right ideals. Hence, $\mathbf{J}(A) = A\mathbf{J}(A) = 0$ since $\mathbf{J}(A)$ annihilates the simple A-module I_i for all i.

Assume now that A is semisimple. First, we prove that the regular A-module A is completely reducible. To do that, we claim that there exist maximal right ideals $M_1, \ldots, M_n$ of A such that

$$\bigcap_{j=1}^{n} M_j = 0 .$$

If this is the case, the map $a \mapsto (a + M_1, \ldots, a + M_n)$ maps A isomorphically into a submodule of the completely reducible A-module $A/M_1 \oplus \ldots \oplus A/M_n$. Then, by Corollary (1.12), A is completely reducible. To prove the claim, among the subspaces L of A which are intersections of a finite number of maximal right ideals, we choose L of minimal dimension. If $L \neq 0$, then L is not contained in $\mathbf{J}(A) = 0$. Since $\mathbf{J}(A)$ is the intersection of all the maximal right ideals of A (Theorem (1.5)), we have that there exists a maximal right ideal M such that $L \cap M < M$. This contradicts the choice of L and proves the claim.

Now, write $A = \sum_i I_i$ as a sum of minimal right ideals of A. If V is an A-module and $\mathcal{B}$ is an F-basis of V, we have that

$$V = \sum_{v \in \mathcal{B}, i} v I_i .$$

Since the map $I_i \to v I_i$ given by $x \mapsto vx$ is a surjective A-homomorphism and I_i is a minimal right ideal, it follows that the kernel of the map is I_i or zero. Hence, $v I_i$ is isomorphic to I_i or 0. Therefore, V is a sum of simple A-submodules, as required. ∎

(1.14) COROLLARY. *If A is a semisimple F-algebra and B is an ideal of A, then the F-algebra A/B is semisimple.*

Proof. If V is an (A/B)-module, then V is an A-module with $va = v(a+B)$ for $v \in V$ and $a \in A$. Hence, V is a sum of simple A-submodules. Since $VB = 0$, these are also simple (A/B)-submodules of V. Now, Theorem (1.13) applies. ∎

(1.15) COROLLARY. *If A is a semisimple F-algebra, then every simple A-module is isomorphic to a minimal right ideal of A.*

Proof. If V is a simple A-module, by Theorem (1.5) we know that V is isomorphic to A/I, where I is a maximal right ideal of A. Since A is completely reducible, by Theorem (1.11) there exists a right ideal J such that $A = I \oplus J$. Since $J \cong A/I$, we have that J is a minimal right ideal and the result follows. ∎

(A word on notation is appropriate here. Although we usually write functions on the left, the notation fg indicates that f is applied first and then g.)

If $a \in A$ and V is an A-module, we denote by a_V the right multiplication map $V \to V$ given by $v \mapsto va$. It is easy to check that the map $a \mapsto a_V$ is an algebra homomorphism $A \to \mathrm{End}_F(V)$. Furthermore, the kernel of this map is $\mathrm{ann}(V)$. We also denote its image by $A_V = \{a_V \mid a \in A\}$.

(1.16) THEOREM (Double Centralizer). *Suppose that A is a simple F-algebra and let $I \neq 0$ be a right ideal of A. If $B = \mathrm{End}_A(I)$, then I is a B-module and the natural map $A \to \mathrm{End}_B(I)$ given by $a \mapsto a_I$ is an isomorphism of F-algebras. Consequently, if I is a minimal right ideal of A and F is algebraically closed, then A is isomorphic to $\mathrm{End}_F(I)$.*

Proof (Rieffel). As we said before Schur's lemma, it is straightforward to check that B is an F-algebra. Now, I is a B-module with $xf = f(x)$, for $x \in I$ and $f \in B$. Hence, $\mathrm{End}_B(I)$ is again an F-algebra.

First notice that $a_I \in \mathrm{End}_B(I)$ because if $x \in I$ and $f \in B$, we have that

$$a_I(xf) = (xf)a = f(x)a = f(xa) = a_I(x)f.$$

We wish to prove that the map $a \mapsto a_I$ is an algebra isomorphism. By the comments preceding the statement of this theorem, we have that the map $a \mapsto a_I$ is an algebra homomorphism with kernel $\mathrm{ann}(I)$. Since $\mathrm{ann}(I)$ is a proper two sided ideal of the simple algebra A, we conclude that our map is injective.

What we need to prove is that $A_I = \{a_I \mid a \in A\} = \mathrm{End}_B(I)$. Since the identity lies in A_I, it suffices to show that A_I is a right ideal of the ring $\mathrm{End}_B(I)$. Since $AI = A$ (because $AI \neq 0$ is a two sided ideal of A), it suffices to show that $(AI)_I$ is a right ideal of $\mathrm{End}_B(I)$. Let $a \in A$, $y \in I$ and $f \in \mathrm{End}_B(I)$. We prove that $(ay)_I f \in \mathrm{End}_B(I)$. If $u \in I$, let us denote by $u_l : I \to I$ the map given by $u_l(x) = ux$. Notice that $u_l(xa) = uxa = u_l(x)a$ and hence $u_l \in B$. In particular, since I is a B-module, we may write $zu_l = u_l(z) = uz$ for $u, z \in I$. Now, for $x \in I$, we have that

$$((ay)_I f)(x) = f((ay)_I(x)) = f(xay) = f((xa)y) = f(y(xa)_l)$$

$$= f(y)(xa)_l = xaf(y) = (af(y))_l(x).$$

Since this is for all $x \in I$, we have that

$$(ay)_I f = (af(y))_I \in A_I.$$

This completes the proof of the first part of the theorem. The second follows from the first and Schur's lemma. ∎

There is a good reason why the previous theorem is called the double centralizer theorem. With the same notation, notice that, by definition, $\text{End}_A(I) = \mathbf{C}_{\text{End}_F(I)}(A_I)$. Also, $\text{End}_B(I) = \mathbf{C}_{\text{End}_F(I)}(\text{End}_A(I))$. Hence, the double centralizer theorem proves that

$$A_I = \mathbf{C}_{\text{End}_F(I)}(\mathbf{C}_{\text{End}_F(I)}(A_I)).$$

An **idempotent** of A (any ring) is a nonzero element $e \in A$ such that $e^2 = e$. Two idempotents e and f are **orthogonal** if $ef = fe = 0$. An idempotent e is **primitive** if it is not the sum of two orthogonal idempotents.

The following key result classifies the semisimple algebras over algebraically closed fields: they are direct sums of matrix algebras.

(1.17) THEOREM (Wedderburn). *Suppose that A is a semisimple algebra over F, an algebraically closed field.*

(a) There is only a finite set $\{B_1, \ldots, B_n\}$ of distinct minimal ideals of A. Also,

$$A = \bigoplus_{j=1}^{n} B_j.$$

(b) If I_j is a minimal right ideal of A contained in B_j, then $\{I_1, \ldots, I_n\}$ is a complete set of representatives of pairwise nonisomorphic minimal right ideals of A. In particular, $\{I_1, \ldots, I_n\}$ is a complete set of representatives of pairwise nonisomorphic simple A-modules. Moreover,

$$\text{ann}(I_i) = \sum_{j \neq i} B_j.$$

(c) If we write $1 = e_1 + \ldots + e_n$ for $e_i \in B_i$, then every central idempotent of A is a sum of some e_i's. Hence, the complete set of distinct primitive idempotents of $\mathbf{Z}(A)$ is $\{e_1, \ldots, e_n\}$. Also, $B_j = e_j A$ and B_j is a simple F-algebra with identity e_j.

(d) *The natural map induced by right multiplication yields an isomorphism* $B_j \cong \mathrm{End}_F(I_j)$. *In particular,*

$$\dim_F(A) = \sum_{j=1}^{n} (\dim_F(I_j))^2 .$$

(e) *The ideal* B_j *is the direct sum of* $\dim_F(I_j)$ *minimal right ideals of* A *isomorphic to* I_j.

Proof. We prove part (a) by induction on $\dim_F(A)$.

Let B be a minimal ideal of A. Since A is semisimple, we know that the regular A-module is completely reducible by Theorem (1.13). Therefore, there exists a right ideal I such that $A = B \oplus I$. Hence, $IB \subseteq I \cap B = 0$ and thus $I \subseteq X = \{a \in A \mid aB = 0\}$. Since B is an ideal, it is clear that X is an ideal of A. Now, $B^2 \neq 0$ (because $0 = \mathbf{J}(A)$ contains every nilpotent ideal of A) and it follows that B is not contained in X. Since B is a minimal ideal of A, we have that $B \cap X = 0$. However, since $X = I \oplus (B \cap X)$, it follows that $I = X$ is a two sided ideal of A. Also, $IB = BI = 0$ since $BI \subseteq B \cap I = 0$. Notice now that $J \subseteq I$ is an ideal of I if and only if J is an ideal of A (again, because $IB = BI = 0$). Furthermore, if $C \neq B$ is a minimal ideal of A, by minimality we have that $C \cap B = 0$. Thus, $CB = BC \subseteq B \cap C = 0$ and hence $C \subseteq X = I$. We know by Corollary (1.14) that A/B is semisimple. It is clear that the projection map $A \to I$ is a linear map with kernel B, which is onto, multiplicative and mapping the identity of A to what is the identity of I. Hence, I is a semisimple F-algebra and the proof of part (a) clearly follows by induction.

Since $\dim_F(A)$ is finite, we may find a minimal right ideal I_j of A contained in B_j for every j. Now, let J be any minimal right ideal of A. We prove that J is isomorphic to some I_i. Clearly,

$$J = JA = JB_1 + \ldots + JB_n ,$$

and thus there exists an i such that $JB_i \neq 0$. Since JB_i is a right ideal of A, we have that $JB_i = J$ by the minimality of J. In particular, $J \subseteq B_i$ since B_i is an ideal of A. Also, $JB_j = JB_iB_j = 0$ for all $j \neq i$. Now, $\mathrm{ann}(J)$ is an ideal of A which contains B_j for every $j \neq i$. Hence, B_i is not contained in $\mathrm{ann}(J)$ and thus $\mathrm{ann}(J) \cap B_i = 0$ (because B_i is a minimal ideal). Therefore, $JI_i \neq 0$. This implies that there exists an $x \in J$ such that $xI_i \neq 0$. Now, since xI_i is a nonzero right ideal of A contained in the minimal right ideal J, we have that $xI_i = J$. Since the map $y \mapsto xy$ from $I_i \to J$ is a nonzero A-homomorphism between simple modules, we have that $y \mapsto xy$ is an isomorphism by Schur's lemma. This proves that every minimal right ideal of A (and, by Corollary (1.15), every simple A-module) is isomorphic to some I_i.

Now, we wish to prove that I_i and I_j are not isomorphic if $i \neq j$. Since two isomorphic A-modules have the same annihilator, it suffices to prove the last part of (b). It is clear that $\sum_{j \neq i} B_j \subseteq \mathrm{ann}(I_i)$ since $I_i \subseteq B_i$ and $B_i B_j = 0$. On the other hand, since $\mathrm{ann}(I_i)$ is an ideal of A which does not contain B_i (otherwise $I_i = 0$), it follows that $\mathrm{ann}(I_i) \cap B_i = 0$. Since

$$\mathrm{ann}(I_i) = \sum_{j \neq i} B_j + (\mathrm{ann}(I_i) \cap B_i)$$

the proof of part (b) is complete.

As a consequence of part (b), note that if J_1 and J_2 are two minimal right ideals of A contained in B_j, then J_1 and J_2 are isomorphic as A-modules.

For part (c), write $1 = e_1 + \ldots + e_n$, where $e_i \in B_i$. For every $a \in A$, we have that

$$ae_1 + \ldots + ae_n = a1 = a = 1a = e_1 a + \ldots + e_n a.$$

Since the B_i's are two sided ideals whose sum is direct, it follows that $ae_i = e_i a$ for every i. Hence, $e_i \in Z(A)$. Furthermore, using again that the sum of the B_i's is direct, it follows that $ae_i = e_i a = a$ for every $a \in B_i$. Hence, e_i is the identity in B_i and therefore

$$e_i e_j = \delta_{ij} e_i.$$

Clearly, $e_j A = B_j$. Also, since B_j is an F-algebra whose ideals are the ideals of A contained in B_j, we have that B_j is a simple algebra. Now, let e be any central idempotent of A. One can easily check that

$$eB_i \oplus (1 - e)B_i = B_i.$$

Since eB_i and $(1-e)B_i$ are ideals of A, it follows that $eB_i = 0$ or $(1-e)B_i = 0$. Therefore, $ee_i = 0$ or $e_i = ee_i$. Since $e = ee_1 + \ldots + ee_n$, we see that e is a sum of certain e_i's. From this, it is straightforward to complete the proof of part (c).

Now, since I_j is a minimal right ideal of B_j and F is algebraically closed, by the double centralizer theorem applied to the simple algebra B_j, we have that right multiplication defines an isomorphism $B_j \to \mathrm{End}_F(I_j)$. In particular, we see that $\dim_F(B_j) = (\dim_F(I_j))^2$, and this proves part (d).

Finally, since A is semisimple, the A-module B_j is the direct sum of n_j minimal right ideals of A. By part (b), every minimal right ideal of A contained in B_j is isomorphic to I_j. Now, by computing dimensions, we see that $\dim_F(B_j) = n_j \dim_F(I_j)$. Since we know that $\dim_F(B_j) = (\dim_F(I_j))^2$ by part (d), it follows that $n_j = \dim_F(I_j)$. This completes the proof of the theorem. ∎

(1.18) DEFINITION. A **representation** of an F-algebra A is an algebra homomorphism

$$\mathcal{X} : A \to \mathrm{Mat}(n, F).$$

The **degree** of the representation $\mathcal{X}$ is n. The **kernel** of $\mathcal{X}$ is $\ker(\mathcal{X}) = \{a \in A \mid \mathcal{X}(a) = 0\}$.

If $V = F^n$ and $\mathcal{X}$ is a representation of A of degree n, then we may define

$$va = v\mathcal{X}(a)$$

for $v \in V$ and $a \in A$, and V becomes an A-module. Conversely, suppose that V is an A-module and let $\{v_1, \ldots, v_n\}$ be a basis of V. If for every $a \in A$ we write

$$v_i a = \sum_{j=1}^{n} a_{ij} v_j ,$$

then the map $\mathcal{X}(a) = (a_{ij})$ is a representation of A. Because of this correspondence, the study of A-modules is equivalent to the study of representations of A.

It is easy to check that two A-modules V_1 and V_2 are isomorphic if and only if any two associated representations $\mathcal{X}_1$ and $\mathcal{X}_2$ are **similar**; which means that there exists an invertible matrix $P \in \mathrm{GL}(n, F)$ such that

$$\mathcal{X}_1(a)P = P\mathcal{X}_2(a)$$

for all $a \in A$. Hence, a representation of A uniquely determines an A-module up to isomorphism and an A-module uniquely determines a representation of A up to similarity.

A representation $\mathcal{X}$ is **irreducible** if its corresponding module is simple.

Sometimes, it is convenient to define the representations of A as the algebra homomorphisms $A \to \mathrm{End}_F(V)$, where V is a vector space over F of finite dimension. Via the "fix a basis"-isomorphism between $\mathrm{Mat}(n, F)$ and $\mathrm{End}_F(V)$, this new definition is entirely equivalent to the old one. If $\mathcal{X} : A \to \mathrm{End}_F(V)$ is such a representation, then V becomes an A-module with $va = \mathcal{X}(a)(v)$. Conversely, if V is an A-module, the map $\mathcal{X} : A \to \mathrm{End}_F(V)$ given by $\mathcal{X}(a) = a_V$, where $a_V(v) = va$, defines an algebra homomorphism.

(1.19) THEOREM. *Let A be an F-algebra, where F is algebraically closed. Then there is only a finite number $\mathcal{X}_1, \ldots, \mathcal{X}_n$ of irreducible nonsimilar representations of A. Each $\mathcal{X}_i$ is surjective. Furthermore, the trace functions $\chi_i : A \to F$ afforded by the $\mathcal{X}_i$'s are linearly independent.*

Proof. Since the irreducible representations for A and $A/\mathbf{J}(A)$ are the same, we may assume that $\mathbf{J}(A) = 0$. By Wedderburn's theorem, we know that there exists a finite number of nonisomorphic minimal right ideals of A, say $\{I_1, \ldots, I_n\}$, and hence a finite number of nonsimilar representations $\mathcal{X}_i : A \to \operatorname{End}_F(I_i)$ given by right multiplication. In the notation of Wedderburn's theorem, we have that $\ker(\mathcal{X}_i) = \sum_{j \neq i} B_j$. Also, by part (d) of the same theorem, we have that $\mathcal{X}_i$ is onto. Therefore, there is $x_i \in B_i$ such that $\chi_i(x_i) = 1$. Hence, $\chi_i(x_j) = \delta_{ij}$, which proves that the trace functions are linearly independent. ∎

If V is an A-module and A is semisimple, then V is a direct sum of simple A-submodules. Hence, every representation $\mathcal{X}$ of A is similar to a diagonal representation

$$\begin{pmatrix} \mathcal{X}_1 & \cdots & 0 \\ \vdots & \ddots & \vdots \\ 0 & \cdots & \mathcal{X}_s \end{pmatrix},$$

where the $\mathcal{X}_i$'s are irreducible representations of A.

In the nonsemisimple case, it is still possible to do something. If we find a **composition series**

$$V = V_0 > V_1 > \ldots > V_{s-1} > V_s = 0$$

(that is, with V_{i-1}/V_i a simple A-module), then by choosing an appropriate basis of V, we find that every representation $\mathcal{X}$ is similar to an upper triangular representation in block form

$$\begin{pmatrix} \mathcal{X}_1 & \cdots & * \\ \vdots & \ddots & \vdots \\ 0 & \cdots & \mathcal{X}_s \end{pmatrix},$$

where the $\mathcal{X}_i$'s are irreducible representations of A corresponding to the modules V_{i-1}/V_i. This may be done by choosing a basis of V_{s-1}, extending it to a basis of V_{s-2} and so on. (Afterwards, number the basis in the reverse order.) By a Jordan-Hölder theorem for A-modules, it follows that the irreducible representations $\mathcal{X}_i$ are uniquely determined by V (and hence by $\mathcal{X}$) up to similarity. We say in this case that $\mathcal{X}_i$ is an **irreducible constituent** of $\mathcal{X}$ or that V_{i-1}/V_i is a **simple constituent** of V.

Note that if V is completely reducible, then V_{i-1} is completely reducible and thus, we may write $V_{i-1} = V_i \oplus U_i$ for some A-submodule U_i of V. Therefore, for completely reducible modules V, we deduce that each simple constituent of V is isomorphic to a simple submodule of V.

Finally, note that a representation $\mathcal{X} : A \to \mathrm{Mat}(n, F)$ is irreducible if and only if $\mathcal{X}$ is not similar to a representation in block form

$$\begin{pmatrix} * & * \\ 0 & * \end{pmatrix}.$$

An easy fact that will be used in these notes is the following.

(1.20) THEOREM. *Suppose that $\mathcal{X} : A \to \mathrm{Mat}(n, F)$ is an irreducible representation of an algebra A over an algebraically closed field F. Suppose that $M \in \mathrm{Mat}(n, F)$ is such that $M\mathcal{X}(a) = \mathcal{X}(a)M$ for every $a \in A$. Then M is a scalar matrix. In particular, if $z \in \mathbf{Z}(A)$, then $\mathcal{X}(z)$ is a scalar matrix. Consequently, if A is commutative, then $\mathcal{X}$ is one dimensional.*

Proof. To prove that M is a scalar matrix, we could use Theorem (1.19) which asserts that $\mathcal{X}(A) = \mathrm{Mat}(n, F)$. There is another argument, which directly follows from Schur's lemma. Write $V = F^n$. We know that V is an A-module via $va = v\mathcal{X}(a)$. If f is the linear map $v \mapsto vM$, by hypothesis, we easily see that $f \in \mathrm{End}_A(V)$. By Schur's lemma, we have that f (and therefore M) is scalar. Now, if $z \in \mathbf{Z}(A)$, then $\mathcal{X}(z)$ commutes with the elements of $\mathcal{X}(a)$ for all $a \in A$, and it follows from the first part that $\mathcal{X}(z)$ is a scalar matrix. If A is commutative, then $\mathbf{Z}(A) = A$ and thus, every F-subspace of V is A-invariant. Since V is simple, necessarily V has dimension one. ∎

If G is a finite abelian group, it follows from Theorem (1.20) that the irreducible representations of FG, where F is algebraically closed, are one dimensional.

We have not forgotten the main object of these notes: the group algebra. Maschke's theorem is easy but fundamental.

(1.21) THEOREM (Maschke). *If G is a finite group and F is a field such that $\mathrm{char}(F)$ does not divide $|G|$, then FG is semisimple.*

Proof. We prove that if V is an FG-module and U is a submodule of V, then there is an FG-submodule W such that $V = U \oplus W$. To see this, write $V = U \oplus U_0$, where U_0 is a subspace of V, and let $f : V \to U$ be the projection. We modify f as follows. Let $\tilde{f} : V \to U$ be given by

$$\tilde{f}(v) = \frac{1}{|G|} \sum_{x \in G} f(vx)x^{-1}.$$

Now,

$$\tilde{f}(vy) = \frac{1}{|G|} \sum_{x \in G} f(vyx)(yx)^{-1}y = \tilde{f}(v)y$$

for $y \in G$ and thus $\tilde{f}$ is an FG-homomorphism. Since $\tilde{f}(u) = u$ for all $u \in U$, we have that $(\tilde{f})^2 = \tilde{f}$. Also, it is clear that $U \cap \ker(\tilde{f}) = 0$. Now, if $v \in V$, we have that $v = (v - \tilde{f}(v)) + \tilde{f}(v)$, where $\tilde{f}(v) \in U$ and $v - \tilde{f}(v) \in \ker(\tilde{f})$. Hence, $V = U \oplus \ker(\tilde{f})$, and we are done. ∎

A representation of FG determines (and is uniquely determined by) a group homomorphism $\mathcal{X} : G \rightarrow \mathrm{GL}(n, F)$. We will also refer to the latter as representations of G.

PROBLEMS

(1.1) Let V be an A-module. Prove that V is completely reducible if and only if $V \mathbf{J}(A) = 0$.

(1.2) Let A be an F-algebra. If $z \in \mathbf{Z}(A)$ is **nilpotent** (that is, there exists an n such that $z^n = 0$), prove that $z \in \mathbf{J}(A)$.

(1.3) Suppose that A is a ring and let e be an idempotent of A. Show that the following conditions are equivalent:

(a) e is primitive;

(b) eA is **indecomposable** as right A-module (that is, it is not the direct sum of two proper right ideals of A);

(c) e is the unique idempotent of eAe.

(1.4) Suppose that A is a ring and let e be a central idempotent. Show that the following conditions are equivalent:

(a) e is a central primitive idempotent;

(b) eA is not the direct sum of two proper (two sided) ideals of A.

(1.5) If G is a group and F is any field, show that there is a unique F-linear algebra homomorphism $\delta : FG \rightarrow F$ such that $\delta(g) = 1$ for all $g \in G$. The map δ is usually called the **augmentation map**. Show that $\{1 - g \mid g \in G - \{1\}\}$ is a basis of $\ker(\delta)$.

(1.6) If G is a group and $\mathrm{char}(F)$ divides $|G|$, show that $(\sum_{g \in G} g)^2 = 0$. Deduce that FG is not semisimple.

(1.7) (D.G. Higman) Let G be a group and let $H \subseteq G$ with index not divisible by $\mathrm{char}(F)$, where F is a field. If V is an FG-module such that V considered as an FH-module is completely reducible, show that V is completely reducible as an FG-module.

(HINT: Mimic the proof of Maschke's theorem.)

2. Brauer Characters

Let $\mathbf{R}$ be the ring of algebraic integers in $\mathbb{C}$. Throughout these notes, we fix a prime p and we choose a maximal ideal M of $\mathbf{R}$ containing $p\mathbf{R}$. Let $F = \mathbf{R}/M$ be a field of characteristic p, and let

$$^* : \mathbf{R} \to F$$

be the natural ring homomorphism. We will need to work in a larger ring than $\mathbf{R}$. We define

$$S = \{\frac{r}{s} \mid r \in \mathbf{R}, \, s \in \mathbf{R} - M\}.$$

It is clear that S is a ring and that

$$\mathcal{P} = \{\frac{r}{s} \mid r \in M, \, s \in \mathbf{R} - M\}$$

is an ideal of S such that every $\alpha \in S - \mathcal{P}$ has an inverse in S. Then $\mathcal{P}$ contains every ideal of S and therefore $\mathbf{J}(S) = \mathcal{P}$.

Notice that the map $^* : \mathbf{R} \to F$ can be extended to S in a natural way. If $r \in \mathbf{R}$ and $s \in \mathbf{R} - M$, then

$$(\frac{r}{s})^* = r^*(s^*)^{-1}$$

defines a ring homomorphism $S \to F$ extending * with kernel $\mathcal{P}$. In particular, $\mathbf{R} \cap \mathcal{P} = M$.

The following lemma tells us all the information that we need about F. First, let us set $\mathbf{U} = \{\xi \in \mathbb{C} \mid \xi^m = 1 \text{ for some integer } m \text{ not divisible by } p\} \subseteq \mathbf{R}$, the multiplicative group of p'-roots of unity. Also, we denote by $\mathbb{Z}_p = \mathbb{Z}/p\mathbb{Z}$ the field of p elements.

(2.1) LEMMA. *The restriction of * to* **U** *defines an isomorphism* **U** $\to F^\times$ *of multiplicative groups. Also, F is the algebraic closure of its prime field* $\mathbb{Z}^* \cong \mathbb{Z}_p$.

Proof. First of all, we claim that $M \cap \mathbb{Z} = p\mathbb{Z}$. It is clear that $p\mathbb{Z} \subseteq M \cap \mathbb{Z}$. If $m \in M \cap \mathbb{Z}$ is not divisible by p, then there are integers $a, b \in \mathbb{Z}$ such that $ap + bm = 1$. Hence $1 \in M$, and this is a contradiction which proves the claim. In particular, we see that $\mathbb{Z}^* \cong \mathbb{Z}/p\mathbb{Z} = \mathbb{Z}_p$.

Now, if $\xi \in \mathbf{U}$ has order m, then

$$1 + x + \ldots + x^{m-1} = \frac{x^m - 1}{x - 1} = \prod_{i=1}^{m-1} (x - \xi^i).$$

Setting $x = 1$, we see that $1 - \xi$ divides m in **R**. Therefore, if $\xi^* = 1$, then $m^* = 0$ and thus $m \in M$. This is not possible since $M \cap \mathbb{Z} = p\mathbb{Z}$. This shows that * is one to one from **U**. Now, suppose that $f \in F^\times$ and let $r \in \mathbf{R}$ with $r^* = f$. Since r is an algebraic integer, we have that

$$r^n + a_{n-1}r^{n-1} + \ldots + a_1 r + a_0 = 0$$

for some integers a_i. Then

$$0 = f^n + a_{n-1}^* f^{n-1} + \ldots + a_1^* f + a_0^*$$

and thus f is algebraic over $\mathbb{Z}^*$. To complete the proof, it suffices to show that if K/F is an algebraic extension, then $K^\times \subseteq \mathbf{U}^*$. This will prove that * is onto and that F is algebraically closed. If $k \in K$, then k is algebraic over $\mathbb{Z}^*$ by the transitivity of the algebraic extensions. Hence, the field $\mathbb{Z}_p[k]$ is finite and thus $k^m = 1$, where $m = |\mathbb{Z}_p[k]| - 1$ is not divisible by p. Since we already know m distinct roots in $\mathbf{U}^*$ of the polynomial $x^m - 1$ (the different powers of ξ^*), it follows that k should be one of those. ∎

We are ready to define Brauer characters. Let us denote by G^0 the set of **p-regular elements** of a finite group G (that is, the set of elements $g \in G$ such that the order of g is not divisible by p). Suppose that $\mathcal{X} : G \to \mathrm{GL}(n, F)$ is a representation of G. If $g \in G^0$, by Lemma (2.1) we have that all the eigenvalues of $\mathcal{X}(g)$ (which lie in $F^\times$ since F is algebraically closed and $\mathcal{X}(g)$ is an invertible matrix) are of the form $\xi_1^*, \ldots, \xi_n^*$ for uniquely determined $\xi_1, \ldots, \xi_n \in \mathbf{U}$. We say that $\varphi : G^0 \to \mathbb{C}$ defined for $g \in G^0$ by

$$\varphi(g) = \xi_1 + \ldots + \xi_n$$

is the **Brauer character** of G afforded by the representation $\mathcal{X}$. Notice that φ is uniquely determined (once M has been chosen) by the equivalence class of the representation $\mathcal{X}$. We say that φ is **irreducible** if $\mathcal{X}$ is irreducible, and we denote the set of irreducible Brauer characters of G by $\mathrm{IBr}(G)$.

(Brauer characters are also called **modular characters**.)

If $\mathcal{X} : G \rightarrow \mathrm{GL}(n, F)$ is a representation of G affording the Brauer character φ, the **degree** of φ is $n = \varphi(1)$.

The **trivial representation** $\mathcal{X} : G \rightarrow F$ given by $\mathcal{X}(g) = 1$ for $g \in G$ affords the **principal** Brauer character 1_{G^0}, where $1_{G^0}(g) = 1$ for $g \in G^0$. The corresponding FG-module (called the **trivial module**) is just F with the trivial action $fg = f$ for $f \in F$ and $g \in G$.

If $\mathcal{X}$ is an F-representation of G and $g \in G$ is p-regular, then $\mathcal{X}_{\langle g \rangle}$ is completely reducible by Maschke's theorem. Since the irreducible representations of $\langle g \rangle$ are one dimensional (by Theorem (1.20)), it follows that $\mathcal{X}(g)$ is similar to a diagonal matrix. This is a useful fact which will be used later on.

Next, we collect a few elementary results on Brauer characters so that we familiarize ourselves with its definition. We denote the space of complex class functions defined on G^0 by $\mathrm{cf}(G^0)$.

(2.2) LEMMA. *Suppose that φ is a Brauer character of G.*
(a) $\varphi \in \mathrm{cf}(G^0)$.
(b) *If $g \in G^0$, then $\varphi(g^{-1}) = \overline{\varphi(g)}$.*
(c) *The function $\bar{\varphi} : G^0 \rightarrow \mathbb{C}$ defined by $\bar{\varphi}(g) = \varphi(g^{-1})$ is a Brauer character of G. In fact, if V is an FG-module affording φ, then $V^* = \mathrm{Hom}_F(V, F)$ is an FG-module affording $\bar{\varphi}$.*
(d) *If H is a subgroup of G, then $\varphi_H = (\varphi)_{H^0}$ is a Brauer character of H.*

Proof. (a) If $g, h \in G$, then the matrices $\mathcal{X}(g)$ and $\mathcal{X}(g^h)$ are similar. Therefore, they have the same set of eigenvalues. This proves (a).

(b) By the comments preceding the statement of this lemma, we know that $\mathcal{X}(g)$ is similar to a diagonal matrix $\mathrm{diag}(\xi_1^*, \ldots, \xi_n^*)$, where $\xi_1, \ldots, \xi_n \in \mathbf{U}$. Hence, $\mathcal{X}(g^{-1}) = \mathcal{X}(g)^{-1}$ is similar to $\mathrm{diag}((\xi_1^{-1})^*, \ldots, (\xi_n^{-1})^*)$. Since

$$\xi_i^{-1} = \bar{\xi}_i \, ,$$

part (b) follows.

(c) Although to prove (c) it suffices to prove the second part, it is interesting to notice that if $\mathcal{X}$ affords φ, then by part (b), the representation

$$\mathcal{Y}(g) = \mathcal{X}(g^{-1})^{\mathrm{t}}$$

affords $\bar{\varphi}$.

Now, if V is an FG-module, $\theta \in V^* = \mathrm{Hom}_F(V, F)$ and $g \in G$, we may define an action of G on V^* by

$$(\theta g)(v) = \theta(vg^{-1}).$$

It is easy to see that this action (extended linearly to FG) makes V^* an FG-module (called the **dual** module). If, in a certain basis $\mathcal{B}$, the module V affords the representation $\mathcal{X}$, then in the dual basis of $\mathcal{B}$, the module V^* affords the representation $\mathcal{Y}(g) = \mathcal{X}(g^{-1})^{\mathrm{t}}$.

(d) If $\mathcal{X}$ is a representation of FG affording φ, then the restriction $\mathcal{X}_{FH}$ is a representation of FH affording φ_H. ∎

It is important to notice that if two FG-modules (representations) have the same set of irreducible constituents (counting multiplicities), then they afford the same Brauer character. This follows from the fact that every representation $\mathcal{X}$ of G is similar to some upper triangular representation in block form with the representations of the composition factors appearing on the diagonal. In this case, it is not difficult to check that the Brauer character afforded by $\mathcal{X}$ is the sum of the Brauer characters afforded by the irreducible representations appearing on the diagonal.

Since it is clear that sums of Brauer characters are Brauer characters, this shows the following.

(2.3) THEOREM. *A class function $\varphi \in \mathrm{cf}(G^0)$ is a Brauer character if and only if φ is a nonnegative integer linear combination of $\mathrm{IBr}(G)$.*

Proof. This easily follows from the above discussion. ∎

The next lemma tells us why Brauer characters are only defined on p-regular elements.

(2.4) LEMMA. *If $\mathcal{X}$ is a representation of FG with character ψ and $g \in G$, then $\psi(g) = \psi(g_{p'})$. In particular, if $\mathcal{X}$ affords the Brauer character μ, then $\psi(g) = \mu(g_{p'})^*$. Furthermore, if we define $\varphi^*(g) = \varphi(g_{p'})^*$ for $g \in G$ and $\varphi \in \mathrm{IBr}(G)$, then $\{\varphi^* \mid \varphi \in \mathrm{IBr}(G)\}$ is the set of the trace functions of the irreducible representations of FG.*

Proof. By writing $\mathcal{X}$ in the upper triangular block form, we may certainly assume that $\mathcal{X}$ is irreducible. There is no loss if we also assume that G is generated by the element g. In this case, by Theorem (1.20), the irreducible representations of G are one dimensional. Hence, $\mathcal{X} : G \to F^\times$ is just a homomorphism of groups. Since $\mathcal{X}(g_p)$ has p-power order in $F^\times$, it follows that $\mathcal{X}(g_p) = 1$. Then $\mathcal{X}(g) = \mathcal{X}(g_{p'})$, as desired. The second and the third parts easily follow from the first by using the definition of Brauer characters. ∎

If χ is an ordinary character of a group G, by arguing as in Lemma (2.4), note that

$$\chi(g)^* = \chi(g_{p'})^*$$

for every $g \in G$.

To develop the basic properties and relationships between blocks and characters, we need the following fact whose proof uses elementary properties of Dedekind domains.

(2.5) LEMMA. *Suppose that I is a proper ideal of* **R** *and let* $\alpha_1, \ldots, \alpha_n$ *be algebraic over* $\mathbb{Q}$ *and not all zero. Then there is a* $\beta \in \mathbb{Q}(\alpha_1, \ldots, \alpha_n)$ *such that* $\beta\alpha_i \in$ **R** *for all i, but not all $\beta\alpha_i$ lie in I.*

Proof. Let $K = \mathbb{Q}(\alpha_1, \ldots, \alpha_n)$, a finite degree extension over $\mathbb{Q}$. Then $D = K \cap$ **R** is a Dedekind domain (see Corollary (28.22) of [Isaacs, *Algebra*], for instance). Let A be the D-module generated by $\{\alpha_1, \ldots, \alpha_n\}$. Since D is Dedekind, there is a finitely generated D-module B inside K such that $AB = D$. Since AB is not contained in I, there is a $\beta \in B$ such that βA is not contained in I. ∎

We will denote the algebraic closure of $\mathbb{Q}$ in $\mathbb{C}$ by **K**. It is clear that **R** $\subseteq$ **K**.

(2.6) THEOREM. *If G is a finite group, then* $\mathrm{IBr}(G)$ *is linearly independent over* $\mathbb{C}$.

Proof. Let $\{\mathcal{X}_1, \ldots, \mathcal{X}_h\}$ be a complete set of representatives of the irreducible representations of FG. Since the Brauer characters have their values in **R** $\subseteq$ **K**, by elementary linear algebra it suffices to show that the irreducible Brauer characters are linearly independent over **K**. Suppose that

$$\sum_{\varphi \in \mathrm{IBr}(G)} a_\varphi \varphi = 0$$

for $a_\varphi \in$ **K** and assume that there is some $a_\varphi \neq 0$. By Lemma (2.5), we know that there is a $\beta \in$ **K** such that $\beta a_\varphi \in$ **R** for all $\varphi \in \mathrm{IBr}(G)$, but not all $\beta\alpha_\varphi$ are in M. If we define

$$\varphi^*(g) = \varphi(g_{p'})^*,$$

by Lemma (2.4) we have that $\{\varphi^* \mid \varphi \in \mathrm{IBr}(G)\}$ is the set of trace functions of the irreducible representations of FG. Now,

$$\sum_{\varphi \in \mathrm{IBr}(G)} (\beta a_\varphi)^* \varphi^* = 0$$

and this contradicts the fact (Theorem (1.19)) that the trace functions are F-linearly independent. ∎

If ψ is a Brauer character of G, we may write

$$\psi = \sum_{\varphi \in \mathrm{IBr}(G)} a_\varphi \varphi$$

for uniquely determined nonnegative integers a_φ. We say that a_φ is the **multiplicity** of φ in ψ. If V and W are FG-modules affording ψ and φ, respectively, notice that the multiplicity of φ in ψ is the number of simple constituents of V isomorphic to W.

By the linear independence of $\mathrm{IBr}(G)$ (and Theorem (2.3)), a Brauer character φ of G is irreducible if and only if φ is not of the form $\alpha + \beta$ for Brauer characters α and β of G.

Since complex conjugates of Brauer characters are Brauer characters, also observe that $\varphi \in \mathrm{IBr}(G)$ if and only if $\bar{\varphi} \in \mathrm{IBr}(G)$.

To prove the next theorem, we will use a well-known fact about irreducible complex characters which easily follows from Theorem (1.17); namely, every $\chi \in \mathrm{Irr}(G)$ can be afforded by a representation with entries in $\mathbf{K}$. (See page 22 of [Isaacs, *Characters*].) In fact, more is going on.

(2.7) THEOREM. *Let $\mathcal{X}$ be a complex representation of G. Then $\mathcal{X}$ is similar to some representation $\mathcal{Y}$ of G with matrix entries in S.*

To prove Theorem (2.7), we need a lemma.

(2.8) LEMMA. *Suppose that V is a vector space over $\mathbf{K}$ and let W be a finitely generated S-submodule of V. Then there exist $w_1, \ldots, w_n \in W$ which are linearly independent over $\mathbf{K}$ and such that $W = Sw_1 + \ldots + Sw_n$.*

Proof. We have that $W/\mathcal{P}W$ becomes a vector space over F by defining

$$\alpha^*(w + \mathcal{P}W) = \alpha w + \mathcal{P}W$$

for $\alpha \in S$. (Notice that if $\alpha^* = \beta^*$ for $\alpha, \beta \in S$, then $\alpha - \beta \in \mathcal{P}$ and thus $\alpha w - \beta w \in \mathcal{P}W$.) Since W is finitely generated over S, it follows that $\dim_F(W/\mathcal{P}W)$ is finite. Then we may find $w_1, \ldots, w_n$ in W such that $\{w_1 + \mathcal{P}W, \ldots, w_n + \mathcal{P}W\}$ is a basis of $W/\mathcal{P}W$. If $U = Sw_1 + \ldots + Sw_n$ is the S-span of $\{w_1, \ldots, w_n\}$, we see that $W = U + \mathcal{P}W$. Then, by Nakayama's lemma (1.7), we conclude that $U = W$. Now, it suffices to show that $\{w_1, \ldots, w_n\}$ is $\mathbf{K}$-linearly independent. If

$$\sum_{i=1}^{n} \alpha_i w_i = 0$$

for some $\alpha_i \in \mathbf{K}$ that are not all zero, then, by Lemma (2.5), there is some $\beta \in \mathbf{K}$ such that $\beta\alpha_i \in \mathbf{R}$ for every i, but not all $\beta\alpha_i$ are in M. Since V is a $\mathbf{K}$-module, it follows that

$$0 = \beta \sum_{i=1}^{n} \alpha_i w_i = \sum_{i=1}^{n} (\beta\alpha_i) w_i \,.$$

Hence,

$$\sum_{i=1}^{n} (\beta\alpha_i)^* (w_i + \mathcal{P}W) = 0 \,,$$

and this contradicts the fact that $\{w_1 + \mathcal{P}W, \ldots, w_n + \mathcal{P}W\}$ is F-linearly independent. ∎

Proof of Theorem (2.7). We may assume that $\mathcal{X}$ is a $\mathbf{K}$-representation of G. Let V be a $\mathbf{K}G$-module affording $\mathcal{X}$ and let $\mathcal{B}$ be a basis of V. Let us denote by W the S-span of $\{vg \mid v \in \mathcal{B}, g \in G\}$. Note that W is G-invariant. Now, W is a finitely generated S-submodule of V, and by Lemma (2.8) we conclude that there exist $w_1, \ldots, w_n \in W$ which are linearly independent over $\mathbf{K}$ and such that $W = Sw_1 + \ldots + Sw_n$. Now, since $\mathcal{B} \subseteq W$, it follows that every element of $\mathcal{B}$ is an S-linear combination of $\{w_1, \ldots, w_n\}$. Hence, the $\mathbf{K}$-span of $\{w_1, \ldots, w_n\}$ is V. This proves that $\{w_1, \ldots, w_n\}$ is a $\mathbf{K}$-basis of V. Also, since $w_i g \in W$, we have that $w_i g$ is an S-linear combination of $\{w_1, \ldots, w_n\}$. This proves the theorem. ∎

If A is a matrix with entries in the ring S, we denote by A^* the matrix in F which results from applying * to every entry of A. It is clear that

$$* : \mathrm{Mat}_n(S) \to \mathrm{Mat}_n(F)$$

is a ring homomorphism. Also,

$$\det(A)^* = \det(A^*)$$

for $A \in \mathrm{Mat}_n(S)$.

Notice furthermore that the ring homomorphism $* : S \to F$ can also be extended to the polynomial rings

$$* : S[x] \to F[x] \,.$$

In fact, if all the roots $\{\xi_1, \ldots, \xi_n\}$ of $p(x) \in S[x]$ are in S, then the roots of the polynomial $p(x)^*$ are $\{\xi_1^*, \ldots, \xi_n^*\}$.

If $\chi \in \mathrm{cf}(G)$ is a complex class function of G, we denote the restriction of χ to G^0 by χ^0.

The next corollary is of fundamental importance since it relates ordinary and Brauer characters.

(2.9) COROLLARY. *If $\mathcal{X}$ is a representation of G with matrix entries in S affording the character χ, then the F-representation $\mathcal{X}^*$, defined by*

$$\mathcal{X}^*(g) = \mathcal{X}(g)^*$$

for $g \in G$, affords the Brauer character χ^0. In particular, if χ is an ordinary character of G, then χ^0 is a Brauer character of G.

Proof. Suppose that $g \in G^0$ and let $\xi_1, \ldots, \xi_n \in \mathbf{U}$ be the eigenvalues of $\mathcal{X}(g)$ (which lie in $\mathbf{U}$ since the order of g is a p'-number). Then $\xi_1^*, \ldots, \xi_n^*$ are the eigenvalues of $\mathcal{X}^*(g)$ because they are the roots of the polynomial

$$\det(xI - \mathcal{X}(g))^* = \det(xI - \mathcal{X}^*(g)).$$

Now, it is clear that $\mathcal{X}^*$ is a representation of FG (because $* : \mathrm{Mat}_n(S) \to \mathrm{Mat}_n(F)$ is a ring homomorphism) which affords the Brauer character χ^0. Finally, if χ is an ordinary character of G, by Theorem (2.7) we may find an S-representation of G affording χ. By the first part, the theorem is proved. ∎

Using the above corollary and the fact that the elements of $\mathrm{IBr}(G)$ are linearly independent, it follows that if $\chi \in \mathrm{Irr}(G)$, then we may write

$$\chi^0 = \sum_{\varphi \in \mathrm{IBr}(G)} d_{\chi\varphi}\varphi$$

for uniquely determined nonnegative integers $d_{\chi\varphi}$ called the **decomposition numbers**.

There is an important consequence of Corollary (2.9).

(2.10) COROLLARY. *The set $\mathrm{IBr}(G)$ is a basis of $\mathrm{cf}(G^0)$. Therefore, $|\mathrm{IBr}(G)|$ is the number of conjugacy classes of p-regular elements of G.*

Proof. If $\xi \in \mathrm{cf}(G^0)$, it suffices to show that ξ is a $\mathbb{C}$-linear combination of $\mathrm{IBr}(G)$. Now, let $\delta \in \mathrm{cf}(G)$ be any extension of ξ. Then

$$\delta = \sum_{\chi \in \mathrm{Irr}(G)} a_\chi \chi$$

for some $a_\chi \in \mathbb{C}$. Therefore,

$$\xi = \sum_{\chi \in \mathrm{Irr}(G)} a_\chi \chi^0$$

is a linear combination of $\mathrm{IBr}(G)$, as required. ∎

The matrix

$$D = (d_{\chi\varphi})_{\chi \in \mathrm{Irr}(G), \varphi \in \mathrm{IBr}(G)}$$

is called the **decomposition matrix**, and we prove next that it has maximum rank.

(2.11) COROLLARY. *The decomposition matrix D has rank $|\mathrm{IBr}(G)|$.*

Proof. Since $\{\chi^0 \mid \chi \in \mathrm{Irr}(G)\}$ spans $\mathrm{cf}(G^0)$, by elementary linear algebra there is a subset $\mathcal{B}$ of $\mathrm{Irr}(G)$ such that $\{\chi^0 \mid \chi \in \mathcal{B}\}$ is a basis of $\mathrm{cf}(G^0)$. This easily implies that the rows of the square matrix $(d_{\chi\varphi})_{\chi \in \mathcal{B}, \varphi \in \mathrm{IBr}(G)}$ are linearly independent. Hence, there is an invertible submatrix of D of order $|\mathrm{IBr}(G)|$. ∎

Note that this implies that for every $\varphi \in \mathrm{IBr}(G)$, there is a $\chi \in \mathrm{Irr}(G)$ such that $d_{\chi\varphi} \neq 0$ (since D cannot have a column consisting of zeros). This was not clear until now.

Next, we see why the interesting case for modular representation theory is when p divides $|G|$.

(2.12) THEOREM. *If p does not divide $|G|$, then $\mathrm{IBr}(G) = \mathrm{Irr}(G)$.*

Proof. By Maschke's theorem, we know that FG is semisimple. Also, since F is algebraically closed, by Wedderburn's theorem (1.17) we have that

$$\sum_{\varphi \in \mathrm{IBr}(G)} \varphi(1)^2 = |G|.$$

Now,

$$|G| = \sum_{\chi \in \mathrm{Irr}(G)} \chi(1)^2 = \sum_{\chi \in \mathrm{Irr}(G)} \left(\sum_{\varphi \in \mathrm{IBr}(G)} d_{\chi\varphi}\varphi(1) \right)^2$$

$$= \sum_{\chi \in \mathrm{Irr}(G)} \sum_{\varphi, \mu \in \mathrm{IBr}(G)} d_{\chi\varphi} d_{\chi\mu}\varphi(1)\mu(1) \geq \sum_{\chi \in \mathrm{Irr}(G)} \sum_{\varphi \in \mathrm{IBr}(G)} (d_{\chi\varphi})^2 \varphi(1)^2$$

$$= \sum_{\varphi \in \mathrm{IBr}(G)} \left(\sum_{\chi \in \mathrm{Irr}(G)} (d_{\chi\varphi})^2 \right) \varphi(1)^2 \geq \sum_{\varphi \in \mathrm{IBr}(G)} \varphi(1)^2 = |G|,$$

where the last inequality follows from the fact that for a given $\varphi \in \mathrm{IBr}(G)$, there is a $\chi \in \mathrm{Irr}(G)$ with $d_{\chi\varphi} \neq 0$. Now, we conclude that $d_{\chi\varphi} d_{\chi\mu} = 0$ for $\varphi \neq \mu$ and that for every $\varphi \in \mathrm{IBr}(G)$ there exists a unique $\chi \in \mathrm{Irr}(G)$ such that $0 \neq d_{\chi\varphi}$. In fact, $d_{\chi\varphi} = 1$. This proves the theorem. ∎

Notice that if φ is a Brauer character of G and H is a p'-subgroup of G, then φ_H is an ordinary character of H. This follows from the fact that φ_H is a Brauer character of H (Lemma (2.2.d)) and Theorems (2.3) and (2.12) (applied in H). We will use this fact without further reference.

We now introduce two fundamental concepts in modular representation theory. If D is the decomposition matrix, we say that $C = D^t D$ is the **Cartan matrix**. Since D has maximum rank, we know that

$$C = (c_{\varphi\mu})_{\varphi,\mu \in \mathrm{IBr}(G)}$$

is a positive definite symmetric matrix with nonnegative integer coefficients. Notice that

$$c_{\varphi\mu} = \sum_{\chi \in \mathrm{Irr}(G)} d_{\chi\varphi} d_{\chi\mu}.$$

For $\varphi \in \mathrm{IBr}(G)$, we define the **projective indecomposable** character Φ_φ associated with φ as

$$\Phi_\varphi = \sum_{\chi \in \mathrm{Irr}(G)} d_{\chi\varphi} \chi.$$

If $\varphi \in \mathrm{IBr}(G)$, notice that

$$(\Phi_\varphi)^0 = \sum_{\chi \in \mathrm{Irr}(G)} d_{\chi\varphi} \chi^0 = \sum_{\mu \in \mathrm{IBr}(G)} c_{\varphi\mu} \mu.$$

We next prove a few properties of the projective indecomposable characters. But first, if φ and θ are complex class functions defined on G or G^0, we write

$$[\varphi, \theta]^0 = \frac{1}{|G|} \sum_{x \in G^0} \varphi(x)\overline{\theta(x)} = \overline{[\theta, \varphi]^0}.$$

We denote the space of complex class functions of G vanishing off G^0 by $\mathrm{vcf}(G)$. Also, we write $\mathrm{cl}(G^0)$ for the set of conjugacy classes of G which consist of p-regular elements.

(2.13) THEOREM. *The set $\{\Phi_\varphi \mid \varphi \in \mathrm{IBr}(G)\}$ is a basis of $\mathrm{vcf}(G)$. Also,*

$$[\theta, \Phi_\varphi]^0 = \delta_{\theta\varphi} = [\Phi_\varphi, \theta]^0$$

for $\theta, \varphi \in \mathrm{IBr}(G)$. In particular, $C^{-1} = ([\theta, \varphi]^0)_{\theta,\varphi \in \mathrm{IBr}(G)}$.

Proof. Let $y \in G$ be p-regular and let $x \in G$. By the second orthogonality relation, we have that

$$\sum_{\chi \in \mathrm{Irr}(G)} \overline{\chi(x)}\chi(y) = \delta_{\mathrm{cl}(x),\mathrm{cl}(y)}|\mathbf{C}_G(x)|,$$

where $\mathrm{cl}(x)$ is the conjugacy class in G of x. Hence, writing

$$\chi(y) = \sum_{\varphi \in \mathrm{IBr}(G)} d_{\chi\varphi}\varphi(y),$$

we have that

$$\sum_{\varphi \in \mathrm{IBr}(G)} \overline{\Phi_\varphi(x)} \varphi(y) = \delta_{\mathrm{cl}(x),\mathrm{cl}(y)} |\mathbf{C}_G(x)| \,.$$

If we read this equation for elements x of order divisible by p, it follows that

$$\sum_{\varphi \in \mathrm{IBr}(G)} \overline{\Phi_\varphi(x)} \varphi = 0 \,.$$

Since the irreducible Brauer characters are linearly independent, we conclude that $\Phi_\varphi(x) = 0$ if x is not p-regular.

We now write the above equation in a convenient matrix form. If for every conjugacy class of p-regular elements $K \in \mathrm{cl}(G^0)$, we choose $x_K \in K$ and let

$$\Phi = (\Phi_\varphi(x_K))_{\varphi \in \mathrm{IBr}(G), K \in \mathrm{cl}(G^0)}$$

and

$$Y = (\varphi(x_K))_{\varphi \in \mathrm{IBr}(G), K \in \mathrm{cl}(G^0)} \,,$$

then

$$\overline{\Phi}^t Y = \mathrm{diag}_{K \in \mathrm{cl}(G^0)} \{|\mathbf{C}_G(x_K)|\} \,. \tag{1}$$

If $E = \mathrm{diag}_{K \in \mathrm{cl}(G^0)} \{|\mathbf{C}_G(x_K)|\}$, then

$$\overline{\Phi}^t Y E^{-1} = I$$

and thus

$$I = (Y E^{-1}) \overline{\Phi}^t \,.$$

Now, if $\theta, \varphi \in \mathrm{IBr}(G)$, we have

$$[\theta, \Phi_\varphi]^0 = \sum_{K \in \mathrm{cl}(G^0)} \theta(x_K) \frac{1}{|\mathbf{C}_G(x_K)|} \Phi_\varphi(x_K^{-1}) = \delta_{\theta\varphi} \,,$$

as desired.

From this formula, it easily follows that the characters $\{\Phi_\varphi\}_{\varphi \in \mathrm{IBr}(G)}$ are linearly independent. Since it is clear that $\dim_{\mathbb{C}}(\mathrm{vcf}(G)) = |\mathrm{IBr}(G)|$, we conclude that $\{\Phi_\varphi\}_{\varphi \in \mathrm{IBr}(G)}$ is a basis of $\mathrm{vcf}(G)$. Finally,

$$(\Phi_\varphi)^0 = \sum_{\chi \in \mathrm{Irr}(G)} d_{\chi\varphi} \chi^0 = \sum_{\mu \in \mathrm{IBr}(G)} c_{\mu\varphi} \mu$$

and consequently

$$\delta_{\theta\varphi} = [\theta, \Phi_\varphi]^0 = [\theta, (\Phi_\varphi)^0]^0 = \sum_{\mu \in \mathrm{IBr}(G)} c_{\mu\varphi}[\theta, \mu]^0.$$

This proves that the matrix $\left([\theta, \varphi]^0\right)_{\theta,\varphi \in \mathrm{IBr}(G)}$ is the inverse of the Cartan matrix. ∎

If $\varphi, \mu \in \mathrm{IBr}(G)$, by Theorem (2.13), we have that

$$[\Phi_\varphi, \Phi_\mu] = [(\Phi_\varphi)^0, \Phi_\mu]^0 = \sum_{\theta \in \mathrm{IBr}(G)} c_{\varphi\theta}[\theta, \Phi_\mu]^0 = c_{\varphi\mu}.$$

We will use this fact without further reference.

(2.14) COROLLARY (Dickson). *If $\varphi \in \mathrm{IBr}(G)$, then $|G|_p$ divides $\Phi_\varphi(1)$.*

Proof. If $P \in \mathrm{Syl}_p(G)$, we have that $[(\Phi_\varphi)_P, 1_P] = \frac{\Phi_\varphi(1)}{|P|}$. Since this is an integer, the corollary follows. ∎

We will prove in Chapter 10 that $\Phi_\varphi(1)_p = |G|_p$ if the group G is p-solvable.

To get more information on the Cartan matrix C, we need to appeal to Brauer's characterization of characters. We denote by

$$\mathbb{Z}[\mathrm{Irr}(G)] = \left\{ \sum_{\chi \in \mathrm{Irr}(G)} a_\chi \chi \,|\, a_\chi \in \mathbb{Z} \right\},$$

the ring of generalized characters of G. Also,

$$\mathbb{Z}[\mathrm{IBr}(G)] = \left\{ \sum_{\varphi \in \mathrm{IBr}(G)} a_\varphi \varphi \,|\, a_\varphi \in \mathbb{Z} \right\}.$$

If $\theta \in \mathrm{cf}(G) \cup \mathrm{cf}(G^0)$ and $|G|_p = p^a$, we define two new useful functions, $\tilde{\theta} \in \mathrm{vcf}(G)$ and $\hat{\theta} \in \mathrm{cf}(G)$, as follows. We set

$$\tilde{\theta}(x) = p^a \theta(x)$$

for $x \in G^0$ and $\tilde{\theta}(x) = 0$ otherwise. We define

$$\hat{\theta}(x) = \theta(x_{p'})$$

for $x \in G$.

(2.15) LEMMA. *If $\theta \in \mathbb{Z}[\mathrm{Irr}(G)] \cup \mathbb{Z}[\mathrm{IBr}(G)]$, then $\hat{\theta}$ and $\tilde{\theta}$ are generalized characters of G.*

Proof. By Brauer's characterization of characters, it suffices to show that if $P \times Q \subseteq G$, where P is a p-group and Q is a p'-group, then $\hat{\theta}_{P\times Q}$ and $\tilde{\theta}_{P\times Q}$ are generalized characters. (Recall that p is our fixed prime.) If ρ_P is the regular character of P, notice that

$$\tilde{\theta}_{P\times Q} = \frac{|G|_p}{|P|}(\rho_P \times \theta_Q)$$

and

$$\hat{\theta}_{P\times Q} = 1_P \times \theta_Q.$$

Since Q is a p'-group, by Lemma (2.2.d) and Theorem (2.12) we have that θ_Q is a generalized character of Q, and the proof of the lemma is complete. ∎

As a consequence of the fact that $\hat{\theta}$ is a generalized character, we can prove the following two results. The fact that $\tilde{\theta}$ is a generalized character will be exploited later on.

(2.16) COROLLARY. *If $\varphi \in \mathrm{IBr}(G)$, then φ is a $\mathbb{Z}$-linear combination of $\{\chi^0 \,|\, \chi \in \mathrm{Irr}(G)\}$.*

Proof. Since $(\hat{\varphi})^0 = \varphi$, the result follows from Lemma (2.15). ∎

(2.17) COROLLARY. *If a character χ of G vanishes off G^0, then χ is a $\mathbb{Z}$-linear combination of $\{\Phi_\varphi \,|\, \varphi \in \mathrm{IBr}(G)\}$.*

Proof. By Theorem (2.13), we may write

$$\chi = \sum_{\varphi \in \mathrm{IBr}(G)} a_\varphi \Phi_\varphi$$

for some $a_\varphi \in \mathbb{C}$. Also,

$$a_\theta = [\chi, \theta]^0$$

for $\theta \in \mathrm{IBr}(G)$ by Theorem (2.13). Since

$$[\chi, \theta]^0 = [\chi, \hat{\theta}]$$

because θ vanishes off G^0, it follows that

$$a_\theta = [\chi, \hat{\theta}] \in \mathbb{Z}$$

by Lemma (2.15). ∎

As we will see in Chapter 10, it is a fact that, in p-solvable groups, a character $\chi \in \mathrm{vcf}(G)$ is a nonnegative integer combination of $\{\Phi_\varphi \,|\, \varphi \in \mathrm{IBr}(G)\}$.

Now, we give more information on the Cartan matrix which depends on the class functions $\tilde{\theta}$.

(2.18) COROLLARY. *The determinant of the Cartan matrix C is a power of p. In fact, if $x_K \in K$ for $K \in \mathrm{cl}(G^0)$, then*

$$\det(C) = \prod_{K \in \mathrm{cl}(G^0)} |\mathbf{C}_G(x_K)|_p \,.$$

Proof. If $\theta, \varphi \in \mathrm{IBr}(G)$, then

$$[\tilde{\theta}, \tilde{\varphi}] = |G|_p^2 [\theta, \varphi]^0 \,.$$

Hence, by using the fact that $([\theta, \varphi]^0)_{\theta, \varphi \in \mathrm{IBr}(G)}$ is the inverse of the Cartan matrix C and that $[\tilde{\theta}, \tilde{\varphi}] \in \mathbb{Z}$ by Lemma (2.15), we get that C times some integer matrix is $|G|_p^2 I$ (where I is the identity matrix). This proves that the determinant of C is a power of p.

Now, let

$$\Phi = (\Phi_\varphi(x_K))_{\varphi \in \mathrm{IBr}(G), K \in \mathrm{cl}(G^0)}$$

and

$$Y = (\varphi(x_K))_{\varphi \in \mathrm{IBr}(G), K \in \mathrm{cl}(G^0)}$$

be as in Theorem (2.13). In that theorem, we obtained the equation (1):

$$\overline{\Phi}^{\,t} Y = \mathrm{diag}_{K \in \mathrm{cl}(G^0)} \{|\mathbf{C}_G(x_K)|\} \,.$$

Now,

$$\Phi_\varphi = \sum_{\mu \in \mathrm{IBr}(G)} c_{\varphi\mu} \mu \,,$$

and we have that $\Phi = CY$. Thus,

$$\bar{Y}^t C Y = \mathrm{diag}_{K \in \mathrm{cl}(G^0)} \{|\mathbf{C}_G(x_K)|\} \,.$$

Since $\mathrm{IBr}(G) = \{\bar{\varphi} \mid \varphi \in \mathrm{IBr}(G)\}$, we deduce that

$$\det(\bar{Y}^t) = \det(\bar{Y}) = \pm\det(Y) \,.$$

Hence,

$$\det(Y)^2 = \pm \frac{\prod_{K \in \mathrm{cl}(G^0)} |\mathbf{C}_G(x_K)|}{\det(C)} \,.$$

Now, since $\det(Y) \in \mathbf{R}$ (recall that Y has its entries in $\mathbf{R}$), we have that

$$\det(Y)^2 \in \mathbf{R} \cap \mathbb{Q} = \mathbb{Z} \,.$$

By Lemma (2.4), we know that the set $\{\varphi^* \mid \varphi \in \mathrm{IBr}(G)\}$, where $\varphi^*(g) = \varphi(g_{p'})^*$, is the set of trace functions of all the irreducible representations of FG. By Theorem (1.19), this set is linearly independent and it follows that the matrix Y^* is invertible. Thus,

$$\det(Y)^* = \det(Y^*) \neq 0.$$

Hence,

$$(\det(Y)^2)^* = (\det(Y)^*)^2 \neq 0.$$

Therefore,

$$\det(Y)^2 \in \mathbb{Z} - p\mathbb{Z}.$$

We deduce that the integers $\prod_{K \in \mathrm{cl}(G^0)} |\mathbf{C}_G(x_K)|$ and $\det(C)$ have the same p-part, as desired.   ∎

We now discuss p'-sections of a group. If G is a group, a p'-**section** of G is the set of all elements of G whose p'-part is conjugate to some fixed $x \in G^0$. This is denoted by $S_{p'}(x)$. Hence, there are as many p'-sections as conjugacy classes of p-regular elements. The most important p'-section of G is the set G_p of the elements of G which have p-power order.

(2.19) THEOREM. *The characteristic function of a p'-section of G is an S-linear combination of $\mathrm{Irr}(G)$.*

To prove Theorem (2.19), we will need some lemmas.

(2.20) LEMMA. *The set $\{\hat{\varphi} \mid \varphi \in \mathrm{IBr}(G)\}$ is a basis of the complex space of class functions of G which are constant on p'-sections.*

Proof. By definition, each $\hat{\varphi}$ is constant on p'-sections. Since the dimension of the space of class functions of G which are constant on p'-sections is the number of p-regular classes of G, it suffices to show that $\{\hat{\varphi} \mid \varphi \in \mathrm{IBr}(G)\}$ is linearly independent. Since $(\hat{\varphi})^0 = \varphi$, this follows from the fact that the elements of $\mathrm{IBr}(G)$ are linearly independent.   ∎

(2.21) LEMMA. *If $x \in G$ and $\varphi \in \mathrm{IBr}(G)$, then*

$$\frac{\Phi_\varphi(x)}{|\mathbf{C}_G(x)|_p} \in \mathbf{R}.$$

Proof. We may assume that x is p-regular; otherwise, $\Phi_\varphi(x) = 0$. Write $H = \mathbf{C}_G(x)$. Since the character $(\Phi_\varphi)_H$ is zero off H^0, by Corollary (2.17) we have that

$$(\Phi_\varphi)_H = \sum_{\mu \in \mathrm{IBr}(H)} a_\mu \Phi_\mu$$

for some integers a_μ. Hence, we see that it is no loss if we assume that $H = G$. Let $\chi \in \mathrm{Irr}(G)$. Since $x \in \mathbf{Z}(G)$, it follows from Clifford's theorem that

$$\chi_{\langle x \rangle} = \chi(1)\lambda_\chi$$

for some $\lambda_\chi \in \mathrm{Irr}(\langle x \rangle)$. We claim that λ_χ is always the same character whenever $d_{\chi\varphi} \neq 0$. If $d_{\chi\varphi} \neq 0$, then, since $\langle x \rangle \subseteq G^0$ and $\chi^0 = \varphi + \psi$ for some Brauer character ψ, it follows that $\chi_{\langle x \rangle}$ contains the ordinary character $\varphi_{\langle x \rangle}$ (see the remark after the proof of Theorem (2.12)). Hence, λ_χ only depends on φ whenever $d_{\chi\varphi} \neq 0$, as claimed. Call this constant character λ_φ. Now,

$$\Phi_\varphi(x) = \sum_{\substack{\chi \in \mathrm{Irr}(G) \\ d_{\chi\varphi} \neq 0}} d_{\chi\varphi} \chi(x)$$

$$= \sum_{\substack{\chi \in \mathrm{Irr}(G) \\ d_{\chi\varphi} \neq 0}} d_{\chi\varphi} \chi(1)\lambda_\varphi(x) = \Phi_\varphi(1)\lambda_\varphi(x).$$

Since $|G|_p$ divides $\Phi_\varphi(1)$ by Corollary (2.14), the result follows. ∎

Proof of Theorem (2.19). If $S_{p'}(x)$ is the p'-section of $x \in G^0$ and γ is the characteristic function of $S_{p'}(x)$, by Lemma (2.20) we may write

$$\gamma = \sum_{\varphi \in \mathrm{IBr}(G)} a_\varphi \hat{\varphi}$$

for some complex numbers a_φ. Notice that if $\theta \in \mathrm{IBr}(G)$, then

$$[\gamma, \Phi_\theta]^0 = \sum_{\varphi \in \mathrm{IBr}(G)} a_\varphi [\hat{\varphi}, \Phi_\theta]^0 = \sum_{\varphi \in \mathrm{IBr}(G)} a_\varphi [\varphi, \Phi_\theta]^0 = a_\theta,$$

where the last equality follows from Theorem (2.13). Also,

$$[\gamma, \Phi_\theta]^0 = \frac{1}{|G|} \sum_{y \in G^0 \cap S_{p'}(x)} \Phi_\theta(y^{-1}) = \frac{1}{|G|} \sum_{y \in \mathrm{cl}(x)} \Phi_\theta(y^{-1}) = \frac{\Phi_\theta(x^{-1})}{|\mathbf{C}_G(x)|}.$$

Since $\mathbf{C}_G(x) = \mathbf{C}_G(x^{-1})$, we have that $a_\theta \in S$ by Lemma (2.21). Now, by Lemma (2.15), the class function $\hat{\varphi}$ is a $\mathbb{Z}$-linear combination of the elements of $\mathrm{Irr}(G)$, and this proves the theorem. ∎

Our next objective is to give some bounds for the Cartan invariants. For notational reasons, we will write $\Phi_{1_{G^0}} = \Phi_1$, $c_{\varphi 1_{G^0}} = c_{\varphi 1}$ and $d_{\chi 1_{G^0}} = d_{\chi 1}$.

(2.22) THEOREM. *If $\varphi, \mu \in \mathrm{IBr}(G)$, then*

$$c_{\varphi\mu} \leq \Phi_1(1).$$

First, we prove that the product of Brauer characters is a Brauer character.

In general, if V and W are FG-modules, we construct the **tensor product** $V \otimes W$ in the following way. Let $\{v_1, \ldots, v_n\}$ be a basis of V, let $\{w_1, \ldots, w_m\}$ be a basis of W, and let $V \otimes W$ be the F-vector space spanned by the nm symbols $v_i \otimes w_j$. If $v = \sum_{i=1}^{n} a_i v_i$ and $w = \sum_{j=1}^{m} b_j w_j$, we define

$$v \otimes w = \sum_{\substack{1 \leq i \leq n \\ 1 \leq j \leq m}} a_i b_j (v_i \otimes w_j) \in V \otimes W.$$

Now, for $g \in G$, we define $(v_i \otimes w_j)g = v_i g \otimes w_j g$ and we extend this linearly to $V \otimes W$. It is straightforward to check that

$$(v \otimes w)g = vg \otimes wg.$$

This action of G makes $V \otimes W$ an FG-module. Furthermore, it is an easy exercise to check that the FG-module $V \otimes W$ is uniquely determined (up to isomorphism) by V and W.

Before proving that products of Brauer characters are Brauer characters, we remind the reader of some basic facts on tensor products of matrices (which will be used in the proof).

Recall that if $A \in \mathrm{Mat}(n, F)$ and $B \in \mathrm{Mat}(m, F)$, then $A \otimes B \in \mathrm{Mat}(nm, F)$ is defined as

$$A \otimes B = \begin{pmatrix} a_{11}B & \cdots & a_{1n}B \\ \vdots & \ddots & \vdots \\ a_{n1}B & \cdots & a_{nn}B \end{pmatrix},$$

where $A = (a_{ij})$. It is easy to check that

$$(A \otimes B)(C \otimes D) = AC \otimes BD$$

for $C \in \mathrm{Mat}(n, F)$ and $D \in \mathrm{Mat}(m, F)$. In fact, $A \otimes B$ is regular if and only if A and B are regular. In this case,

$$(A \otimes B)^{-1} = A^{-1} \otimes B^{-1}.$$

(2.23) THEOREM. *If α and β are Brauer characters of G, then $\alpha\beta$ is a Brauer character of G.*

Proof. Suppose that V is an FG-module affording α and that W is an FG-module affording β. Write

$$v_i g = \sum_{l=1}^{n} x_{il}(g)v_l \quad \text{and} \quad w_j g = \sum_{k=1}^{m} y_{jk}(g)w_k \,,$$

so that $\mathcal{X}(g) = (x_{ij}(g))$ and $\mathcal{Y}(g) = (y_{ij}(g))$ are the representations afforded by V and W, respectively. Then the representation afforded by $V \otimes W$ (with respect to the basis $\{v_i \otimes w_j\}_{i,j}$) is $\mathcal{X} \otimes \mathcal{Y}$, where

$$(\mathcal{X} \otimes \mathcal{Y})(g) = \mathcal{X}(g) \otimes \mathcal{Y}(g)$$

is the tensor product of matrices.

Let $g \in G^0$. If $\lambda_1, \ldots, \lambda_n$ are the eigenvalues of $\mathcal{X}(g)$ and $\mu_1, \ldots, \mu_m$ are the eigenvalues of $\mathcal{Y}(g)$, we have that $\lambda_i \mu_j$ for $1 \le i \le n$ and $1 \le j \le m$ are the eigenvalues of $\mathcal{X}(g) \otimes \mathcal{Y}(g)$ by elementary linear algebra. (The reader who is not familiar with the tensor product of matrices should first convince himself that it is no loss to assume that $\mathcal{X}(g)$ and $\mathcal{Y}(g)$ are diagonal by using the remarks before the statement of this theorem and the fact that $\mathcal{X}(g)$ and $\mathcal{Y}(g)$ are similar to diagonal matrices. In this case, $\mathcal{X}(g) \otimes \mathcal{Y}(g)$ is diagonal with diagonal $\{\lambda_i \mu_j\}_{i,j}$.) Now, by applying the definition of Brauer characters, we easily check that the representation $\mathcal{X} \otimes \mathcal{Y}$ affords $\alpha\beta$. ∎

(2.24) LEMMA. *If φ is a Brauer character, then 1_{G^0} is a constituent of $\varphi\bar{\varphi}$.*

Proof. If V is an FG-module affording φ, we know by Lemma (2.2) that V^* is an FG-module affording $\bar{\varphi}$. Now, there is a unique map $\theta : V \otimes V^* \to F$ such that

$$\theta(v \otimes f) = f(v)$$

for $v \in V$ and $f \in V^*$. Notice that this is an FG-homomorphism from the FG-module $V \otimes V^*$ onto the trivial FG-module F since

$$\theta(vg \otimes fg) = (fg)(vg) = f(vgg^{-1}) = f(v) = \theta(v \otimes f) = \theta(v \otimes f)g \,.$$

Now, the trivial module is a simple constituent of $V \otimes V^*$ and this proves the lemma. ∎

If $\alpha \in \mathrm{vcf}(G)$ and $\beta \in \mathrm{cf}(G^0)$, we denote the unique class function in $\mathrm{vcf}(G)$ extending $\alpha^0\beta$ by $\alpha\beta$.

(2.25) LEMMA. *If* $\varphi, \mu \in \mathrm{IBr}(G)$, *then*

$$\Phi_\varphi \mu = \sum_{\gamma \in \mathrm{IBr}(G)} a_\gamma \Phi_\gamma \,,$$

where a_γ *is the multiplicity of* φ *in the Brauer character* $\gamma\bar{\mu}$. *In particular,* $\Phi_\varphi \mu$ *is an ordinary character of* G.

Proof. Since $\Phi_\varphi \mu \in \mathrm{vcf}(G)$, by Theorem (2.13) we may write

$$\Phi_\varphi \mu = \sum_{\gamma \in \mathrm{IBr}(G)} a_\gamma \Phi_\gamma$$

for some complex numbers a_γ. Now, again by Theorem (2.13), we have that

$$a_\gamma = [\Phi_\varphi \mu, \gamma]^0 = [\Phi_\varphi, \gamma\bar{\mu}]^0 \,.$$

If we write

$$\gamma\bar{\mu} = \sum_{\tau \in \mathrm{IBr}(G)} b_\tau \tau \,,$$

then

$$[\Phi_\varphi, \gamma\bar{\mu}]^0 = [\Phi_\varphi, \sum_{\tau \in \mathrm{IBr}(G)} b_\tau \tau]^0 = b_\varphi \,,$$

which is the multiplicity of φ in $\gamma\bar{\mu}$, as desired. ∎

(2.26) COROLLARY. *If* $\varphi \in \mathrm{IBr}(G)$, *then*

$$\frac{\Phi_1(1)}{\varphi(1)} \le \Phi_\varphi(1) \le \Phi_1(1)\varphi(1) \,.$$

Proof. Since 1 is the multiplicity of φ in φ, then Φ_1 is a constituent of $\Phi_\varphi\bar{\varphi}$ by Lemma (2.25). Also, since the multiplicity of 1_{G^0} in $\varphi\bar{\varphi}$ is nonzero (by Lemma (2.24)), it follows that Φ_φ is a constituent of $\Phi_1\varphi$, again by Lemma (2.25). Now, compute degrees. ∎

We are now ready to prove Theorem (2.22).

Proof of Theorem (2.22). Since

$$(\Phi_\mu)^0 = \sum_{\varphi \in \mathrm{IBr}(G)} c_{\varphi\mu} \varphi,$$

it follows that

$$\Phi_\mu(1) = \sum_{\varphi \in \mathrm{IBr}(G)} c_{\varphi\mu} \varphi(1).$$

First we prove that $c_{\mu\mu} \leq \Phi_1(1)$. By the above equation, we have that $\Phi_\mu(1) \geq c_{\mu\mu}\mu(1)$. Hence, by using Corollary (2.26), we get

$$\Phi_1(1) \geq \frac{\Phi_\mu(1)}{\mu(1)} \geq c_{\mu\mu}.$$

Finally, by using the Cauchy-Schwarz inequality, we have

$$(c_{\varphi\mu})^2 = \left(\sum_{\chi \in \mathrm{Irr}(G)} d_{\chi\varphi} d_{\chi\mu} \right)^2 \leq \sum_{\chi \in \mathrm{Irr}(G)} (d_{\chi\varphi})^2 \sum_{\chi \in \mathrm{Irr}(G)} (d_{\chi\mu})^2$$

$$= c_{\varphi\varphi} c_{\mu\mu} \leq \Phi_1(1)^2.$$

This proves the theorem. ∎

The projective indecomposable character Φ_1 is particulary easy to understand in groups with a p-**complement** (that is, in groups G which have a subgroup $H \subseteq G$ such that $|H| = |G|_{p'}$).

(2.27) THEOREM. *Suppose that H is a p'-subgroup of G. Then the character Φ_1 is a constituent of the character $(1_H)^G$. If H is a p-complement of G, then $\Phi_1 = (1_H)^G$.*

Proof. Since $H \subseteq G^0$, it follows that

$$\chi_H = \sum_{\varphi \in \mathrm{IBr}(G)} d_{\chi\varphi} \varphi_H$$

for $\chi \in \mathrm{Irr}(G)$. Now, since φ_H is an ordinary character of H (see the remark after the proof of Theorem (2.12)), we have that

$$[\chi, (1_H)^G] = [\chi_H, 1_H] \geq d_{\chi 1}.$$

This proves the first part of the theorem. Now, if H is a p-complement, by Corollary (2.14) we have that

$$(1_H)^G(1) = |G|_p \leq \Phi_1(1)$$

and necessarily $\Phi_1 = (1_H)^G$, as required. ∎

It is always pleasant to find group theoretic consequences for character theoretic results. Frobenius conjectured that a group G with exactly $|G|_{p'}$ elements of p'-order has a normal p-complement. (The conjecture is known to be true by using the classification of simple groups.)

(2.28) THEOREM (Brauer-Nesbitt). *Suppose that* $|G| = p^a q^b r^c$ *where* p, q, r *are primes. If* G *has exactly* $q^b r^c$ *elements of* p'-*order, then* G *has a normal* p-*complement.*

We need a lemma.

(2.29) LEMMA. *Suppose that* G *has exactly* $|G|_{p'}$ *elements of* p'-*order. If* $c_{11} \leq |G|_p$, *then* G *has a normal* p-*complement.*

Proof. Consider the complex class function $\tilde{1}_G$ which has the value $|G|_p$ on p'-elements and vanishes off G^0. We know by Lemma (2.15) that $\tilde{1}_G$ is a generalized character of G. Using the fact that $|G^0| = |G|_{p'}$, we get

$$[\tilde{1}_G, \tilde{1}_G] = \frac{|G|_p^2 |G|_{p'}}{|G|} = |G|_p.$$

Also, by Theorem (2.13), we have

$$[\Phi_1, \tilde{1}_G] = |G|_p [\Phi_1, 1_{G^0}]^0 = |G|_p.$$

Since $[\Phi_1, \Phi_1] = c_{11}$ (see the remark after the proof of Theorem (2.13)), by the Cauchy-Schwarz inequality we obtain that

$$|G|_p^2 = [\Phi_1, \tilde{1}_G]^2 \leq [\Phi_1, \Phi_1][\tilde{1}_G, \tilde{1}_G] = c_{11} |G|_p \leq |G|_p^2,$$

where the last inequality follows by hypothesis. Therefore, we have equality throughout and, by the Cauchy-Schwarz equality, we deduce that Φ_1 and $\tilde{1}_G$ are proportional. Hence,

$$\Phi_1 = \frac{\Phi_1(1)}{|G|_p} \tilde{1}_G$$

and thus, $G^0 = \ker(\Phi_1)$ is a normal subgroup of G, as desired. ∎

Proof of Theorem (2.28). Let H be a Sylow q-subgroup of G and let K be a Sylow r-subgroup of G. Let T be a complete set of representatives of the (H, K)-double cosets, so that

$$G = \bigcup_{t \in T} HtK$$

is a disjoint union. Since, by elementary group theory, we have that

$$|HtK| = \frac{|H||K|}{|H^t \cap K|} = |H||K|,$$

it follows that $|T| = |G|_p$. By Theorem (2.27), the character Φ_1 is a constituent of $(1_H)^G$. Hence, we may write $(1_H)^G = \Phi_1 + \psi$, where ψ is a character of G. Since

$$(\Phi_1)^0 = \sum_{\varphi \in \mathrm{IBr}(G)} c_{1\varphi}\varphi$$

and $K \subseteq G^0$, it follows that

$$((1_H)^G)_K = c_{11}1_K + \tau \,,$$

where τ is some ordinary character of K. Now, by using Mackey's theorem (see Problem (5.6) of [Isaacs, *Characters*]), we get

$$c_{11} \leq [((1_H)^G)_K, 1_K] = \sum_{t \in T}[(1_{H^t \cap K})^K, 1_K]$$

$$= \sum_{t \in T}[1_{H^t \cap K}, 1_{H^t \cap K}] = |T| = |G|_p \,.$$

By Lemma (2.29), the theorem is proved. ∎

Our next objective is to prove a remarkable fact on real valued Brauer characters in even characteristic. We know no character theoretic proof of it.

(2.30) THEOREM (Fong). *Suppose that $p = 2$ and let $\varphi \in \mathrm{IBr}(G)$ be real valued. If $\varphi(1)$ is odd, then $\varphi = 1_{G^0}$.*

Recall that if V is an FG-module, then V^* denotes the dual FG-module of V (defined in Lemma (2.2.c)).

(2.31) THEOREM. *Suppose that V is a simple nontrivial FG-module, where $\mathrm{char}(F) = 2$. If V is isomorphic to V^*, then $\dim_F(V)$ is even.*

Proof. Let $\alpha : V \to V^*$ be an FG-isomorphism. Then

$$\langle x, y \rangle = \alpha(x)(y)$$

defines a nonzero bilinear form on V. In fact, since the map α is an FG-homomorphism, then the bilinear form is G-invariant; that is

$$\langle xg, yg \rangle = \langle x, y \rangle$$

for all $g \in G$. We claim that every G-invariant bilinear form on V is a multiple of $\langle \cdot, \cdot \rangle$. If $B(\cdot, \cdot)$ is another G-invariant bilinear form on V, observe that the map $V \to V^*$ defined by $x \mapsto B(x, \cdot)$ is an FG-homomorphism.

Since V is simple, notice that, to within a scalar multiple, there is a unique FG-homomorphism $V \to V^*$ by Schur's lemma. This proves the claim.

Our next goal is to show that $\langle \cdot, \cdot \rangle$ is symmetric. To see this, notice that the new form

$$\ll x, y \gg = \langle y, x \rangle$$

is also G-invariant and thus

$$\langle y, x \rangle = a \langle x, y \rangle$$

for some fixed $a \in F^\times$ and every $x, y \in V$. Hence,

$$\langle x, y \rangle = a^2 \langle x, y \rangle$$

for every $x, y \in V$ and thus $a^2 = 1$. Since char$(F) = 2$, it follows that $a = 1$ and, therefore, $\langle \cdot, \cdot \rangle$ is symmetric.

Now, since α is an isomorphism, we have that $V^\perp = \{x \in V \mid \langle x, y \rangle = 0$ for all $y \in V\} = 0$. Let

$$U = \{x \in V \mid \langle x, x \rangle = 0\}.$$

If $x, y \in U$, notice that

$$\langle x + y, x + y \rangle = \langle x, x \rangle + \langle x, y \rangle + \langle y, x \rangle + \langle y, y \rangle = 2 \langle x, y \rangle = 0.$$

Hence, U is a G-invariant space of V, and since V is simple, it follows that $U = 0$ or $U = V$. If $U = V$, then the bilinear form $\langle \cdot, \cdot \rangle$ is symplectic and since $V^\perp = 0$, it is well known that $\dim_F(V)$ is even. Suppose now that $U = 0$ and assume that $\dim_F(V) > 1$. Let $x, y \in V$ be linearly independent. If $b \in F$, we consider the equation

$$\langle x + by, x + by \rangle = \langle x, x \rangle + b^2 \langle y, y \rangle.$$

Since the square map on the field F is surjective, we can choose b to make $\langle x + by, x + by \rangle = 0$. This is a contradiction which proves that $\dim_F(V) = 1$. Now, if $x \in V$ is nonzero and $g \in G$, then $xg = cx$ for some $c \in F$. Then

$$\langle x, x \rangle = \langle xg, xg \rangle = c^2 \langle x, x \rangle$$

and thus $c^2 = 1$. Hence, $c = 1$ and V is the trivial module. This proves the theorem. ∎

Proof of Theorem (2.30). If V is an FG-module affording φ, then V^* affords the irreducible Brauer character $\bar{\varphi}$ by Lemma (2.2). If $\varphi = \bar{\varphi}$, it follows that V and V^* are FG-isomorphic because different irreducible Brauer characters are afforded by nonisomorphic FG-modules. Now, apply Theorem (2.31). ∎

There is a nice application of Fong's theorem due to T. Okuyama. It has been proven by G. Michler (using the classification of simple groups) that the irreducible Brauer characters of a group G have degrees not divisible by p if and only if G has a normal Sylow p-subgroup. We will give Okuyama's proof of this result for $p = 2$.

First, we discuss the relationship between the Brauer characters of a group G and the Brauer characters of its factor groups.

If $\varphi \in \mathrm{IBr}(G)$, we know that φ uniquely determines an irreducible F-representation $\mathcal{X}$ up to similarity. Note, therefore, that the kernel of the group homomorphism $\mathcal{X} : G \to \mathrm{GL}(n, F)$ is uniquely determined by φ. We define the **kernel** of the irreducible Brauer character φ as $\ker(\varphi) = \{g \in G \,|\, \mathcal{X}(g) = I_n\}$. We say that φ is **faithful** if $\ker(\varphi) = 1$.

If $N \triangleleft G$ and $\eta \in \mathrm{IBr}(G/N)$, then it is easy to check that the class function $\varphi \in \mathrm{cf}(G^0)$ defined by $\varphi(g) = \eta(gN)$ is an irreducible Brauer character of G with $N \subseteq \ker(\varphi)$. (If $\mathcal{X} : G/N \to \mathrm{GL}(n, F)$ affords η, then the irreducible representation $\mathcal{Y} : G \to \mathrm{GL}(n, F)$ defined by $\mathcal{Y}(g) = \mathcal{X}(gN)$ affords the Brauer character φ.) Also, if $\varphi \in \mathrm{IBr}(G)$ and $N \subseteq \ker(\varphi)$, the well-defined function η on $(G/N)^0$ given by $\eta(gN) = \varphi(g_{p'})$ is an irreducible Brauer character of G/N. Usually, as happens with ordinary characters, we identify the functions φ and η and view $\mathrm{IBr}(G/N)$ as a subset of $\mathrm{IBr}(G)$.

(2.32) LEMMA. *If $\mathcal{X}$ is an irreducible F-representation of G, then*

$$\mathbf{O}_p(G) \subseteq \ker(\mathcal{X}).$$

Proof. Let V be any FG-module affording $\mathcal{X}$ and consider V as FP-module, where $P = \mathbf{O}_p(G)$. Since P has a unique p-regular class, by Corollary (2.10) we have that the only simple FP-module is the trivial one. Hence, if $W > 0$ is a simple FP-submodule of V, then $0 < W \subseteq \mathbf{C}_V(P)$. Since $\mathbf{C}_V(P)$ is G-invariant (because $P \triangleleft G$) and V is a simple FG-module, it follows that $\mathbf{C}_V(P) = V$, as required. ∎

In the proof of Okuyama's theorem, we will use a standard group theoretic fact due to R. Baer: if p is a prime and $x \in G$ is a p-element such that $\langle x, x^g \rangle$ is a p-group for every $g \in G$, then $x \in \mathbf{O}_p(G)$. Also, we remind the reader that if u and v are involutions of G, then $\langle u, v \rangle$ is a dihedral group. This easily follows from the equation

$$(uv)^u = vu = (uv)^{-1}.$$

(2.33) THEOREM (Okuyama). *Suppose that $p = 2$ and let P be a Sylow 2-subgroup of G. Then $\varphi(1)$ is odd for every $\varphi \in \mathrm{IBr}(G)$ if and only if $P \lhd G$.*

Proof. One implication can easily be done for every prime p. If $P \lhd G$, then every irreducible Brauer character φ of G has P contained in its kernel by Lemma (2.32). Hence, $\varphi \in \mathrm{Irr}(G/P)$ by Theorem (2.12). Since G/P is a p'-group and $\varphi(1)$ divides $|G/P|$, we deduce that $\varphi(1)$ is not divisible by p.

Suppose now that $\varphi(1)$ is odd for every $\varphi \in \mathrm{IBr}(G)$. We prove that $P \lhd G$ by induction on $|G|$. We may assume that 2 divides the order of G. If $N \lhd G$, we have that G/N satisfies the hypothesis of the theorem since $\mathrm{IBr}(G/N) \subseteq \mathrm{IBr}(G)$. Hence, we may assume that $\mathbf{O}_2(G) = 1$. Now, by Fong's theorem, we have that 1_{G^0} is the only real valued irreducible Brauer character of G. Next, we use Brauer's lemma on character tables (Theorem (6.32) of [Isaacs, *Characters*]) to deduce that $\{1\}$ is the unique real conjugacy class of G consisting of elements of odd order (that is, the unique conjugacy class $\mathrm{cl}(x)$ with $o(x)$ odd and $\mathrm{cl}(x) = \mathrm{cl}(x^{-1})$). (Although Brauer's lemma is stated for the ordinary character table of G, the only fact which is needed in its proof is that the character table matrix is invertible. This happens with the Brauer character table by Corollary (2.10).) Now, let t be an involution of G. By Baer's theorem, there is a $g \in G$ such that $\langle t, t^g \rangle$ is not a 2-group. By elementary group theory, we have that $\langle t, t^g \rangle$ is a dihedral group whose order is divisible by some odd prime number p. Hence, there is a p-element $1 \neq y \in \langle t, t^g \rangle$ which is G-conjugate to y^{-1}. This contradiction proves the theorem. ∎

As another example in which we use Fong's Theorem (2.30), we calculate the Brauer character table of $G = A_5$ with $p = 2$. We are about to see that this character table is uniquely determined by A_5.

Below is the ordinary character table of A_5. (See the [*Atlas*] or page 288 of [Isaacs, *Characters*].)

$|\langle x \rangle|$	1	2	3	5	5
χ_1	1	1	1	1	1
χ_2	4	0	1	-1	-1
χ_3	5	1	-1	0	0
χ_4	3	-1	0	$\frac{1+\sqrt{5}}{2}$	$\frac{1-\sqrt{5}}{2}$
χ_5	3	-1	0	$\frac{1-\sqrt{5}}{2}$	$\frac{1+\sqrt{5}}{2}$

We know that $|\text{IBr}(G)|$ is the number of p-regular classes of G. There are four such classes. Since G is simple, the only homomorphism $G \to F^\times$ is the trivial one. Therefore, there is a unique Brauer character of degree 1: the trivial Brauer character $\varphi_1 = 1_{G^0}$. Now, G has two real valued ordinary irreducible characters χ_4 and χ_5 of degree 3.

By Fong's theorem, we have that $(\chi_4)^0$ is not an irreducible Brauer character. Hence, $(\chi_4)^0$ decomposes. By degrees, necessarily, $(\chi_4)^0 = 1_{G^0} + \varphi_2$, where $\varphi_2 \in \text{IBr}(G)$. For the same reason, $(\chi_5)^0 = 1_{G^0} + \varphi_3$ for $\varphi_3 \in \text{IBr}(G)$.

Finally, A_5 has an ordinary irreducible character χ_2 of degree $4 = |G|_2$. As we will see soon (Theorem (3.18)), a character $\chi \in \text{Irr}(G)$ with $\chi(1)_p = |G|_p$ **lifts** an irreducible Brauer character of G; that is, $\chi^0 \in \text{IBr}(G)$. Hence, $(\chi_2)^0 = \varphi_4 \in \text{IBr}(G)$. (Notice that $\varphi_4 = \varphi_2\varphi_3$.)

We sumarize this in the following table.

$|\langle x \rangle|$	1	3	5	5
φ_1	1	1	1	1
φ_2	2	-1	$\frac{-1+\sqrt{5}}{2}$	$\frac{-1-\sqrt{5}}{2}$
φ_3	2	-1	$\frac{-1-\sqrt{5}}{2}$	$\frac{-1+\sqrt{5}}{2}$
φ_4	4	1	-1	-1

Perhaps this is a good place to show that although the set $\text{IBr}(G)$ is not uniquely determined by G, the decomposition matrix is (modulo a permutation of the rows and columns).

By definition, the values of the Brauer characters of G lie in $\mathbb{Q}_{|G|_{p'}} \subseteq \mathbb{Q}_{|G|}$ (we use the notation $\mathbb{Q}_n = \mathbb{Q}(e^{\frac{2\pi i}{n}})$).

It is well known by elementary character theory that $\text{Gal}(\mathbb{Q}_{|G|}/\mathbb{Q})$ (the Galois group of the field extension $\mathbb{Q}_{|G|}/\mathbb{Q}$) permutes $\text{Irr}(G)$ via

$$\chi^\sigma(g) = \chi(g)^\sigma$$

for $\sigma \in \text{Gal}(\mathbb{Q}_{|G|}/\mathbb{Q})$.

For $\theta \in \text{cf}(G) \cup \text{cf}(G^0)$ with values in $\mathbb{Q}_{|G|}$ and $\sigma \in \text{Gal}(\mathbb{Q}_{|G|}/\mathbb{Q})$, we denote by θ^σ the class function defined by

$$\theta^\sigma(g) = \theta(g)^\sigma .$$

(2.34) THEOREM. *Suppose that M_1 and M_2 are maximal ideals of* **R** *containing p**R** and write $F_1 = $ **R**$/M_1$ and $F_2 = $ **R**$/M_2$. If $\mathrm{IBr}_{F_1}(G)$ and $\mathrm{IBr}_{F_2}(G)$ are the set of irreducible Brauer characters with respect to F_1 and F_2, respectively, then there is a $\sigma \in \mathrm{Gal}(\mathbb{Q}_{|G|}/\mathbb{Q}_{|G|_p})$ such that $\varphi \mapsto \varphi^\sigma$ is a bijection from $\mathrm{IBr}_{F_1}(G)$ onto $\mathrm{IBr}_{F_2}(G)$. Moreover, if $\chi \in \mathrm{Irr}(G)$, then $d_{\chi^\sigma \varphi^\sigma} = d_{\chi\varphi}$. Therefore, the decomposition matrix is unique up to a permutation of the rows and columns.*

Proof. Since F_i is the algebraic closure of $\{M_i, 1 + M_i, \ldots, (p-1) + M_i\}$ (its prime field), by field extensions theory there is a field isomorphism $\rho : F_1 \to F_2$.

We now consider the multiplicative group isomorphisms $*_1 : \mathbf{U} \to F_1{}^\times$ and $*_2 : \mathbf{U} \to F_2{}^\times$ and let $\tau \in \mathrm{Aut}(\mathbf{U})$ be the automorphism of $\mathbf{U}$ obtained by composing first $*_1$, then ρ and afterwards the inverse of $*_2$.

Let ξ be a primitive $|G|_{p'}$th root of unity. Since $o(\xi) = o(\tau(\xi))$, it follows that $\tau(\xi) = \xi^m$, where m is an integer coprime with $|G|_{p'}$. Hence

$$\rho(\delta^{*_1}) = (\delta^{*_2})^m$$

for every $\delta \in \langle \xi \rangle$.

By Galois theory, there exists a unique $\sigma \in \mathrm{Gal}(\mathbb{Q}_{|G|}/\mathbb{Q}_{|G|_p})$ such that $\sigma(\xi) = \xi^m$. Thus, $\sigma(\delta) = \delta^m$ for every $\delta \in \langle \xi \rangle$.

We claim that if $\mathcal{X}$ is an F_1-representation of G affording the F_1-Brauer character φ, then the F_2-representation $\mathcal{X}^\rho = \mathcal{Y}$ of G obtained by applying ρ to every entry of $\mathcal{X}$ affords the F_2-Brauer character φ^σ. Let $g \in G^0$. If $\epsilon_1, \ldots, \epsilon_h$ are the roots in F_1 of $\det(xI_h - \mathcal{X}(g))$, then it is clear that $\rho(\epsilon_1), \ldots, \rho(\epsilon_h)$ are the eigenvalues in F_2 of $\mathcal{Y}(g)$ since ρ induces a ring isomorphism $F_1[x] \to F_2[x]$. Hence, if $\delta_1, \ldots, \delta_h \in \mathbf{U}$ are such that $(\delta_i)^{*_1} = \epsilon_i$, we have that

$$\varphi(g) = \delta_1 + \ldots + \delta_h.$$

But now, since $g \in G^0$, it follows that $o(g)$ divides $|G|_{p'}$. Since $o(\epsilon_i) = o(\delta_i)$ divides $o(g)$, we have that $\delta_i \in \langle \xi \rangle$. Hence, $((\delta_i)^m)^{*_2} = \rho(\epsilon_i)$. Therefore, if η is the Brauer character afforded by $\mathcal{Y}$, we have that

$$\eta(g) = (\delta_1)^m + \ldots + (\delta_h)^m = \varphi(g)^\sigma = \varphi^\sigma(g),$$

as claimed.

The first part of the proof is complete once we notice that if $\{\mathcal{X}_1, \ldots, \mathcal{X}_k\}$ is a complete set of nonsimilar irreducible F_1-representations of G, then $\{\mathcal{X}_1^\rho, \ldots, \mathcal{X}_k^\rho\}$ is a complete set of nonisomorphic F_2-representations of G. (Observe that two representations $\mathcal{X}$ and $\mathcal{Z}$ of $F_1 G$ are similar if and only if $\mathcal{X}^\rho$ and $\mathcal{Z}^\rho$ are similar. Also, $\mathcal{X}$ is similar to an upper triangular representation in block form if and only if $\mathcal{X}^\rho$ is similar to an upper triangular representation in block form. This proves the observation above.)

Finally, if $\chi \in \operatorname{Irr}(G)$, then $(\chi^0)^\sigma = (\chi^\sigma)^0$ by definition. Now,

$$\sum_{\varphi \in \operatorname{IBr}_{F_1}(G)} d_{\chi\varphi} \varphi^\sigma = \left(\sum_{\varphi \in \operatorname{IBr}_{F_1}(G)} d_{\chi\varphi} \varphi \right)^\sigma$$

$$= (\chi^0)^\sigma = (\chi^\sigma)^0 = \sum_{\varphi \in \operatorname{IBr}_{F_1}(G)} d_{\chi^\sigma \varphi^\sigma} \varphi^\sigma ,$$

and this proves that

$$d_{\chi^\sigma \varphi^\sigma} = d_{\chi\varphi} ,$$

as required. ∎

In particular, notice that rational valued irreducible Brauer characters are uniquely defined.

It is not true, in general, that $\operatorname{Gal}(\mathbb{Q}_{|G|}/\mathbb{Q})$ permutes the set of irreducible Brauer characters $\operatorname{IBr}(G)$. Let $G = J_1$ be the sporadic Janko group and set $p = 11$. Let ξ be a primitive fifth root of unity and let $\sigma \in \operatorname{Gal}(\mathbb{Q}_{|G|}/\mathbb{Q}_{|G|_{5'}})$ be the automorphism which maps ξ to ξ^2. In the notation of [*Atlas*] and [*Atlas of Brauer*], the irreducible character $\chi_2 \in \operatorname{Irr}(G)$ is such that $(\chi_2)^0 = \varphi_6 \in \operatorname{IBr}(G)$. However, if $\chi_3 = (\chi_2)^\sigma$, then $(\chi_3)^0$ is not irreducible.

There are certain elements of $\operatorname{Gal}(\mathbb{Q}_{|G|}/\mathbb{Q})$ which do permute $\operatorname{IBr}(G)$ (Problem (2.10)).

Next, we show that the degrees of the irreducible Brauer characters do not necessarily divide the order of the group.

(2.35) LEMMA. *Suppose that G is a nonabelian simple group and suppose that p is an odd prime number. If $1_{G^0} \neq \varphi \in \operatorname{IBr}(G)$, then $\varphi(1) > 2$.*

Proof. Suppose that $\varphi(1) = 2$ and let $\mathcal{X} : G \to \operatorname{GL}(2, F)$ be a representation affording φ. Let $g \in G$ be an element of order 2. Since p is odd, we have that the representation $\mathcal{X}_{(g)}$ is completely reducible. Therefore, $\mathcal{X}(g)$ is similar to a matrix

$$\begin{pmatrix} \epsilon & 0 \\ 0 & \delta \end{pmatrix}$$

where $\epsilon^2 = \delta^2 = 1$. Since G is simple, we have that $\ker(\mathcal{X}) = 1$ and $\det(\mathcal{X}(g)) = 1$. This forces $\epsilon = \delta = -1$. In this case, $\mathcal{X}(g)$ commutes with $\mathcal{X}(h)$ for all $h \in G$, and then, since $\ker(\mathcal{X}) = 1$, we have that $g \in \mathbf{Z}(G)$, a contradiction. ∎

Let G be the simple group $\operatorname{PSL}(2,7) \cong \operatorname{GL}(3,2)$ of order $168 = 2^3 \cdot 3 \cdot 7$. Set $p = 7$. The character table of G is

$\lvert\langle x\rangle\rvert$	1	2	3	4	7	7
χ_1	1	1	1	1	1	1
χ_2	3	-1	0	1	$\frac{-1+i\sqrt{7}}{2}$	$\frac{-1-i\sqrt{7}}{2}$
χ_3	3	-1	0	1	$\frac{-1-i\sqrt{7}}{2}$	$\frac{-1+i\sqrt{7}}{2}$
χ_4	6	2	0	0	-1	-1
χ_5	7	-1	1	-1	0	0
χ_6	8	0	-1	0	1	1

(See the [*Atlas*] or page 289 of [Isaacs, *Characters*].)

We have that G has four conjugacy classes of p'-elements. Put $\mathrm{IBr}(G) = \{\varphi_1 = 1_{G^0}, \varphi_2, \varphi_3, \varphi_4\}$. Since $\chi_5(1) = 7$, by Theorem (3.18) we will soon know that $(\chi_5)^0 = \varphi_4 \in \mathrm{IBr}(G)$. Now G is simple and thus φ_1 is the only Brauer character of G of degree 1. Since by Lemma (2.35) the irreducible Brauer characters of G have degree at least 3, we deduce that $(\chi_2)^0 = (\chi_3)^0 = \varphi_2 \in \mathrm{IBr}(G)$. It remains to find φ_3. Now, we use a group theoretic fact on G; namely, G has a p-complement H (in fact, H is isomorphic to $\mathrm{Sym}(4)$). (To see this consider the natural action of $G = \mathrm{GL}(3,2)$ on the three dimensional $\mathbb{Z}_2$-space V. Then G acts transitively on the nontrivial elements of V and the stabilizer of any of them is isomorphic to $\mathrm{Sym}(4)$.) Hence,

$$\Phi_1 = \sum_{\chi \in \mathrm{Irr}(G)} d_{\chi 1}\chi = (1_H)^G,$$

by Theorem (2.27). Since $(\chi_2)^0 = (\chi_3)^0 = \varphi_2 \neq 1_{G^0}$, we see that χ_2 and χ_3 are not irreducible constituents of Φ_1. Since Φ_1 contains 1_G with multiplicity 1, by computing degrees, necessarily

$$\Phi_1 = 1_G + \chi_4.$$

In particular, $d_{\chi_4 1} = 1$. If we write

$$(\chi_4)^0 = 1_{G^0} + a\varphi_2 + b\varphi_3 + c\varphi_4,$$

by degrees, we see that $c = 0$. Hence, $6 = 1 + 3a + b\varphi_3(1)$. If $a \geq 1$, then $b \neq 0$ and $\varphi_3(1) \leq 2$. This is not possible by Lemma (2.35). Therefore, $a = 0$ and $(\chi_4)^0 = 1_{G^0} + b\varphi_3$. Thus, $5 = b\varphi_3(1)$ and we conclude that $b = 1$ and $\varphi_3 = (\chi_4)^0 - 1_{G^0}$. The Brauer character table is, therefore, uniquely determined by G and can be written as follows :

$\lvert\langle x\rangle\rvert$	1	2	3	4
φ_1	1	1	1	1
φ_2	3	-1	0	1
φ_3	5	1	-1	-1
φ_4	7	-1	1	-1

Notice that $\varphi_3(1) = 5$ does not divide $|G|$.
The decomposition matrix is

$$\begin{pmatrix} 1 & 0 & 0 & 0 \\ 0 & 1 & 0 & 0 \\ 0 & 1 & 0 & 0 \\ 1 & 0 & 1 & 0 \\ 0 & 0 & 0 & 1 \\ 0 & 1 & 1 & 0 \end{pmatrix}.$$

(In this particular case, there is a deep reason why the decomposition numbers are not greater than one. This will be understood in Chapter 11.)
The Cartan matrix is

$$\begin{pmatrix} 2 & 0 & 1 & 0 \\ 0 & 3 & 1 & 0 \\ 1 & 1 & 2 & 0 \\ 0 & 0 & 0 & 1 \end{pmatrix}$$

and has determinant 7.

PROBLEMS

(2.1) Let $\sigma : G \to H$ be an isomorphism of groups. If ψ is a class function defined on G (or on G^0), we define

$$\psi^\sigma(x) = \psi(x^{\sigma^{-1}})$$

for $x \in H$ (or H^0). Prove that ψ is a Brauer character of G if and only if ψ^σ is a Brauer character of H. Also, show that $d_{\chi\varphi} = d_{\chi^\sigma \varphi^\sigma}$ for $\chi \in \mathrm{Irr}(G)$ and $\varphi \in \mathrm{IBr}(G)$.

(2.2) If H is any p'-subgroup of G, prove that $\Phi_1(1) \le |G : H|$. Deduce that $c_{\varphi\mu} \le |G : H|$ for $\varphi, \mu \in \mathrm{IBr}(G)$.

(2.3) Prove that $|G^0|$ is not divisible by p and that $|G|_{p'}$ divides $|G^0|$. Also, prove that if p divides $|G|$ and G_p is the set of p-elements of G, then $|G|_p$ divides $|G_p|$.

(HINT: If $P \in \mathrm{Syl}_p(G)$, then P acts on G^0. Use the Schur-Zassenhaus theorem in the group $\mathbf{C}_G(P)$. For the other parts, compute $[\tilde{1}_G, 1]$ and use Theorem (2.19).)

(2.4) If $H \subseteq G$ and $\mu \in \mathrm{IBr}(H)$, prove that $(\Phi_\mu)^G$ is a nonnegative integer linear combination of $\{\Phi_\varphi \mid \varphi \in \mathrm{IBr}(G)\}$.

(NOTE: This result is also true for the characters $(\Phi_\varphi)_H$. We will prove this after we discuss induction for Brauer characters.)

(2.5) Prove that if D is the decomposition matrix, then the matrix D^* in F also has rank $|\mathrm{IBr}(G)|$.

(2.6) (Isaacs) Let H_1 and H_2 be p-complements in a finite group G. Show that

$$|H_1 \cap K| = |H_2 \cap K|$$

for every conjugacy class K of G.

(2.7) A Brauer character φ is **linear** if $\varphi(1) = 1$. (Of course, Brauer linear characters are irreducible.) If N is the smallest normal subgroup of G such that G/N is an abelian p'-group (that is, $N = G'\mathbf{O}^{p'}(G)$), show that the map $\mathrm{Irr}(G/N) \to \{\varphi \in \mathrm{IBr}(G) \mid \varphi(1) = 1\}$ given by $\chi \mapsto \chi^0$ is a bijection. (In fact, this is an isomorphism of groups.) Conclude that the number of Brauer linear characters of G is $|G : G'|_{p'}$.

(2.8) If G is a group with a p-complement H and $\lambda \in \mathrm{IBr}(G)$ is linear, prove that

$$\Phi_\lambda = (\lambda_H)^G.$$

(2.9) If ρ_G is the regular character of G, prove that

$$\rho_G = \sum_{\varphi \in \mathrm{IBr}(G)} \varphi(1)\Phi_\varphi$$

and

$$(\rho_G)^0 = \sum_{\varphi \in \mathrm{IBr}(G)} \Phi_\varphi(1)\varphi.$$

Also, prove that $(\rho_G)^0$ is the Brauer character afforded by FG. Deduce that

$$|G| = \sum_{\varphi \in \mathrm{IBr}(G)} \Phi_\varphi(1)\varphi(1).$$

(2.10) Given a power p^f of p, there is a unique $\sigma \in \mathrm{Gal}(\mathbb{Q}_{|G|}/\mathbb{Q}_{|G|_{p'}})$ such that $\sigma(\xi) = \xi^{p^f}$ for every p'-root of unity ξ in $\mathbb{Q}_{|G|}$. Prove that $\varphi \in \mathrm{IBr}(G)$ if and only if $\varphi^\sigma \in \mathrm{IBr}(G)$.

(HINT: Use Frobenius automorphisms in the field F.)

(2.11) Suppose that $\varphi \in \mathrm{cf}(G^0)$. Prove that $\varphi \in \mathbb{Z}[\mathrm{IBr}(G)]$ if and only if $\varphi_H \in \mathbb{Z}[\mathrm{Irr}(H)]$ for every p'-subgroup H of G. Deduce that if $\varphi \in \mathrm{IBr}(G)$ and $\sigma \in \mathrm{Gal}(\mathbb{Q}_{|G|}/\mathbb{Q})$, then $\varphi^\sigma \in \mathbb{Z}[\mathrm{IBr}(G)]$.

(2.12) Suppose that $\mathcal{X}$ is a representation of a group G with entries in L, where L is an algebraic extension of $\mathbb{Q}$. Show that $\mathcal{X}$ is L-similar to some representation with entries in $S_L = \{\frac{\alpha}{\beta} \mid \alpha \in \mathbf{R} \cap L, \beta \in \mathbf{R} \cap L - M\}$.

(HINT: Prove that S_L is a local ring and mimic the proof of Theorem (2.7).)

(2.13) If $\lambda \in \mathrm{IBr}(G)$ is linear and $\varphi \in \mathrm{IBr}(G)$, prove that $\lambda\varphi \in \mathrm{IBr}(G)$. Also, show that

$$\lambda\Phi_\varphi = \Phi_{\lambda\varphi}.$$

(2.14) Show that the Brauer character table of A_5 for $p = 3$ is

$\|\langle x\rangle\|$	1	2	5	5
φ_1	1	1	1	1
φ_2	4	0	-1	-1
φ_3	3	-1	$\frac{1+\sqrt5}{2}$	$\frac{1-\sqrt5}{2}$
φ_4	3	-1	$\frac{1-\sqrt5}{2}$	$\frac{1+\sqrt5}{2}$

Calculate the Cartan and the decomposition matrices. Do the same for $p = 5$.

3. Blocks

If $\chi \in \mathrm{Irr}(G)$, it is well known that χ uniquely determines an algebra homomorphism $\omega_\chi : \mathbf{Z}(\mathbb{C}G) \to \mathbb{C}$. If K is a conjugacy class of G, $x_K \in K$ and $\hat{K} = \sum_{x \in K} x$, then

$$\omega_\chi(\hat{K}) = \frac{|K|\chi(x_K)}{\chi(1)} .$$

(This fact can be proved as follows. Since $\hat{K} \in \mathbf{Z}(\mathbb{C}G)$, we have by Theorem (1.20) that if $\mathcal{X}$ is a representation of G affording χ, then $\mathcal{X}(\hat{K}) = \omega_\chi(\hat{K})I$ is a scalar matrix depending only on χ (not on $\mathcal{X}$). The value $\omega_\chi(\hat{K})$ is easily determined by taking traces. This completely determines the map ω_χ since, for every ring R, the set $\{\hat{K} \mid K \in \mathrm{cl}(G)\}$ is an R-basis of $\mathbf{Z}(RG)$.)

One of the fundamental facts in character theory is that $\omega_\chi(\hat{K})$ is an algebraic integer (Theorem (3.7) of [Isaacs, *Characters*]). Using this result, we may construct a map $\lambda_\chi : \mathbf{Z}(FG) \to F$ by setting

$$\lambda_\chi(\hat{K}) = \omega_\chi(\hat{K})^* .$$

Since

$$\hat{K}\hat{L} = \sum_{M \in \mathrm{cl}(G)} |A_{KLM}|\hat{M} ,$$

where $A_{KLM} = \{(x,y) \in K \times L \mid xy = x_M\}$ and ω_χ is an algebra homomorphism, it easily follows that λ_χ is an algebra homomorphism $\mathbf{Z}(FG) \to F$.

We see that every irreducible character χ has associated an algebra homomorphism $\mathbf{Z}(FG) \to F$. The same is true for irreducible Brauer characters, as we are about to show.

We use the same argument as before. If $\varphi \in \mathrm{IBr}(G)$ and $\mathcal{X} : FG \to \mathrm{Mat}(n, F)$ is an irreducible F-representation of G affording φ, by Theorem (1.20) we know that $\mathcal{X}(\hat{K})$ is a scalar matrix for every $K \in \mathrm{cl}(G)$. This scalar only depends on φ. Hence, the equality

$$\mathcal{X}(\hat{K}) = \lambda_\varphi(\hat{K})I_n$$

defines an algebra homomorphism $\lambda_\varphi : \mathbf{Z}(FG) \to F$.

(3.1) DEFINITION. The *p*-**blocks** of G are the equivalence classes in $\mathrm{Irr}(G) \cup \mathrm{IBr}(G)$ under the relation $\chi \sim \varphi$ if $\lambda_\chi = \lambda_\varphi$ for $\chi, \varphi \in \mathrm{Irr}(G) \cup \mathrm{IBr}(G)$.

By definition, we have that two characters $\chi, \psi \in \mathrm{Irr}(G)$ lie in the same *p*-block of G if and only if

$$\left(\frac{|K|\chi(x_K)}{\chi(1)}\right)^* = \left(\frac{|K|\psi(x_K)}{\psi(1)}\right)^*$$

for all $K \in \mathrm{cl}(G)$.

We introduce some notation. If B is a *p*-block of G, we write $\mathrm{Irr}(B) = B \cap \mathrm{Irr}(G)$ and $\mathrm{IBr}(B) = \mathrm{IBr}(G) \cap B$. Also, if $\psi \in B$, we put $\lambda_B = \lambda_\psi$. It is customary to write $\mathrm{k}(B) = |\mathrm{Irr}(B)|$, $\mathrm{l}(B) = |\mathrm{IBr}(B)|$ and $\mathrm{Bl}(G)$ for the set of *p*-blocks of G.

The **principal block** of G is the unique block of G which contains 1_G.

The first thing that we prove is that the subsets $\mathrm{Irr}(B)$ of $\mathrm{Irr}(G)$ do not depend on the choice of the maximal ideal M. (We will see another proof of this fact in Theorem (3.19) below.)

(3.2) THEOREM. *If $\chi, \psi \in \mathrm{Irr}(G)$, then $\lambda_\chi = \lambda_\psi$ if and only if $\omega_\chi(\hat{K}) - \omega_\psi(\hat{K}) \in J$ for every maximal ideal J of $\mathbf{R}$ containing $p\mathbf{R}$ and every conjugacy class K of G.*

Proof. We wish to prove that if $\alpha(\hat{K}) = \omega_\chi(\hat{K}) - \omega_\psi(\hat{K}) \in M$ for every $K \in \mathrm{cl}(G)$, then $\alpha(\hat{K})$ lies in every maximal ideal of $\mathbf{R}$ containing $p\mathbf{R}$.

Let $\sigma \in \mathcal{G} = \mathrm{Gal}(\mathbb{Q}_{|G|}/\mathbb{Q})$. If ξ is a primitive $|G|$th root of unity, we know that there is an integer i coprime with $|G|$ such that $\sigma(\xi) = \xi^i$. Hence, $\eta(g)^\sigma = \eta(g^i)$ for all $\eta \in \mathrm{Irr}(G)$ and all $g \in G$. Also, $|\mathrm{cl}(g)| = |\mathrm{cl}(g^i)|$. Therefore, if $K = \mathrm{cl}(g)$ and $K^\sigma = \mathrm{cl}(g^i)$, it follows that

$$\alpha(\hat{K})^\sigma = \left(\frac{|K|\chi(x_K)}{\chi(1)} - \frac{|K|\psi(x_K)}{\psi(1)}\right)^\sigma = \alpha(\widehat{K^\sigma}) \in M$$

for all $\sigma \in \mathcal{G}$. Now,

$$\alpha(\hat{K}) = |K| \left(\frac{\chi(x_K)}{\chi(1)} - \frac{\psi(x_K)}{\psi(1)}\right) \in \mathbb{Q}_{|G|},$$

and thus the monic polynomial

$$\prod_{\sigma \in \mathcal{G}} (x - \alpha(\hat{K})^\sigma)$$

has all its coefficients (but the leading one) in $M \cap \mathbb{Q} = M \cap \mathbf{R} \cap \mathbb{Q} = M \cap \mathbb{Z} = p\mathbb{Z}$. Therefore, by putting $x = \alpha(\hat{K})$, we deduce that $\alpha(\hat{K})^m \in p\mathbf{R}$ for some integer m (the degree of the polynomial above). In particular, since maximal ideals are prime, $\alpha(\hat{K})$ lies in every maximal ideal containing $p\mathbf{R}$, as desired.

∎

(3.3) THEOREM. *If $\chi \in \mathrm{Irr}(G)$ and $\varphi \in \mathrm{IBr}(G)$ are such that $d_{\chi\varphi} \neq 0$, then $\lambda_\chi = \lambda_\varphi$. Consequently, $\mathrm{IBr}(B) = \{\varphi \in \mathrm{IBr}(G) \,|\, d_{\chi\varphi} \neq 0$ for some $\chi \in \mathrm{Irr}(B)\}$.*

Proof. By Theorem (2.7), we know that there is an S-representation $\mathcal{X}$ of G affording χ. Also, by Corollary (2.9), $\mathcal{X}^*$ is an F-representation of G which affords the Brauer character χ^0. If $K \in \mathrm{cl}(G)$, then, since $\mathcal{X}(\hat{K}) = \omega_\chi(\hat{K})I$, it follows that $\mathcal{X}^*(\hat{K}) = \lambda_\chi(\hat{K})I$. Now, we know that $\mathcal{X}^*$ is similar to some F-representation $\mathcal{Z}$ of G in block upper triangular form. Notice that

$$\mathcal{X}^*(\hat{K}) = \lambda_\chi(\hat{K})I = \mathcal{Z}(\hat{K}),$$

since $\mathcal{X}^*(\hat{K})$ is a scalar matrix. Now, $d_{\chi\varphi} \neq 0$ and it follows that some representation $\mathcal{Y}$ in the block diagonal of $\mathcal{Z}$ affords φ. Since $\mathcal{Y}(\hat{K}) = \lambda_\varphi(\hat{K})I$ and $\mathcal{Z}(\hat{K}) = \lambda_\chi(\hat{K})I$, this forces $\lambda_\varphi(\hat{K}) = \lambda_\chi(\hat{K})$, as desired.

By the first part it is clear that $\{\varphi \in \mathrm{IBr}(G) \,|\, d_{\chi\varphi} \neq 0$ for some $\chi \in \mathrm{Irr}(B)\} \subseteq \mathrm{IBr}(B)$. Now, let $\varphi \in \mathrm{IBr}(B)$. By the remark after the proof of Corollary (2.11), we know that there is a $\chi \in \mathrm{Irr}(G)$ such that $d_{\chi\varphi} \neq 0$. By the first part, $\chi \in \mathrm{Irr}(B)$ and the proof is complete. ∎

Note that, as a consequence of Theorem (3.3), if we arrange the ordinary and Brauer characters in blocks, we may write the decomposition matrix as

$$D = \begin{pmatrix} D_{B_1} & 0 & \cdots & 0 \\ 0 & D_{B_2} & \cdots & 0 \\ \vdots & \vdots & \ddots & \vdots \\ 0 & 0 & \cdots & D_{B_t} \end{pmatrix},$$

where $\mathrm{Bl}(G) = \{B_1, \ldots, B_t\}$.

Observe that, for $B \in \mathrm{Bl}(G)$, the rank of the submatrix D_B is $\mathrm{l}(B)$ since D has maximum rank. In particular, this shows that $\mathrm{l}(B) \leq \mathrm{k}(B)$. We may also write the Cartan matrix as

$$C = \begin{pmatrix} C_{B_1} & 0 & \cdots & 0 \\ 0 & C_{B_2} & \cdots & 0 \\ \vdots & \vdots & \ddots & \vdots \\ 0 & 0 & \cdots & C_{B_t} \end{pmatrix},$$

where $C_B = (D_B)^t D_B$. Therefore, $\det(C_B) > 0$.

Our next objective is to show that the matrix D_B is not of the form

$$\begin{pmatrix} * & 0 \\ 0 & * \end{pmatrix}.$$

(The same is true for the matrix C_B, as we will see in the problems.)

(3.4) DEFINITION. We say that $\chi, \psi \in \mathrm{Irr}(G)$ are **linked** if there is a $\varphi \in \mathrm{IBr}(G)$ such that $d_{\chi\varphi} \neq 0 \neq d_{\psi\varphi}$. We say that χ and ψ are **connected** if they lie in the same connected component of the graph defined by linking. This graph is called the **Brauer graph**.

(3.5) LEMMA. *If B is a p-block, then $\mathrm{Irr}(B)$ is a union of connected components of the Brauer graph.*

Proof. If two irreducible characters of G are linked (and therefore if they are connected) then, by Theorem (3.3), they lie in the same p-block. ∎

The importance of this definition lies in the fact that if $\mathcal{A}$ is a connected component (inside some p-block B) and $\mathcal{B} = \{\varphi \in \mathrm{IBr}(B) \,|\, d_{\chi\varphi} \neq 0$ for some $\chi \in \mathcal{A}\}$ then, by first arranging the ordinary characters in $\mathcal{A}$ and the Brauer characters in $\mathcal{B}$, we see that the decomposition matrix D_B has the form

$$\begin{pmatrix} * & 0 \\ 0 & * \end{pmatrix}.$$

We need to study the connections between the primitive idempotents of $\mathbf{Z}(\mathbb{C}G)$ and $\mathbf{Z}(FG)$. We remind the reader that the primitive idempotents of $\mathbf{Z}(\mathbb{C}G)$ are

$$\{e_\chi = \frac{\chi(1)}{|G|} \sum_{g \in G} \chi(g^{-1})g \,|\, \chi \in \mathrm{Irr}(G)\}.$$

(It is not difficult to prove this fact by using Theorem (1.17). The reader who wants to review the proof of this result may see Theorem (2.12) of [Isaacs, *Characters*].) Also, for $\chi, \psi \in \mathrm{Irr}(G)$, we have

$$\omega_\chi(e_\psi) = \delta_{\chi\psi}.$$

It will turn out that the primitive idempotents of $\mathbf{Z}(FG)$ are intimately connected with sums of the e_χ's. If $\mathcal{A}$ is any subset of $\mathrm{Irr}(G)$, we write

$$f_\mathcal{A} = \sum_{\chi \in \mathcal{A}} e_\chi.$$

Since $f_A \in \mathbf{Z}(\mathbb{C}G)$, we may write

$$f_A = \sum_{K \in \mathrm{cl}(G)} f_A(\hat{K})\hat{K}$$

for uniquely determined $f_A(\hat{K}) \in \mathbb{C}$. By using the expression for the e_χ's, notice that

$$f_A(\hat{K}) = \frac{1}{|G|} \sum_{\chi \in A} \chi(1)\chi(x_K^{-1}),$$

where x_K is a fixed element in K.

When A is a connected component of the Brauer graph, we have a convenient expression for these kind of sums.

(3.6) THEOREM (Osima). *Let A be a union of connected components of $\mathrm{Irr}(G)$ and let $B = \{\varphi \in \mathrm{IBr}(G) \mid d_{\chi\varphi} \neq 0 \text{ for some } \chi \in A\}$. If $x \in G$ is p-regular and $y \in G$, then*

$$\sum_{\chi \in A} \chi(x)\chi(y) = \sum_{\varphi \in B} \varphi(x)\Phi_\varphi(y).$$

Proof. We have that

$$\sum_{\chi \in A} \chi(x)\chi(y) = \sum_{\chi \in A}\left(\sum_{\varphi \in \mathrm{IBr}(G)} d_{\chi\varphi}\varphi(x)\right)\chi(y)$$

$$= \sum_{\chi \in A}\left(\sum_{\varphi \in B} d_{\chi\varphi}\varphi(x)\right)\chi(y) = \sum_{\varphi \in B}\left(\sum_{\chi \in A} d_{\chi\varphi}\chi(y)\right)\varphi(x).$$

Now, suppose that $\varphi \in B$ and let $\chi \in \mathrm{Irr}(G)$ be such that $d_{\chi\varphi} \neq 0$. Since $\varphi \in B$, there is a $\psi \in A$ such that $d_{\psi\varphi} \neq 0$. In this case, $\chi \in A$ since A is a union of connected components. Therefore,

$$\sum_{\varphi \in B}\left(\sum_{\chi \in A} d_{\chi\varphi}\chi(y)\right)\varphi(x) = \sum_{\varphi \in B}\left(\sum_{\chi \in \mathrm{Irr}(G)} d_{\chi\varphi}\chi(y)\right)\varphi(x)$$

$$= \sum_{\varphi \in B} \varphi(x)\Phi_\varphi(y),$$

as desired. ∎

The next result will be proved in a general (much deeper) form in Chapter 5. Recall that $g \in G$ is p**-singular** if p divides the order of g.

(3.7) COROLLARY (Weak Block Orthogonality). *Suppose that B is a p-block of G and let $x, y \in G$. If x is p-regular and y is p-singular, then*

$$\sum_{\chi \in \mathrm{Irr}(B)} \chi(x)\overline{\chi(y)} = 0.$$

Proof. By Lemma (3.5), we know that if B is a p-block, then $\mathrm{Irr}(B)$ is a union of connected components of the Brauer graph. By Theorem (3.6), we have that

$$\sum_{\chi \in \mathrm{Irr}(B)} \chi(x)\overline{\chi(y)} = \sum_{\chi \in \mathrm{Irr}(B)} \chi(x)\chi(y^{-1}) = \sum_{\varphi \in B} \varphi(x)\Phi_\varphi(y^{-1}).$$

Since p divides the order of y^{-1}, $\Phi_\varphi(y^{-1}) = 0$ and the theorem is proved. ∎

If B is a p-block, it has been conjectured by K. Harada that the sets $\mathrm{Irr}(B)$ are the minimal subsets X of $\mathrm{Irr}(G)$ satisfying

$$\sum_{\chi \in X} \chi(1)\chi(g) = 0$$

for all $g \in G$ with order divisible by p. The conjecture is known to be true for p-solvable groups (as we will see in Chapter 10).

If $B \in \mathrm{Bl}(G)$, we set

$$f_B = f_{\mathrm{Irr}(B)} = \sum_{\chi \in \mathrm{Irr}(B)} e_\chi = \sum_{K \in \mathrm{cl}(G)} f_B(\hat{K})\hat{K}.$$

(3.8) COROLLARY (Osima). *Suppose that $\mathcal{A}$ is a union of connected components of the Brauer graph. Then $f_B \in \mathbf{Z}(SG)$. In other words, $f_{\mathcal{A}}(\hat{K}) \in S$ for all conjugacy class K of G. Furthermore, $f_{\mathcal{A}}(\hat{K}) = 0$ if K does not consist of p-regular elements.*

Proof. If $x \in K$, we know that

$$f_{\mathcal{A}}(\hat{K}) = \frac{1}{|G|}\sum_{\chi \in \mathcal{A}} \chi(1)\chi(x^{-1}).$$

Now, by Theorem (3.6) (with the same notation), we have that

$$\frac{1}{|G|}\sum_{\chi \in \mathcal{A}} \chi(1)\chi(x^{-1}) = \frac{1}{|G|}\sum_{\varphi \in B} \varphi(1)\Phi_\varphi(x^{-1}).$$

Since Φ_φ vanishes off p-regular elements, we see that $f_{\mathcal{A}}(\hat{K})$ is zero if p divides the order of x. Now, assume that x is p-regular. We have that

$$\frac{1}{|G|}\sum_{\chi\in\mathcal{A}}\chi(1)\chi(x^{-1}) = \frac{1}{|G|}\sum_{\varphi\in B}\varphi(x^{-1})\Phi_\varphi(1).$$

Since $|G|_p$ divides $\Phi_\varphi(1)$ by Dickson's theorem, we deduce that $f_{\mathcal{A}}(\hat{K}) \in S$.

Suppose now that B is a p-block. By Lemma (3.5), we know that $\mathrm{Irr}(B)$ is a union of connected components and we deduce that f_B is an S-linear combination of $\hat{K}$ for $K \in \mathrm{cl}(G^0)$ from the first part. ∎

The next theorem gives us two characterizations of the ordinary characters in a p-block: they are the minimal subsets $\mathcal{A} \subseteq \mathrm{Irr}(G)$ for which $f_{\mathcal{A}} \in \mathbf{Z}(SG)$ and they are exactly the connected components of the Brauer graph.

The ring homomorphism $^* : S \to F$ extends to a ring homomorphism $^* : SG \to FG$ by setting

$$\left(\sum_{g\in G} s_g g\right)^* = \sum_{g\in G} s_g^* g.$$

Notice that * maps $\mathbf{Z}(SG)$ onto $\mathbf{Z}(FG)$ via

$$\left(\sum_{K\in\mathrm{cl}(G)} s_K \hat{K}\right)^* = \sum_{K\in\mathrm{cl}(G)} s_K^* \hat{K}.$$

If $z \in \mathbf{Z}(SG)$, then it is clear that

$$\lambda_\chi(z^*) = \omega_\chi(z)^*.$$

(3.9) THEOREM. If $B \in \mathrm{Bl}(G)$, then $\mathrm{Irr}(B)$ is a single connected component of the Brauer graph. Furthermore, if $\mathcal{A}$ is any subset of $\mathrm{Irr}(G)$ such that $f_{\mathcal{A}} \in \mathbf{Z}(SG)$, then $\mathcal{A}$ is a union of $\mathrm{Irr}(B)$ for some p-blocks B of G.

Proof. We already know by Lemma (3.5) that $\mathrm{Irr}(B)$ is a union of connected components of the Brauer graph. Then, by Corollary (3.8), it suffices to show the second part. We know that

$$\omega_\chi(f_{\mathcal{A}}) = 1$$

if and only if $\chi \in \mathcal{A}$, being zero otherwise. By hypothesis, we have that $f_{\mathcal{A}} \in \mathbf{Z}(SG)$. Thus $\lambda_\chi((f_{\mathcal{A}})^*) = \omega_\chi(f_{\mathcal{A}})^*$, and we see that

$$\lambda_\chi((f_{\mathcal{A}})^*) = 1$$

if and only if $\chi \in \mathcal{A}$. Since λ_χ only depends on the block of χ, it follows that $\mathcal{A}$ is a union of $\mathrm{Irr}(B)$ for some p-blocks B of G, as required. ∎

(3.10) COROLLARY. *The decomposition matrix D_B is not of the form*

$$\begin{pmatrix} * & 0 \\ 0 & * \end{pmatrix}.$$

Proof. Otherwise, $\mathrm{Irr}(B)$ would be the disjoint union of some connected components of the Brauer graph. ∎

By Corollary (3.8), we know that $f_B \in \mathbf{Z}(SG)$. We will write

$$e_B = (f_B)^* \in \mathbf{Z}(FG).$$

This is the **block idempotent** of B. Since $* : \mathbf{Z}(SG) \to \mathbf{Z}(FG)$ is a ring homomorphism, notice that

$$e_B e_{B'} = \delta_{BB'} e_B$$

for $B, B' \in \mathrm{Bl}(G)$. Also, since $1 = \sum\limits_{\chi \in \mathrm{Irr}(G)} e_\chi$, it follows that

$$1 = \sum_{B \in \mathrm{Bl}(G)} e_B.$$

(3.11) THEOREM. *Every idempotent of $\mathbf{Z}(FG)$ is a sum of some of the e_B. In particular, the set of all the different primitive idempotents of $\mathbf{Z}(FG)$ is $\{e_B \,|\, B \in \mathrm{Bl}(G)\}$. Also, $\{\lambda_B \,|\, B \in \mathrm{Bl}(G)\}$ is the set of all the algebra homomorphisms $\mathbf{Z}(FG) \to F$ and $\lambda_B(e_{B'}) = \delta_{BB'}$. Furthermore,*

$$\bigcap_{B \in \mathrm{Bl}(G)} \ker(\lambda_B) = \mathbf{J}(\mathbf{Z}(FG)).$$

Proof. Since the Jacobson radical of an algebra is the intersection of the kernels of its irreducible representations, it follows that

$$\mathbf{J}(\mathbf{Z}(FG)) \subseteq \bigcap_{B \in \mathrm{Bl}(G)} \ker(\lambda_B).$$

Now, since for every block B there is a $\varphi \in \mathrm{IBr}(G)$ such that $\lambda_B = \lambda_\varphi$ and conversely, we have that

$$\bigcap_{B \in \mathrm{Bl}(G)} \ker(\lambda_B) = \bigcap_{\varphi \in \mathrm{IBr}(G)} \ker(\lambda_\varphi) \subseteq \mathbf{Z}(FG) \cap \mathbf{J}(FG) = \mathbf{J}(\mathbf{Z}(FG)),$$

where the last equality follows from Theorem (1.8). This proves the last part of the theorem.

Now, if $\chi \in \mathrm{Irr}(B)$ and B' is a block of G, then

$$\lambda_B(e_{B'}) = \omega_\chi(f_{B'})^*$$

because $f_{B'} \in \mathbf{Z}(SG)$ by Corollary (3.8). Since $\omega_\chi(f_{B'}) = 1$ if $\chi \in \mathrm{Irr}(B')$ (and is zero otherwise), it follows that

$$\lambda_B(e_{B'}) = \delta_{BB'}.$$

(In particular, $e_B \neq 0$.)

If e is an idempotent of $\mathbf{Z}(FG)$, we have that

$$e = \sum_{B \in \mathrm{Bl}(G)} e e_B$$

because $\sum_{B \in \mathrm{Bl}(G)} e_B = 1$. To prove that e is a sum of some of the e_B, it suffices to show that if $e e_B \neq 0$, then $e e_B = e_B$. Suppose that $e e_B \neq 0$. Notice, then, that $e e_B$ is an idempotent of $\mathbf{Z}(FG)$. Also, $\lambda_{B'}(e e_B) = \lambda_{B'}(e)\lambda_{B'}(e_B) = 0$ for blocks $B' \neq B$. Since the idempotent $e e_B$ does not lie in the nilpotent ideal $\mathbf{J}(\mathbf{Z}(FG))$ (because all the powers of $e e_B$ are just $e e_B$), by the first part of the proof we deduce that $\lambda_B(e e_B) \neq 0$. Hence, $\lambda_B(e e_B)^2 = \lambda_B((e e_B)^2) = \lambda_B(e e_B)$ and thus $\lambda_B(e e_B) = 1$. Then $\lambda_{B'}(e_B(1 - e)) = 0$ for all blocks B', and it follows that

$$e_B(1 - e) \in \bigcap_{B' \in \mathrm{Bl}(G)} \ker(\lambda_{B'}) = \mathbf{J}(\mathbf{Z}(FG)),$$

a nilpotent ideal. Since $(e_B(1 - e))^2 = e_B(1 - e)$, we have that $e_B(1 - e) = 0$ and therefore $e_B = e e_B$, as desired.

Since $e_B e_{B'} = \delta_{BB'} e_B$, this easily implies that the set of all the different primitive idempotents of $\mathbf{Z}(FG)$ is $\{e_B \mid B \in \mathrm{Bl}(G)\}$.

Now, let λ be an algebra homomorphism $\mathbf{Z}(FG) \to F$. Since λ is F-linear and $\lambda(1) = 1$, it follows that $\mathbf{Z}(FG) = F \oplus \ker(\lambda)$. Hence, we see that λ determines and is uniquely determined by $\ker(\lambda)$. Suppose that $\lambda \neq \lambda_B$ for every $B \in \mathrm{Bl}(G)$. Then $\ker(\lambda_B) \neq \ker(\lambda)$ and thus we may find a $z_B \in \ker(\lambda_B)$ such that $\lambda(z_B) \neq 0$ for all $B \in \mathrm{Bl}(G)$. Now, $z = \prod_{B \in \mathrm{Bl}(G)} z_B$ is nilpotent because

$$z \in \bigcap_{B \in \mathrm{Bl}(G)} \ker(\lambda_B) = \mathbf{J}(\mathbf{Z}(FG)).$$

Hence, there is a positive integer n such that $z^n = 0$. Then $\lambda(z)^n = 0$ and this contradicts the fact that $\lambda(z) \neq 0$. ∎

Since $\sum_{B \in \mathrm{Bl}(G)} e_B = 1$ and $e_B e_{B'} = \delta_{BB'} e_B$, it easily follows that

$$FG = \bigoplus_{B \in \mathrm{Bl}(G)} e_B FG.$$

Notice that $e_B FG$ is an ideal of FG (in fact, it is an algebra with identity e_B). Since e_B is a central primitive idempotent, by Problem (1.4) the ideal $e_B FG$ cannot be decomposed as a direct sum of two ideals of FG. Most authors write $B = e_B FG$. We will also use this notation when we find it convenient.

(3.12) COROLLARY. *If B is a block of G, then $\mathbf{Z}(B) = e_B \mathbf{Z}(FG)$ is a local algebra with $\mathbf{J}(\mathbf{Z}(B)) = e_B \mathbf{J}(\mathbf{Z}(FG))$. The unique representation of $\mathbf{Z}(B)$ is $(\lambda_B)_{\mathbf{Z}(B)}$. Also, $\mathbf{Z}(B) = Fe_B \oplus \mathbf{J}(\mathbf{Z}(B))$ and*

$$\mathbf{Z}(FG) = \bigoplus_{B \in \mathrm{Bl}(G)} \mathbf{Z}(B).$$

Proof. First of all, we claim that if X is any subset of FG with $e_B X \subseteq X$, then $X \cap B = e_B X$. Since $e_B X \subseteq B = e_B FG$, by hypothesis $e_B X \subseteq X \cap B$. On the other hand, if $z \in X \cap B$, then $z = e_B z \in e_B X$ and the claim follows.

Since we have that $BB' = 0 = B'B$ for $B' \neq B$ (recall that $e_B e_{B'} = 0$ if $B \neq B'$), it follows that $\mathbf{Z}(B) \subseteq \mathbf{Z}(FG) \cap B \subseteq \mathbf{Z}(B)$. Hence, $\mathbf{Z}(B) = \mathbf{Z}(FG) \cap B = e_B \mathbf{Z}(FG)$ by our first remark. Also, since the sum $FG = \bigoplus_{B \in \mathrm{Bl}(G)} B$ is direct and every B is an ideal of FG, if $z \in FG$ and we write $z = \sum_{B \in \mathrm{Bl}(B)} z_B$ with $z_B \in B$, we see that $zx = xz$ if and only if $z_B x = x z_B$ for $x \in FG$. Hence, $z \in \mathbf{Z}(FG)$ if and only if $z_B \in \mathbf{Z}(FG) \cap B = \mathbf{Z}(B)$ for all B. Therefore,

$$\mathbf{Z}(FG) = \bigoplus_{B \in \mathrm{Bl}(G)} \mathbf{Z}(B).$$

By using the above direct sum, it follows that every algebra homomorphism $\mathbf{Z}(B) \to F$ may be extended to an algebra homomorphism $\mathbf{Z}(FG) \to F$. By Theorem (3.11), the only algebra homomorphisms of $\mathbf{Z}(FG)$ are of the form $\lambda_{B'}$ for some block B'. Since $\lambda_{B'}(e_B) = 0$, we conclude that $\mathbf{Z}(B)$ has a unique representation of degree 1, namely, the restriction of λ_B to $\mathbf{Z}(B)$. Now, by Theorem (1.20), all the representations of the commutative algebra $\mathbf{Z}(B)$ are one dimensional, and thus $\mathbf{Z}(B)$ is local. In particular,

$\mathbf{J}(\mathbf{Z}(B)) = \ker(\lambda_B) \cap \mathbf{Z}(B)$. Since $\mathbf{Z}(B) = e_B \mathbf{Z}(FG) \subseteq \ker(\lambda_{B'})$ for $B' \neq B$, we have that

$$\mathbf{J}(\mathbf{Z}(B)) = \bigcap_{B' \in \mathrm{Bl}(G)} \ker(\lambda_{B'}) \cap \mathbf{Z}(B) = \mathbf{J}(\mathbf{Z}(FG)) \cap \mathbf{Z}(B) = \mathbf{J}(\mathbf{Z}(FG)) \cap B.$$

By the first remark, $\mathbf{J}(\mathbf{Z}(FG)) \cap B = e_B \mathbf{J}(\mathbf{Z}(FG))$, as required.

Finally, if $z \in \mathbf{Z}(B)$ and we write $z = \lambda_B(z)e_B + (z - \lambda_B(z)e_B)$, we notice that $\lambda_B(z - \lambda_B(z)e_B) = 0$. Hence, $\mathbf{Z}(B) = Fe_B + \mathbf{J}(\mathbf{Z}(B))$. However, if $f \in F$ and $z = fe_B \in \mathbf{J}(\mathbf{Z}(B))$, then $0 = \lambda_B(z) = f$ and thus

$$\mathbf{Z}(B) = Fe_B \oplus \mathbf{J}(\mathbf{Z}(B)),$$

as desired. ∎

So far, we have seen a character theoretic approach to p-blocks. The following lemma relates the character theoretic approach to what some authors take as the definition of when a simple module lies in a block.

(3.13) LEMMA. *Suppose that B is a p-block of G.*

(a) *Let V be a simple $\mathbb{C}G$-module affording $\chi \in \mathrm{Irr}(G)$. Then $\chi \in B$ if and only if $V f_B = V$. Otherwise, $V f_B = 0$.*

(b) *Let M be a simple FG-module affording the Brauer character $\varphi \in \mathrm{IBr}(G)$. Then $\varphi \in B$ if and only if $M e_B = M$. Otherwise, $M e_B = 0$.*

Proof.

(a) If $\mathcal{X}$ is a $\mathbb{C}G$-representation afforded by V, we know that

$$\mathcal{X}(e_\psi) = \omega_\chi(e_\psi)I = \delta_{\chi\psi}I.$$

Hence, e_χ acts like the identity on V and e_ψ annihilates V for $\psi \neq \chi$. Since $\chi \in \mathrm{Irr}(B)$ if and only if $e_\chi f_B = e_\chi$, part (a) is proven.

(b) Since $1 = \sum_{B \in \mathrm{Bl}(G)} e_B$ and $e_B e_{B'} = \delta_{BB'} e_B$, it follows that

$$M = \bigoplus_{B \in \mathrm{Bl}(G)} M e_B.$$

Now, $e_B \in \mathbf{Z}(FG)$ and thus $M e_B$ is an FG-submodule of M. Therefore, there exists a unique block B such that $M e_B = M$ and $M e_{B'} = 0$ for $B' \neq B$. In particular, we see that e_B acts like the identity on M and $e_{B'}$ annihilates M for $B' \neq B$.

If $\mathcal{X}$ is a representation of FG corresponding to M, we know that $\mathcal{X}_{\mathbf{Z}(FG)} = \lambda_{B'}I$, where B' is the unique block to which φ belongs. Since $\mathcal{X}(e_B) = I$, it follows that $\lambda_{B'}(e_B) = 1$ and thus $B = B'$. This completes the proof of the lemma. ∎

It is natural to ask what Brauer character the FG-module $B = e_B FG$ affords.

(3.14) THEOREM. *If B is a p-block, then the FG-module $e_B FG$ affords the Brauer character*

$$\sum_{\chi \in \mathrm{Irr}(B)} \chi(1)\chi^0 \,.$$

Proof. Let θ_B be the Brauer character afforded by $e_B FG = B$. Since e_B acts like the identity on B, it follows that e_B acts like the identity on every composition factor of B. Hence, if $\varphi \in \mathrm{IBr}(G)$ is an irreducible constituent of θ_B, by Lemma (3.13.b) we see that $\varphi \in \mathrm{IBr}(B)$.

Since

$$FG = \bigoplus_{B \in \mathrm{Bl}(G)} B \,,$$

it follows that the Brauer character $\sum_{B \in \mathrm{Bl}(G)} \theta_B$ is the Brauer character afforded by the regular FG-module. We claim that FG affords the Brauer character ρ^0, where ρ is the regular character of G. To see this, consider the representation $\mathcal{X}$ afforded by the $\mathbb{C}G$-module $\mathbb{C}G$ with respect to the basis G. Then $\mathcal{X}(g)$ has its entries in S for all $g \in G$ and affords ρ. Hence, $\mathcal{X}^*$, which is the F-representation of FG with respect to the basis G, affords ρ^0 by Corollary (2.9). This proves the claim. Now,

$$\sum_{B \in \mathrm{Bl}(G)} \theta_B = \rho^0 = \sum_{B \in \mathrm{Bl}(G)} \left(\sum_{\chi \in \mathrm{Irr}(B)} \chi(1)\chi^0 \right) \,.$$

Since every irreducible constituent of $\sum_{\chi \in \mathrm{Irr}(B)} \chi(1)\chi^0$ lies in $\mathrm{IBr}(B)$ (by Theorem (3.3)), and every irreducible constituent of θ_B lies in $\mathrm{IBr}(B)$ (by the first paragraph), by the linear independence of $\mathrm{IBr}(G)$, we deduce that

$$\theta_B = \sum_{\chi \in \mathrm{Irr}(B)} \chi(1)\chi^0 \,,$$

as required. ∎

Notice that

$$\sum_{\chi \in \mathrm{Irr}(B)} \chi(1)\chi^0 = \sum_{\chi \in \mathrm{Irr}(B)} \chi(1) \left(\sum_{\varphi \in \mathrm{IBr}(B)} d_{\chi\varphi}\varphi \right) = \sum_{\varphi \in \mathrm{IBr}(B)} \Phi_\varphi(1)\varphi \,.$$

(3.15) DEFINITION. If B is a p-block of G, we define the **defect** $\mathrm{d}(B)$ of the block B as the integer satisfying

$$p^{a-\mathrm{d}(B)} = \min\left\{\chi(1)_p \mid \chi \in \mathrm{Irr}(B)\right\},$$

where $|G|_p = p^a$. If $\chi \in \mathrm{Irr}(B)$, notice that we may write

$$\chi(1)_p = p^{a-\mathrm{d}(B)+h},$$

for some $h \geq 0$. We say that h is the **height** of χ and write

$$\mathrm{height}(\chi) = h.$$

It is clear that every block has at least one irreducible character of height zero.

If we define the defect $\mathrm{d}(B)$ by using Brauer characters, we get the same answer. We need to strengthen Corollary (2.16).

(3.16) LEMMA. *If $\varphi \in \mathrm{IBr}(B)$, then φ is a $\mathbb{Z}$-linear combination of $\{\chi^0 \mid \chi \in \mathrm{Irr}(B)\}$.*

Proof. By Corollary (2.16), we may write

$$\varphi = \sum_{\chi \in \mathrm{Irr}(G)} a_\chi \chi^0$$

for some integers a_χ. Now,

$$\varphi - \sum_{\chi \in \mathrm{Irr}(B)} a_\chi \chi^0 = \sum_{\chi \in \mathrm{Irr}(G) - \mathrm{Irr}(B)} a_\chi \chi^0.$$

By Theorem (3.3), we know that the class function

$$\varphi - \sum_{\chi \in \mathrm{Irr}(B)} a_\chi \chi^0$$

only involves elements of $\mathrm{IBr}(B)$. For the same reason, the class function

$$\sum_{\chi \in \mathrm{Irr}(G) - \mathrm{Irr}(B)} a_\chi \chi^0$$

only involves elements of $\mathrm{IBr}(G) - \mathrm{IBr}(B)$. By the linear independence of $\mathrm{IBr}(G)$, we see that this class function is zero. This proves the lemma. ∎

(3.17) COROLLARY. *If* $d(B)$ *is the defect of* B, *then*

$$p^{a-d(B)} = \min \{\varphi(1)_p \mid \varphi \in \mathrm{IBr}(B)\},$$

where $|G|_p = p^a$.

Proof. For every $\chi \in \mathrm{Irr}(B)$ and $\varphi \in \mathrm{IBr}(B)$, we have

$$\chi(1) = \sum_{\varphi \in \mathrm{IBr}(B)} d_{\chi\varphi}\varphi(1).$$

By Lemma (3.16), we see that

$$\varphi(1) = \sum_{\chi \in \mathrm{Irr}(B)} a_{\chi\varphi}\chi(1)$$

for some integers $a_{\chi\varphi}$. The corollary clearly follows. ∎

If B is a block of G, $\varphi \in \mathrm{IBr}(B)$ and $|G|_p = p^a$, it follows that we may write

$$\varphi(1)_p = p^{a-d(B)+h}$$

for a uniquely determined nonnegative integer h. We say that h is the **height** of the irreducible Brauer character φ.

If $\varphi \in \mathrm{IBr}(G)$, it is not true that $\varphi(1)_p \leq |G|_p$. If $p = 2$, the McLaughlin simple group M^{c}L has an irreducible Brauer character of degree $2^9 \cdot 7$ while the order of the group M^{c}L is $2^7 \cdot 3^6 \cdot 5^3 \cdot 7 \cdot 11$ (see [*Atlas of Brauer*]). As we will see in Chapter 10, this does not happen in p-solvable groups.

The blocks of defect zero, though very interesting, are easy to describe.

(3.18) THEOREM. *Suppose that* B *is a* p-*block of* G. *Then the following conditions are equivalent:*
  (a) $k(B) = l(B)$;
  (b) *if* $\chi \in \mathrm{Irr}(B)$, *then* $\chi(x) = 0$ *for all* p-*singular* $x \in G$;
  (c) $d(B) = 0$;
  (d) *there is a* $\chi \in \mathrm{Irr}(B)$ *such that* $\chi(1)_p = |G|_p$;
  (e) $k(B) = 1$.
  In this case, if $\mathrm{Irr}(B) = \{\chi\}$, *then* $\mathrm{IBr}(B) = \{\chi^0\}$. *Consequently, if* $\chi \in \mathrm{Irr}(G)$ *is such that* $\chi(1)_p = |G|_p$, *then* χ *vanishes off* p-*regular elements, and* $\chi^0 \in \mathrm{IBr}(G)$.

Proof. If we assume (a), then we have that the decomposition matrix D_B is invertible since it has maximum rank (see the remark after the proof of Theorem (3.3)). If we write $(D_B)^{-1} = (b_{\varphi\chi})$, then for every $\chi \in \mathrm{Irr}(B)$, we have that

$$\sum_{\varphi \in \mathrm{IBr}(B)} b_{\varphi\chi} \Phi_\varphi = \sum_{\varphi \in \mathrm{IBr}(B)} b_{\varphi\chi} \left(\sum_{\psi \in \mathrm{Irr}(B)} d_{\psi\varphi} \psi \right)$$

$$= \sum_{\psi \in \mathrm{Irr}(B)} \left(\sum_{\varphi \in \mathrm{IBr}(B)} d_{\psi\varphi} b_{\varphi\chi} \right) \psi = \chi .$$

Hence, χ vanishes on p-singular elements (because the characters Φ_φ do), and we see that (a) implies (b).

If we assume (b) and compute $[\chi_P, 1_P]$ for a Sylow p-subgroup P of G and any $\chi \in \mathrm{Irr}(B)$, we find that $\frac{\chi(1)}{|P|} = [\chi_P, 1_P]$ is an integer. Hence, $\chi(1)_p = |G|_p$ and thus $\mathrm{d}(B) = 0$. This proves that (b) implies (c).

It is clear that (c) implies (d) by choosing in $\mathrm{Irr}(B)$ a character with height zero.

Suppose now that $|G|_p = \chi(1)_p$ for some $\chi \in \mathrm{Irr}(B)$. Then we have that

$$e_\chi = \frac{\chi(1)}{|G|} \sum_{g \in G} \chi(g^{-1}) g \in \mathbf{Z}(SG) .$$

By Theorem (3.9), $\mathrm{Irr}(B) = \{\chi\}$, and this shows that (d) implies (e). Since $\mathrm{k}(B) \geq \mathrm{l}(B) \geq 1$, we see that (e) implies (a), and the proof of the equivalence is completed.

Now, if $\mathrm{Irr}(B) = \{\chi\}$ and $\mathrm{IBr}(B) = \{\varphi\}$, then $\chi^0 = d_{\chi\varphi}\varphi$ and, by Lemma (3.16), $\varphi = a\chi^0$ for some integer a. By comparing degrees, we see that $a = 1$, as required.

Finally, if $\chi(1)_p = |G|_p$ for some $\chi \in \mathrm{Irr}(G)$, then the defect of the block which contains χ is necessarily 0 (by using part (d)) and the proof of the theorem is complete. ∎

An irreducible character $\chi \in \mathrm{Irr}(G)$ is said to have p-**defect zero** if $\chi(1)_p = |G|_p$.

Now, we work toward another characterization of p-blocks. If χ and $\psi \in \mathrm{Irr}(G)$, we will temporarily write $\chi \leftrightarrow \psi$ if

$$\sum_{x \in G^0} \chi(x)\psi(x^{-1}) \neq 0 .$$

(3.19) THEOREM. *The connected components of the graph in* $\mathrm{Irr}(G)$ *defined by* $\leftrightarrow$ *are exactly the sets* $\mathrm{Irr}(B)$ *for* $B \in \mathrm{Bl}(G)$.

The advantage of the present characterization of the ordinary irreducible characters in a p-block is that there is no need to choose a maximal ideal of **R**, compute in F or calculate $\mathrm{IBr}(G)$; the p-blocks are directly determined by reading the ordinary character table of the group. In particular, this gives another proof that the irreducible ordinary characters in a p-block are independent of the maximal ideal M. (In addition, from this approach, it is very easy to give a definition of π-blocks for a set of primes π.)

If $\chi, \psi \in \mathrm{Irr}(G)$, notice that

$$\frac{1}{|G|} \sum_{x \in G^0} \chi(x)\psi(x^{-1}) = [\chi, \psi]^0 = [\chi^0, \psi^0]^0$$

$$= [\sum_{\varphi \in \mathrm{IBr}(G)} d_{\chi\varphi}\varphi, \sum_{\mu \in \mathrm{IBr}(G)} d_{\psi\mu}\mu]^0 = \sum_{\varphi, \mu \in \mathrm{IBr}(G)} [\varphi, \mu]^0 d_{\chi\varphi} d_{\psi\mu} .$$

Since $\Gamma = ([\varphi, \mu]^0)_{\varphi, \mu \in \mathrm{IBr}(G)}$ is the inverse of the Cartan matrix (by Theorem (2.13)), it follows that Γ has a block form

$$\Gamma = \begin{pmatrix} \Gamma_{B_1} & \cdots & 0 \\ \vdots & \ddots & \vdots \\ 0 & \cdots & \Gamma_{B_t} \end{pmatrix},$$

where $\mathrm{Bl}(G) = \{B_1, \ldots, B_t\}$. Now, suppose that $\chi \leftrightarrow \psi$ (which happens if and only if $[\chi, \psi]^0 \neq 0$). Then there exist $\varphi, \mu \in \mathrm{IBr}(G)$ such that

$$[\varphi, \mu]^0 d_{\chi\varphi} d_{\chi\mu} \neq 0 .$$

Since $[\varphi, \mu]^0 \neq 0$, we deduce that φ and μ lie in the same block of G. Also, $d_{\chi\varphi} d_{\chi\mu} \neq 0$ and we conclude that χ and ψ lie in the same block of G.

(3.20) LEMMA. *If* $\chi \in \mathrm{Irr}(B)$, *then the class function* $\tilde{\chi}$ *is a* $\mathbb{Z}$-*linear combination of* $\mathrm{Irr}(B)$.

Proof. By Lemma (2.15), we know that $\tilde{\chi}$ is a generalized character of G. Since $[\tilde{\chi}, \psi] = |G|_p [\chi, \psi]^0$, the proof of the lemma follows from the above discussion. ∎

To prove Theorem (3.19), we will show something remarkable: if $\chi \in \mathrm{Irr}(B)$ has height zero, then the generalized character $\tilde{\chi}$ exactly involves all the irreducible characters in the block B. If this is the case, since $[\tilde{\chi}, \psi] = |G|_p [\chi, \psi]^0$, we will have that $\chi \leftrightarrow \psi$ for all $\psi \in \mathrm{Irr}(B)$.

We find it convenient to introduce a "valuation" function. If $0 \neq n \in \mathbb{Z}$ and $|n|_p = p^a$, we write

$$\nu(n) = a \,.$$

Clearly, $\nu(nm) = \nu(n) + \nu(m)$, and thus, if $\frac{n}{m} = \frac{r}{s}$ then

$$\nu(n) + \nu(s) = \nu(ns) = \nu(mr) = \nu(m) + \nu(r) \,.$$

If we set

$$\nu(\frac{r}{s}) = \nu(r) - \nu(s) \,,$$

we see that the map $\nu : \mathbb{Q}^\times \to \mathbb{Z}$ is well defined. Moreover,

$$\nu(q_1 q_2) = \nu(q_1) + \nu(q_2)$$

for $q_1, q_2 \in \mathbb{Q}^\times$. We write $\nu(0) = \infty$.

(3.21) LEMMA. $S \cap \mathbb{Q} = \{q \in \mathbb{Q} \mid \nu(q) \geq 0\}$, $\mathcal{P} \cap \mathbb{Q} = \{q \in \mathbb{Q} \mid \nu(q) > 0\}$ and $\{q \in \mathbb{Q} \mid \nu(q) = 0\}$ is the set of units of the ring $S \cap \mathbb{Q}$.

Proof. Let $q \in \mathbb{Q}^\times$ with $\nu(q) \geq 0$. If we write $q = \frac{u}{v}$, where $u, v \in \mathbb{Z}$ are coprime, then $\nu(q) = \nu(u) - \nu(v) \geq 0$ and thus $\nu(u) \geq \nu(v)$. Also, since $(u, v) = 1$, it follows that p does not divide v. In other words, $\nu(v) = 0$. Thus, $q \in S \cap \mathbb{Q}$. Also, since $\nu(q) = \nu(u)$, we have that

$$\{q \in \mathbb{Q} \mid \nu(q) \geq 0\} \subseteq S \cap \mathbb{Q}$$

and

$$\{q \in \mathbb{Q} \mid \nu(q) > 0\} \subseteq \mathcal{P} \cap \mathbb{Q} \,.$$

Moreover, if $\nu(q) = 0$, then p does not divide u and therefore q is a unit of the ring $S \cap \mathbb{Q}$.

Now, suppose that $q = \frac{u}{v} \in S \cap \mathbb{Q}$, where again $u, v \in \mathbb{Z}$ are coprime, and write

$$\frac{u}{v} = \frac{\alpha}{\beta}$$

for $\alpha \in \mathbf{R}$ and $\beta \in \mathbf{R} - M$. Suppose that p divides v. Then

$$u\beta = \alpha v \in p\mathbf{R} \,.$$

Now, since p does not divide u, we may find integers a, b with $au + bp = 1$. Then

$$\beta = au\beta + bp\beta \in p\mathbf{R} \subseteq M$$

and this is a contradiction. This proves that p does not divide v and therefore $S \cap \mathbb{Q} = \{q \in \mathbb{Q} \mid \nu(q) \geq 0\}$. Since $\nu(q^{-1}) = -\nu(q)$, we see that the units of the ring $S \cap \mathbb{Q}$ are exactly the set $\{q \in \mathbb{Q} \mid \nu(q) = 0\}$. Since no element of the ideal $\mathcal{P}$ can have an inverse in S, we have that $\{q \in \mathbb{Q} \mid \nu(q) > 0\} = \mathcal{P} \cap \mathbb{Q}$, as desired. ∎

(3.22) LEMMA. Let $\eta \in \mathrm{cf}(G)$ with values in $p^a \mathbf{R}$, where $a = \nu(|G|)$.
(a) If $\chi \in \mathrm{Irr}(G)$, then

$$\frac{[\chi, \eta]}{\chi(1)} \in S.$$

Hence,

$$\frac{[\chi, \tilde{\psi}]}{\chi(1)} \in S \cap \mathbb{Q}$$

for $\chi, \psi \in \mathrm{Irr}(G)$.
(b) If $B \in \mathrm{Bl}(G)$ and $\chi, \psi \in \mathrm{Irr}(B)$, then

$$\frac{[\chi, \eta]}{\chi(1)} \equiv \frac{[\psi, \eta]}{\psi(1)} \bmod \mathcal{P}.$$

Proof. Write $\eta = p^a \theta$, where $\theta \in \mathrm{cf}(G)$ has its values in $\mathbf{R}$, and let $\chi \in \mathrm{Irr}(G)$. If we choose x_K in every conjugacy class K of G, we obtain that

$$\frac{[\chi, \eta]}{\chi(1)} = \frac{1}{|G|_{p'}} \sum_{K \in \mathrm{cl}(G)} \omega_\chi(\hat{K}) \overline{\theta(x_K)} \in S.$$

Now, if $\psi \in \mathrm{Irr}(G)$, we have that $\tilde{\psi}$ is a generalized character (Lemma (2.15)) with values in $p^a \mathbf{R}$. By the foregoing, we deduce that

$$\frac{[\chi, \tilde{\psi}]}{\chi(1)} \in S \cap \mathbb{Q},$$

proving (a).

Suppose now that B is a block of G with $\chi, \psi \in \mathrm{Irr}(B)$. Since

$$\omega_\chi(\hat{K}) - \omega_\psi(\hat{K}) \equiv 0 \bmod M$$

(by the definition of blocks) and $\frac{\overline{\theta(x_K)}}{|G|_{p'}} \in S$, it follows that

$$\frac{\overline{\theta(x_K)}}{|G|_{p'}} \left(\omega_\chi(\hat{K}) - \omega_\psi(\hat{K}) \right) \equiv 0 \bmod \mathcal{P}$$

for all $K \in \mathrm{cl}(G)$. Hence,

$$\frac{[\chi, \eta]}{\chi(1)} = \frac{1}{|G|_{p'}} \sum_{K \in \mathrm{cl}(G)} \omega_\chi(\hat{K}) \overline{\theta(x_K)}$$

$$\equiv \frac{1}{|G|_{p'}} \sum_{K \in \mathrm{cl}(G)} \omega_\psi(\hat{K}) \overline{\theta(x_K)} \equiv \frac{[\psi, \eta]}{\psi(1)} \bmod \mathcal{P} ,$$

as required. ∎

It is not important to remember that we are writing $[\chi, \eta]$ (and not $[\eta, \chi]$) in Lemma (3.22). This is, of course, clear if $[\chi, \eta]$ is a real number. By using the fact that

$$\overline{\omega_\chi(\hat{K})} = \omega_\chi(\hat{L}),$$

where $L = \mathrm{cl}(x_K^{-1})$, the reader can easily check this assertion.

(3.23) LEMMA. *If $\chi \in \mathrm{Irr}(G)$ and $b = \nu(\chi(1))$, then $\frac{1}{p^b}\tilde{\chi}$ is a generalized character, while $\frac{1}{p^{b+1}}\tilde{\chi}$ is not.*

Proof. Write

$$\tilde{\chi} = \sum_{\xi \in \mathrm{Irr}(G)} [\tilde{\chi}, \xi]\xi,$$

where $[\tilde{\chi}, \xi] \in \mathbb{Z}$ by Lemma (2.15). By the definition of the functions $\tilde{\chi}$, we have that

$$[\tilde{\chi}, \xi] = [\chi, \tilde{\xi}].$$

Write $\nu(|G|) = a$. By Lemma (3.22.a), it follows that $\frac{[\chi, \tilde{\xi}]}{\chi(1)} \in S \cap \mathbb{Q}$. Then, by Lemma (3.21), we see that $\nu([\tilde{\chi}, \xi]) \geq \nu(\chi(1)) = b$. Hence,

$$\frac{1}{p^b}\tilde{\chi} = \sum_{\xi \in \mathrm{Irr}(G)} \frac{[\tilde{\chi}, \xi]}{p^b}\xi$$

is a generalized character. Now, assume that $\frac{1}{p^{b+1}}\tilde{\chi}$ is a generalized character and let P be a Sylow p-subgroup of G. Then

$$[\frac{1}{p^{b+1}}(\tilde{\chi})_P, 1_P] = \frac{\chi(1)}{p^{b+1}}$$

is not an integer and this is a contradiction. ∎

(3.24) THEOREM. *Let B be a block and let $\chi \in \mathrm{Irr}(B)$. Set $a = \nu(|G|)$. Then the following conditions are equivalent:*
    (a) $\nu(\chi(1)) = a - \mathrm{d}(B)$ (that is, χ has height zero),
    (b) there exists a $\xi \in \mathrm{Irr}(B)$ such that $\nu([\tilde{\chi}, \xi]) = \nu(\xi(1))$ and
    (c) $\nu([\tilde{\chi}, \xi]) = \nu(\xi(1))$ for every $\xi \in \mathrm{Irr}(B)$.

Proof. Write $d = \mathrm{d}(B)$. By Lemma (3.20), we already know that $[\tilde{\chi}, \xi] = 0$ if $\xi \in \mathrm{Irr}(G)$ does not lie in B. Therefore, we may write

$$\tilde{\chi} = \sum_{\xi \in \mathrm{Irr}(B)} [\tilde{\chi}, \xi] \xi \,.$$

By Lemma (3.22.a) and Lemma (3.21) on "valuations", we have that

$$\nu([\tilde{\chi}, \xi]) \geq \nu(\xi(1)) \geq a - d$$

for every $\xi \in \mathrm{Irr}(B)$. In particular,

$$\frac{1}{p^{a-d}} \tilde{\chi} = \sum_{\xi \in \mathrm{Irr}(B)} \frac{[\tilde{\chi}, \xi]}{p^{a-d}} \xi$$

is a generalized character. (Of course, this latter fact also follows from Lemma (3.23).)

Suppose first that $\nu(\chi(1)) = a - d$. By Lemma (3.23), we have that

$$\frac{1}{p^{a-d+1}} \tilde{\chi} = \sum_{\xi \in \mathrm{Irr}(B)} \frac{[\tilde{\chi}, \xi]}{p^{a-d+1}} \xi$$

is not a generalized character. Hence, there is a $\xi \in \mathrm{Irr}(B)$ such that

$$\nu([\tilde{\chi}, \xi]) = a - d \,.$$

Since $\nu([\tilde{\chi}, \xi]) \geq \nu(\xi(1)) \geq a - d$, it follows that $\nu([\tilde{\chi}, \xi]) = \nu(\xi(1))$. This shows that (a) implies (b).

Suppose now that (b) holds. If $\psi \in \mathrm{Irr}(B)$, by Lemma (3.22.b) we have that

$$\frac{[\tilde{\chi}, \xi]}{\xi(1)} \equiv \frac{[\tilde{\chi}, \psi]}{\psi(1)} \bmod \mathcal{P} \,.$$

Since $\frac{[\tilde{\chi}, \xi]}{\xi(1)}$ is a unit of S, it follows that $\frac{[\tilde{\chi}, \psi]}{\psi(1)} \notin \mathcal{P}$. Hence, $\nu([\tilde{\chi}, \psi]) = \nu(\psi(1))$ (by Lemma (3.21) on "valuations" and using the fact that $\nu([\tilde{\chi}, \psi]) \geq \nu(\psi(1))$). This proves that (b) implies (c).

Finally, suppose that for every $\xi \in \mathrm{Irr}(B)$ we have that $\nu([\tilde{\chi}, \xi]) = \nu(\xi(1))$. Since

$$\nu([\tilde{\chi}, \xi]) = \nu([\chi, \tilde{\xi}]) \geq \nu(\chi(1)) \,,$$

(where the last inequality follows from Lemma (3.22.a)), it follows that

$$\nu(\xi(1)) \geq \nu(\chi(1))$$

for all $\xi \in \mathrm{Irr}(B)$. If we choose $\xi \in \mathrm{Irr}(B)$ of height zero, the proof is complete. ∎

The following includes Theorem (3.19).

(3.25) COROLLARY. *Let B be a block and let $\chi \in \mathrm{Irr}(B)$ be of height zero. If $\psi \in \mathrm{Irr}(B)$, then $\psi \leftrightarrow \chi$.*

Proof. Since $\chi \leftrightarrow \psi$ if and only if $[\tilde{\chi}, \psi] \neq 0$, the corollary follows from Theorem (3.24). ∎

Now, we can give further information about the Cartan matrices associated to blocks.

(3.26) THEOREM. *If C_B is the Cartan matrix of the block B, then $V = p^{\mathrm{d}(B)}(C_B)^{-1}$ is a matrix with integer coefficients. Furthermore, the matrix V^* in F has rank one. As a consequence, the greatest elementary divisor of C_B is $p^{\mathrm{d}(B)}$ and it appears only once.*

Proof. Write $a = \nu(|G|)$ and $d = \mathrm{d}(B)$. If $\varphi \in \mathrm{IBr}(B)$, by Lemma (3.16) we may write $\varphi = \sum_{\chi \in \mathrm{Irr}(B)} l_{\varphi\chi} \chi^0$ for some integers $l_{\varphi\chi}$. Since $\tilde{}$ is a linear map, notice that

$$\tilde{\varphi} = \sum_{\chi \in \mathrm{Irr}(B)} l_{\varphi\chi} \widetilde{\chi^0}.$$

Now, $\widetilde{\chi^0} = \tilde{\chi}$ and we see that $\tilde{\varphi}$ is an integer linear combination of $\{\tilde{\chi} \mid \chi \in \mathrm{Irr}(B)\}$. By Lemma (3.23), each $\frac{1}{p^{a-d}}\tilde{\chi}$ is a generalized character (because $a - d \leq \nu(\chi(1))$). Thus,

$$\frac{1}{p^{a-d}}\tilde{\varphi} = \sum_{\chi \in \mathrm{Irr}(B)} l_{\varphi\chi} \frac{1}{p^{a-d}}\tilde{\chi}$$

$$= \sum_{\chi \in \mathrm{Irr}(B)} l_{\varphi\chi} \left(\sum_{\psi \in \mathrm{Irr}(B)} \frac{[\tilde{\chi}, \psi]}{p^{a-d}}\psi \right) = \sum_{\psi \in \mathrm{Irr}(B)} \left(\sum_{\chi \in \mathrm{Irr}(B)} l_{\varphi\chi} \frac{[\tilde{\chi}, \psi]}{p^{a-d}} \right) \psi$$

is a generalized character. Write

$$a_{\varphi\psi} = \sum_{\chi \in \mathrm{Irr}(B)} l_{\varphi\chi} \frac{[\tilde{\chi}, \psi]}{p^{a-d}} \in \mathbb{Z}$$

and define the matrices

$$A = (a_{\varphi\chi})_{\varphi \in \mathrm{IBr}(B), \chi \in \mathrm{Irr}(B)}$$

and

$$L = (l_{\varphi\chi})_{\varphi \in \mathrm{IBr}(B), \chi \in \mathrm{Irr}(B)} .$$

Notice that $A = LU$, where U is the matrix with integer coefficients defined by

$$U = (\frac{[\tilde{\chi}, \psi]}{p^{a-d}})_{\chi, \psi \in \mathrm{Irr}(B)} .$$

Next, we prove that

$$V = p^d(C_B)^{-1} = AL^t.$$

In particular, this will show the first part of the theorem. Now, if $\varphi, \theta \in \mathrm{IBr}(B)$, then

$$p^d[\varphi, \theta]^0 = \frac{1}{p^{a-d}}[\tilde{\varphi}, \theta]^0 = \frac{1}{p^{a-d}}[\tilde{\varphi}, \sum_{\chi \in \mathrm{Irr}(B)} l_{\theta\chi}\chi^0]^0$$

$$= \sum_{\chi \in \mathrm{Irr}(B)} l_{\theta\chi}[\frac{1}{p^{a-d}}\tilde{\varphi}, \chi^0]^0 = \sum_{\chi \in \mathrm{Irr}(B)} l_{\theta\chi}[\frac{1}{p^{a-d}}\tilde{\varphi}, \chi] = \sum_{\chi \in \mathrm{Irr}(B)} a_{\varphi\chi}l_{\theta\chi},$$

as desired.

Our next objective is to show that the rank of V^* is 1. First, we will prove that the rank of U^* is 1. Suppose that $(\frac{[\tilde{\chi},\psi]}{p^{a-d}})^* \neq 0$ for some $\chi, \psi \in \mathrm{Irr}(B)$. We claim that χ and ψ have height zero. By assumption, we have that $\nu([\tilde{\chi},\psi]) = a - d$. Since $\nu([\tilde{\chi},\psi]) \geq \nu(\psi(1)) \geq a - d$ by Lemma (3.22.a), it follows that $\nu([\tilde{\chi},\psi]) = \nu(\psi(1))$ and that ψ has height zero. Thus, χ has height zero by Theorem (3.24). This proves the claim. By Lemma (3.22.b), note that

$$(\frac{[\tilde{\chi},\psi]}{\psi(1)})^* = (\frac{[\tilde{\chi},\xi]}{\xi(1)})^*$$

for $\chi \in \mathrm{Irr}(G)$ and $\xi, \psi \in \mathrm{Irr}(B)$. Now, if $\mathrm{Irr}_0(B)$ is the set of height zero characters in $\mathrm{Irr}(B)$, we have that

$$\mathrm{rank}(U^*) = \mathrm{rank}\left((\frac{[\tilde{\chi},\psi]}{p^{a-d}})^*\right)_{\chi,\psi \in \mathrm{Irr}_0(B)}$$

$$= \mathrm{rank}\left((\frac{1}{\psi(1)_{p'}})^*(\frac{[\tilde{\chi},\psi]}{p^{a-d}})^*\right)_{\chi,\psi \in \mathrm{Irr}_0(B)} = \mathrm{rank}\left((\frac{[\tilde{\chi},\psi]}{\psi(1)})^*\right)_{\chi,\psi \in \mathrm{Irr}_0(B)} = 1,$$

by Theorem (3.24).

Now, since $V = AL^t = LUL^t$, then $V^* = L^*U^*L^{*t}$ and by elementary linear algebra, it follows that $\mathrm{rank}(V^*) \leq \mathrm{rank}(U^*) = 1$. Suppose that $V^* = 0$. Then p divides the integers $p^d[\varphi, \theta]^0$ for $\varphi, \theta \in \mathrm{IBr}(B)$. Now, let $\varphi \in \mathrm{IBr}(B)$ be of height zero. Since $\tilde{\varphi}$ vanishes off p-regular elements, notice that we may write $\tilde{\varphi} = \sum_{\theta \in \mathrm{IBr}(B)} p^a[\varphi, \theta]^0\Phi_\theta$, by applying Theorem (2.13) and the fact that $[\varphi, \theta]^0 = 0$ for $\theta \in \mathrm{IBr}(G) - B$. We deduce that

$$\frac{1}{p^{a-d+1}}\tilde{\varphi} = \sum_{\theta \in \mathrm{IBr}(B)} \frac{p^d[\varphi, \theta]^0}{p}\Phi_\theta.$$

By Dickson's theorem, we conclude that p^a divides $\frac{\tilde{\varphi}(1)}{p^{a-d+1}} = p^{a-1}$. This is a contradiction which proves that $V^* \neq 0$. Hence, $\mathrm{rank}(V^*) = 1$.

To complete the proof of the theorem, let $Q, P \in \mathrm{GL}(l, \mathbb{Z})$ be such that

$$Q C_B P = \mathrm{diag}(d_1, \ldots, d_l),$$

where $l = \mathrm{l}(B)$ and d_i divides d_j for $i \leq j$. Then

$$\mathrm{diag}(\frac{p^d}{d_1}, \ldots, \frac{p^d}{d_l}) = P^{-1} V Q^{-1} \in \mathrm{Mat}(l, \mathbb{Z}).$$

Since $\mathrm{rank}((P^{-1}VQ^{-1})^*) = \mathrm{rank}(((P^{-1})^*)V^*(Q^{-1})^*) = \mathrm{rank}(V^*) = 1$ (by using the fact that $P^*, Q^* \in \mathrm{GL}(l, F)$), it follows that $d_l = p^d$ and $d_i < p^d$ for $i < l$. ∎

If B is a p-block, it was conjectured by Brauer that $\mathrm{k}(B) \leq p^{\mathrm{d}(B)}$. (This is known as **Brauer's k(B)-conjecture**.) This very deep problem has been recently proved by G. Robinson and J. Thompson for p-solvable G and large p. (See [Robinson-Thompson].) The next result gives a bound for $\mathrm{k}(B)$ in terms of $\mathrm{d}(B)$.

(3.27) THEOREM (Brauer-Feit). *If B is a p-block of defect d, then $\mathrm{k}(B) \leq \frac{1}{4}p^{2d} + 1$.*

Proof. Let $\chi \in \mathrm{Irr}(B)$ be of height zero. Then we have that $b = a - d$, where $a = \nu(|G|)$ and $b = \nu(\chi(1))$.

By Lemma (3.23) and Corollary (3.25), we may write

$$\frac{1}{p^{a-d}}\tilde{\chi} = \sum_{\psi \in \mathrm{Irr}(B)} c_\psi \psi$$

for some integers $c_\psi \neq 0$. Now,

$$[\frac{1}{p^{a-d}}\tilde{\chi}, \frac{1}{p^{a-d}}\tilde{\chi}] = \sum_{\psi \in \mathrm{Irr}(B)} c_\psi^{\,2}.$$

If $c = c_\chi$, we have that

$$c^2 + (\mathrm{k}(B) - 1) \leq \sum_{\psi \in \mathrm{Irr}(B)} c_\psi^{\,2} = \frac{p^{2d}}{p^{2a}}[\tilde{\chi}, \tilde{\chi}] = \frac{p^{2d}}{p^a}[\tilde{\chi}, \chi] = \frac{p^{2d}}{p^a}p^{a-d}c = p^d c.$$

Hence, $\mathrm{k}(B) - 1 \leq p^d c - c^2$. Since the function $p^d c - c^2$ in c takes its maximum value in $c = \frac{1}{2}p^d$, it follows that

$$\mathrm{k}(B) \leq \frac{1}{2}p^{2d} - \frac{1}{4}p^{2d} + 1,$$

as required. ∎

Our next objective in this chapter is to prove the following theorem of Brauer. The treatment we use is essentially due to M. Broué.

(3.28) THEOREM (Brauer). *If B is a p-block of defect d and $|G|_p = p^a$, then*

$$\nu\Big(\sum_{\chi \in \mathrm{Irr}(B)} \chi(1)^2 \Big) = 2a - d.$$

We already know that

$$\sum_{\chi \in \mathrm{Irr}(B)} \chi(1)^2 = \sum_{\varphi \in \mathrm{IBr}(B)} \varphi(1) \Phi_\varphi(1).$$

(See the remark after the proof of Theorem (3.14).) Since p^a divides $\Phi_\varphi(1)$ (by Dickson's theorem) and p^{a-d} divides $\varphi(1)$ for all $\varphi \in \mathrm{IBr}(B)$ (by Corollary (3.17)), we see that p^{2a-d} divides $\sum_{\chi \in \mathrm{Irr}(B)} \chi(1)^2$. The difficult part is to show that p^{2a-d} is the exact power of p dividing this number.

We need some preliminary results.

If θ is a class function of G with values in a ring R, we usually extend θ to a function $\theta : RG \to R$ by setting

$$\theta\Big(\sum_{g \in G} r_g g \Big) = \sum_{g \in G} r_g \theta(g).$$

Suppose now that θ has its values in our ring S. Then θ uniquely determines a function $\theta^* : FG \to F$ defined by

$$\theta^*\Big(\sum_{g \in G} f_g g \Big) = \sum_{g \in G} f_g \theta(g)^*.$$

In fact, if $* : SG \to FG$ is the natural ring homomorphism, then

$$\theta(u)^* = \theta^*(u^*)$$

for $u \in SG$. This also may be done (as we already did for the algebra homomorphisms ω_χ) for functions $\omega : \mathbf{Z}(SG) \to S$. If $\omega : \mathbf{Z}(SG) \to S$, we will denote by ω^* the unique function $\omega^* : \mathbf{Z}(FG) \to F$ satisfying

$$\omega^*(u^*) = \omega(u)^*$$

for $u \in \mathbf{Z}(SG)$.

We denote the ring of S-linear combinations of $\mathrm{Irr}(B)$ by $S[\mathrm{Irr}(B)]$.

(3.29) LEMMA. Let $\theta \in S[\mathrm{Irr}(B)]$, let $d = \mathrm{d}(B)$ and $a = \nu(|G|)$. If $u \in \mathbf{Z}(SG)$, then $\frac{\theta(u)}{p^{a-d}} \in S$. If $u^* \in \mathbf{J}(\mathbf{Z}(FG))e_B$, then $\big(\frac{\theta(u)}{p^{a-d}} \big)^* = 0$.

Proof. Since we may write $\theta = \sum\limits_{\chi \in \mathrm{Irr}(B)} [\theta, \chi]\chi$ with $[\theta, \chi] \in S$, we have that

$\theta(u) = \sum\limits_{\chi \in \mathrm{Irr}(B)} [\theta, \chi]\chi(u)$, and therefore it suffices to prove the lemma when $\theta = \chi \in \mathrm{Irr}(B)$. In this case, since $u \in \mathbf{Z}(SG)$, it follows that

$$\frac{\chi(u)}{\chi(1)} = \omega_\chi(u) \in S.$$

Then

$$\frac{\chi(u)}{p^{a-d}} = \frac{\chi(1)}{p^{a-d}}\omega_\chi(u) \in S.$$

Finally, if $u^* \in \mathbf{J}(\mathbf{Z}(FG))e_B$, then $\omega_\chi(u)^* = \lambda_B(u^*) = 0$ by Corollary (3.12). Therefore,

$$\left(\frac{\chi(u)}{p^{a-d}}\right)^* = 0,$$

as required. $\blacksquare$

(3.30) THEOREM. *Let $\theta \in S[\mathrm{Irr}(B)]$, let $d = \mathrm{d}(B)$ and let $\nu(|G|) = a$. Then $\frac{\theta(f_B)}{p^{a-d}}$ is a unit of S if and only if there exists a $u \in \mathbf{Z}(SG)f_B$ such that $\frac{\theta(u)}{p^{a-d}}$ is a unit of S.*

Proof. For $z \in \mathbf{Z}(SG)$, we let $\eta(z) = \frac{\theta(z)}{p^{a-d}}$. Using Lemma (3.29), we have that $\eta : \mathbf{Z}(SG) \to S$.

Suppose that there exists a $u \in \mathbf{Z}(SG)f_B$ such that $\eta(u)$ is a unit of S. Hence, $\eta(u)^* \neq 0$. By the decomposition in Corollary (3.12), we may write $u^* = ce_B + j$, where $c \in F$ and $j \in \mathbf{J}(\mathbf{Z}(FG))e_B$. Let $v \in \mathbf{Z}(SG)f_B$ be such that $j = v^*$. Now, we have

$$0 \neq \eta(u)^* = \eta^*(u^*) = \eta^*(ce_B) + \eta^*(j) = \eta^*(ce_B) + \eta(v)^* = c\eta^*(e_B)$$

by applying Lemma (3.29) to θ and v. Therefore, $\eta^*(e_B) = \eta(f_B)^* \neq 0$ and thus $\eta(f_B) = \frac{\theta(f_B)}{p^{a-d}}$ is a unit of S. $\blacksquare$

Next, we prove an easy result (which, in fact, has nothing to do with blocks).

If $\theta = \sum\limits_{\chi \in \mathrm{Irr}(G)} [\theta, \chi]\chi$ is a class function and $B \in \mathrm{Bl}(G)$, we let

$$\theta_B = \sum\limits_{\chi \in \mathrm{Irr}(B)} [\theta, \chi]\chi.$$

We say that θ_B is the **B-part** of θ.

(3.31) LEMMA. *If $\theta \in \mathrm{cf}(G)$, $z \in \mathbb{C}G$ and B is a p-block, then*

$$\theta_B(z) = \theta(f_B z).$$

Proof. We may clearly assume that $\theta \in \mathrm{Irr}(G)$. Suppose that V is a $\mathbb{C}G$-module affording θ and let $\mathcal{X}$ be a representation afforded by V. By Lemma (3.13.a), we know that $V = V f_B$ if $\theta \in \mathrm{Irr}(B)$, while $V f_B = 0$ if $\theta \notin \mathrm{Irr}(B)$. Hence, f_B acts like the identity if $\theta \in \mathrm{Irr}(B)$ or else, f_B acts like zero. It follows that

$$\theta(f_B z) = \mathrm{Trace}(\mathcal{X}(f_B z)) = \mathrm{Trace}(\mathcal{X}(f_B)\mathcal{X}(z)) = \theta(z)$$

if $\theta \in B$ and is zero otherwise. This proves the lemma. ∎

(3.32) LEMMA. *If $\chi \in \mathrm{cf}(G)$, $u_\chi = \sum\limits_{g \in G^0} \chi(g^{-1})g$ and $B \in \mathrm{Bl}(G)$, then*

$$u_\chi f_B = |G|_{p'} \sum_{\psi \in \mathrm{Irr}(B)} \frac{[\tilde{\chi}, \psi]}{\psi(1)} e_\psi = u_{\chi_B}.$$

Proof. Write

$$\tilde{\chi} = \sum_{\psi \in \mathrm{Irr}(G)} [\tilde{\chi}, \psi]\psi.$$

We have

$$u_\chi = \frac{1}{|G|_p} \sum_{g \in G} \tilde{\chi}(g^{-1})g = \frac{1}{|G|_p} \sum_{g \in G} \left(\sum_{\psi \in \mathrm{Irr}(G)} [\tilde{\chi}, \psi]\psi(g^{-1}) \right) g$$

$$= \frac{1}{|G|_p} \sum_{\psi \in \mathrm{Irr}(G)} [\tilde{\chi}, \psi] \left(\sum_{g \in G} \psi(g^{-1})g \right) = |G|_{p'} \sum_{\psi \in \mathrm{Irr}(G)} \frac{[\tilde{\chi}, \psi]}{\psi(1)} e_\psi.$$

Now, since ˜ is a linear map, observe that

$$\tilde{\chi} = \sum_{B \in \mathrm{Bl}(G)} \widetilde{\chi_B}.$$

If $\psi \in \mathrm{Irr}(B)$, we know by Lemma (3.20) that $\tilde{\psi}$ only involves irreducible characters in $\mathrm{Irr}(B)$. Therefore,

$$\widetilde{\chi_B} = (\tilde{\chi})_B.$$

Now,

$$u_\chi f_B = |G|_{p'} \sum_{\psi \in \mathrm{Irr}(G)} \frac{[\tilde{\chi}, \psi]}{\psi(1)} e_\psi f_B$$

$$= |G|_{p'} \sum_{\psi \in \mathrm{Irr}(B)} \frac{[\tilde{\chi}, \psi]}{\psi(1)} e_\psi = |G|_{p'} \sum_{\psi \in \mathrm{Irr}(G)} \frac{[(\tilde{\chi})_B, \psi]}{\psi(1)} e_\psi$$

$$= |G|_{p'} \sum_{\psi \in \mathrm{Irr}(G)} \frac{[\widetilde{\chi_B}, \psi]}{\psi(1)} e_\psi = u_{\chi_B},$$

as desired. ∎

Proof of Theorem (3.28). Let $\chi \in \mathrm{Irr}(B)$ be of height zero. By Lemma (3.32), we know that $u f_B = u$ where

$$u = \sum_{g \in G^0} \chi(g^{-1}) g.$$

Hence, $u \in \mathbf{Z}(SG) f_B$.

Now, let θ be the characteristic function of the p'-section of the identity; that is, $\theta(x) = 1$ if x is a p-element of G, and $\theta(x) = 0$ otherwise. By Theorem (2.19), we know that $\theta \in S[\mathrm{Irr}(G)]$ and hence $\theta_B \in S[\mathrm{Irr}(B)]$. Now, by applying Lemma (3.31), we have

$$\theta_B(u) = \theta(u f_B) = \theta(u) = \chi(1).$$

Since χ has height zero, we conclude that

$$\frac{\theta_B(u)}{p^{a-d}}$$

is a unit of S. By Theorem (3.30), we deduce that

$$\frac{\theta_B(f_B)}{p^{a-d}}$$

is also a unit of S. Now, recall that, by Corollary (3.8), the coefficient

$$f_B(\hat{K}) = \frac{1}{|G|} \sum_{\chi \in \mathrm{Irr}(B)} \chi(1) \chi(x_K^{-1})$$

associated to the conjugacy class $K \in \mathrm{cl}(G)$ is zero whenever K does not consist of p-regular elements. Again by Lemma (3.31), it follows that

$$\theta_B(f_B) = \theta(f_B f_B) = \theta(f_B) = f_B(1) = \frac{1}{|G|} \sum_{\chi \in \mathrm{Irr}(B)} \chi(1)^2.$$

Since $\frac{\theta_B(f_B)}{p^{a-d}}$ is a unit of S, we have that

$$\nu(\frac{1}{|G|} \sum_{\chi\in\mathrm{Irr}(B)} \chi(1)^2) = a - d.$$

Hence,

$$\nu(\sum_{\chi\in\mathrm{Irr}(B)} \chi(1)^2) = 2a - d,$$

as desired. ∎

To end this chapter, we discuss selfdual blocks.

If $\chi \in \mathrm{Irr}(G)$ and $\chi^0 = \sum_{\varphi\in\mathrm{IBr}(G)} d_{\chi\varphi}\varphi$, then, since the $d_{\chi\varphi}$'s are integers, we see that

$$(\bar{\chi})^0 = \overline{\chi^0} = \sum_{\varphi\in\mathrm{IBr}(G)} d_{\chi\varphi}\bar{\varphi}.$$

Therefore,

$$d_{\chi\varphi} = d_{\bar{\chi}\bar{\varphi}}.$$

Furthermore, since $[\chi,\psi]^0 \neq 0$ if and only if $[\bar{\chi},\bar{\psi}]^0 = \overline{[\chi,\psi]^0} \neq 0$ for $\chi, \psi \in \mathrm{Irr}(G)$, by using Theorems (3.3) and (3.19), for instance, we easily see that for every block B of G, the set $\bar{B} = \{\bar{\chi} \mid \chi \in B\}$ is another block of G. The block $\bar{B}$ is called the **dual** block of B. We also say that B is **selfdual** if $\bar{B} = B$.

It is clear that a block which contains a real valued (ordinary or modular) character is necessarily selfdual. In general, for odd characteristic, a selfdual block need not contain any real irreducible (ordinary or modular) character.

(3.33) THEOREM (Gow-Willems). *Suppose that $p = 2$ and let B be a p-block of G. Then the following conditions are equivalent.*
   (a) $B = \bar{B}$.
   (b) *There exists a $\varphi \in \mathrm{IBr}(B)$ such that $\bar{\varphi} = \varphi$, φ has height zero and $\Phi_\varphi(1)_2 = |G|_2$.*
   (c) *There exists a $\chi \in \mathrm{Irr}(B)$ such that $\bar{\chi} = \chi$.*

Proof. It is clear that (c) implies (a).

Write $|G|_2 = 2^a$ and let $d = \mathrm{d}(B)$ be the defect of B. Suppose that $B = \bar{B}$. By Brauer's theorem (3.28), we know that

$$(\sum_{\chi\in\mathrm{Irr}(B)} \chi(1)^2)_2 = 2^{2a-d}.$$

By the remark after the proof of Theorem (3.14), we have that

$$2^{2a-d} = (\sum_{\varphi \in \mathrm{IBr}(B)} \varphi(1) \Phi_\varphi(1))_2 .$$

We know that 2^{a-d} divides $\varphi(1)$ and that 2^a divides $\Phi_\varphi(1)$ for every $\varphi \in \mathrm{IBr}(B)$ (by Dickson's theorem). Write $X = \{\varphi \in \mathrm{IBr}(B) \,|\, \bar\varphi = \varphi\}$ for the real valued Brauer characters in B. Since

$$\overline{\Phi_\varphi} = \sum_{\chi \in \mathrm{Irr}(B)} d_{\chi\varphi}\bar\chi = \sum_{\chi \in \mathrm{Irr}(B)} d_{\bar\chi\bar\varphi}\bar\chi = \Phi_{\bar\varphi} ,$$

we see that 2^{2a-d+1} divides

$$\sum_{\varphi \in \mathrm{IBr}(B) - X} \varphi(1) \Phi_\varphi(1) .$$

Since 2^{2a-d} divides

$$\sum_{\varphi \in X} \varphi(1) \Phi_\varphi(1) ,$$

we conclude that

$$(\sum_{\varphi \in X} \varphi(1) \Phi_\varphi(1))_2 = 2^{2a-d} .$$

Hence, there exists an irreducible real valued Brauer character in B such that $\varphi(1)_2 = 2^{a-d}$ and $\Phi_\varphi(1)_2 = |G|_2$. This shows that (a) implies (b).

Finally assume (b). If P is a Sylow 2-subgroup of G, notice that

$$[(\Phi_\varphi)_P, 1_P] = \frac{\Phi_\varphi(1)}{|P|} = \Phi_\varphi(1)_{2'} .$$

Since $d_{\chi\varphi} = d_{\bar\chi\varphi}$ (because φ is real valued) and $[\chi_P, 1_P] = [\bar\chi_P, 1_P]$, we deduce that

$$1 \equiv [(\Phi_\varphi)_P, 1_P] \equiv \sum_{\substack{\chi \in \mathrm{Irr}(B) \\ \bar\chi = \chi}} d_{\chi\varphi}[\chi_P, 1_P] \bmod 2 .$$

This proves that (b) implies (c). ∎

(3.34) COROLLARY. If $p = 2$, then $\Phi_1(1)_2 = |G|_2$.

Proof. If B_0 is the principal block of G, we have that B_0 is selfdual since $1_G = \overline{1_G}$. By Theorem (3.33), we deduce that there exists a real valued $\varphi \in \mathrm{IBr}(B)$ of height zero such that $\Phi_\varphi(1)_2 = |G|_2$. Since B_0 has defect $\nu(|G|)$, it follows that $\varphi(1)$ is odd. Now, by Fong's theorem (2.30), we have that $\varphi = 1_{G^0}$. This proves the corollary. ∎

In general, it is not true that if a block B contains a real valued ordinary irreducible character, then it contains a real valued irreducible Brauer character ($\mathrm{GL}(2,3)$ for $p = 3$ is a counterexample). However, the following result is true.

(3.35) THEOREM. *If a p-block B contains a real valued irreducible Brauer character, then $\mathrm{Irr}(B)$ contains a real valued character.*

We will deduce Theorem (3.35) from an argument due to W. Willems.

(3.36) THEOREM (Willems). *Suppose that $\alpha \in \mathrm{IBr}(G)$ is real valued. Then there exists a real valued $\chi \in \mathrm{Irr}(G)$ such that $d_{\chi\alpha}$ is odd.*

Proof. Consider the generalized character $\hat{\alpha} \in \mathbb{Z}[\mathrm{Irr}(G)]$. Since $\hat{\alpha}(g) = \alpha(g_{p'})$, it follows that $\hat{\alpha}$ is real valued. Hence, $[\hat{\alpha}, \chi] = [\hat{\alpha}, \bar{\chi}]$ for $\chi \in \mathrm{Irr}(G)$. For each pair of nonreal irreducible characters $\{\chi, \bar{\chi}\}$ choose one of them and form a subset $\mathcal{U}$ of $\mathrm{Irr}(G)$. We may write

$$\hat{\alpha} = \sum_{\chi \in \mathcal{U}} [\hat{\alpha}, \chi](\chi + \bar{\chi}) + \sum_{\substack{\chi \in \mathrm{Irr}(G) \\ \bar{\chi} = \chi}} [\hat{\alpha}, \chi]\chi \, .$$

Since $(\hat{\alpha})^0 = \alpha$, we have that

$$\alpha = \sum_{\chi \in \mathcal{U}} [\hat{\alpha}, \chi](d_{\chi\alpha} + d_{\bar{\chi}\alpha})\alpha + \sum_{\substack{\chi \in \mathrm{Irr}(G) \\ \bar{\chi} = \chi}} [\hat{\alpha}, \chi]d_{\chi\alpha}\alpha \, .$$

Now, since α is real valued, we have that

$$d_{\chi\alpha} = d_{\bar{\chi}\alpha}$$

for $\chi \in \mathrm{Irr}(G)$. Hence,

$$1 \equiv \sum_{\substack{\chi \in \mathrm{Irr}(G) \\ \bar{\chi} = \chi}} [\hat{\alpha}, \chi]d_{\chi\alpha} \bmod 2$$

and the theorem easily follows. ∎

Proof of Theorem (3.35). Suppose that $\alpha \in \mathrm{IBr}(B)$ is real valued. By Theorem (3.36), there exists a $\chi \in \mathrm{Irr}(G)$ real with $d_{\chi\alpha}$ odd. In particular, $d_{\chi\alpha} \neq 0$ and $\chi \in \mathrm{Irr}(B)$, as desired.   ∎

PROBLEMS

(3.1) If G is a p-group, show that G has a unique p-block. If G is a p'-group, show that the p-blocks of G are the sets $\{\chi\}$, where $\chi \in \mathrm{Irr}(G)$.

(3.2) Suppose that B is a p-block of a group G. Prove that

$$\dim_F(\mathbf{Z}(B)) = \mathrm{k}(B) \quad \text{and} \quad \dim_F(\mathbf{Z}(B/\mathbf{J}(B))) = \mathrm{l}(B).$$

Also, prove that the FG-module $B/\mathbf{J}(B)$ affords the Brauer character

$$\sum_{\varphi \in \mathrm{IBr}(B)} \varphi(1)\varphi.$$

(HINT: Use Wedderburn's theorem in the F-algebra $B/\mathbf{J}(B)$.)

(3.3) Prove that $\chi, \psi \in \mathrm{Irr}(G)$ are in the same block if and only if $\omega_\chi(\hat{K})^* = \omega_\psi(\hat{K})^*$ for every p-regular class K of G.

(3.4) Prove that the Cartan matrix associated with B is not of the form

$$\begin{pmatrix} * & 0 \\ 0 & * \end{pmatrix}.$$

If $\mu, \varphi \in \mathrm{IBr}(G)$, we say that μ and φ are **linked** if there exists a $\chi \in \mathrm{Irr}(G)$ such that $d_{\chi\mu} \neq 0 \neq d_{\chi\varphi}$. Write $\mu \leftrightarrow \varphi$ in this case. Prove that $\mu, \varphi \in \mathrm{IBr}(G)$ lie in the same p-block of G if and only if there exists a sequence of irreducible Brauer characters $\theta_1, \ldots, \theta_n \in \mathrm{IBr}(G)$ such that $\mu \leftrightarrow \theta_1$, $\theta_i \leftrightarrow \theta_{i+1}$ for $i = 1, \ldots, n-1$ and $\theta_n \leftrightarrow \varphi$.

(HINT: Suppose $\mathcal{B} \subseteq \mathrm{IBr}(B)$ is such that $c_{\varphi\mu} = 0$ for every $\varphi \in \mathcal{B}$ and $\mu \in \mathrm{IBr}(B) - \mathcal{B}$. If $\mathcal{A} = \{\chi \in \mathrm{Irr}(B) \mid \text{there exists } \varphi \in \mathcal{B} \text{ such that } d_{\chi\varphi} \neq 0\}$, prove that every character linked to $\chi \in \mathcal{A}$ lies in $\mathcal{A}$.)

(3.5) If B is a block with positive defect, show that $c_{\varphi\varphi} \geq 2$ for all $\varphi \in \mathrm{IBr}(B)$.

(HINT: Otherwise, prove that there exists a unique $\psi \in \mathrm{Irr}(G)$ such that $d_{\psi\varphi} = 1$ and $d_{\chi\varphi} = 0$ for $\chi \neq \psi$. Consider Φ_φ and use Theorem (3.18).)

(3.6) If $\varphi \in \mathrm{IBr}(B)$, prove that $p^{\mathrm{d}(B)}\varphi$ is a $\mathbb{Z}$-linear combination of $\{(\Phi_\varphi)^0 \mid \varphi \in \mathrm{IBr}(B)\}$. If $\varphi \in \mathrm{IBr}(B)$ has height zero, prove that $p^{\mathrm{d}(B)-1}\varphi$ is not a $\mathbb{Z}$-linear combination of $\{(\Phi_\varphi)^0 \mid \varphi \in \mathrm{IBr}(B)\}$.

(3.7) Let $\alpha \in \mathbb{Z}[\mathrm{Irr}(B)]$ be such that $\nu(\alpha(1)) = a - d$, where $a = \nu(|G|)$ and $d = \mathrm{d}(B)$. Prove that $\frac{1}{p^{a-d}}\tilde{\alpha}$ is a generalized character of G while $\frac{1}{p^{a-d+1}}\tilde{\alpha}$ is not.

(3.8) (Brauer) Prove that the principal 2-block is the only selfdual 2-block of G with maximal defect.

(3.9) Suppose that $\chi \in \mathrm{Irr}(G)$ has p-defect zero and let $N \lhd G$. If θ is an irreducible constituent of χ_N, prove that θ has p-defect zero.

(HINT: Recall that $\frac{\chi(1)}{\theta(1)}$ divides $|G : N|$.)

(3.10) Let $N \lhd G$ and suppose that $\theta \in \mathrm{Irr}(N)$ is G-invariant and has p-defect zero. If G/N is a p-group, prove that θ extends to G.

(HINT: If P is a Sylow p-subgroup of G, prove that $(\theta^G)_P = a\rho_P$, where ρ_P is the regular character of P and a is a p'-number. Deduce that there exists an irreducible constituent χ of θ^G with multiplicity not divisible by p.)

(3.11) If B is a block of positive defect, prove that $\mathrm{Irr}(B)$ contains at least two characters with height zero.

(HINT: Use that $\nu(\sum_{\chi \in \mathrm{Irr}(B)} \chi(1)^2) \geq 2\nu(|G|) - \mathrm{d}(B)$.)

(NOTE: This result is due to **G. H. Cliff**, **W. Plesken** and **A. Weiss**.)

(3.12) (Passman) Suppose that B is a p-block. Prove that the coefficient of 1 in the idempotent e_B is nonzero if and only if B has maximal defect.

(HINT: Use Brauer's theorem (3.28).)

(3.13) (Blau) Let B be a block of G and let $\varphi \in \mathrm{IBr}(B)$ be of height zero. If K is a conjugacy class of p-regular elements of G and $x \in K$, prove that

$$\frac{|K|\varphi(x)}{\varphi(1)} \in S.$$

Also, show that

$$\left(\frac{|K|\varphi(x)}{\varphi(1)}\right)^* = \lambda_B(\hat{K}).$$

(HINT: Use Lemma (3.16).)

4. The First Main Theorem

Our main objective in this chapter is to associate to each p-block a uniquely determined conjugacy class of p-subgroups of G. After this is done, we will be able to relate the blocks of G with the blocks of certain local subgroups of G.

If $K = \mathrm{cl}(x)$ is a conjugacy class of G, we denote the set of all the G-conjugates of any Sylow p-subgroup of $\mathbf{C}_G(x)$ by $\delta(K)$. Each element of $\delta(K)$ is called a **defect group** of K. Sometimes, it is convenient to choose $D_K \in \delta(K)$ and $x_K \in K$ such that $D_K \in \mathrm{Syl}_p(\mathbf{C}_G(x_K))$.

If X and Y are subsets of G, we indicate that X is contained in some G-conjugate of Y by $X \subseteq_G Y$. Also, $X =_G Y$ means that X and Y are G-conjugate.

If D is a p-subgroup of G, we denote by $\mathbf{Z}_D(FG)$ the F-span of all the elements $\hat{K}$ such that K is a conjugacy class of G and some defect group of K is contained in a G-conjugate of D. In other words,

$$\mathbf{Z}_D(FG) = \sum_{D_K \subseteq_G D} F\hat{K} \, .$$

Notice that $\mathbf{Z}_D(FG) = \mathbf{Z}_{D^g}(FG)$ for $g \in G$.

(4.1) THEOREM (Osima). *If D is any p-subgroup of G, then $\mathbf{Z}_D(FG)$ is an ideal of $\mathbf{Z}(FG)$.*

Proof. If $K, L \in \mathrm{cl}(G)$, we know that in $\mathbf{Z}(FG)$,

$$\hat{K}\hat{L} = \sum_{M \in \mathrm{cl}(G)} a_{KLM} \hat{M} \, ,$$

where $a_{KLM} = |A_{KLM}|^*$ and $A_{KLM} = \{(x, y) \in K \times L \,|\, xy = x_M\}$.

We claim that if $a_{KLM} \neq 0$, then $D_M \subseteq_G D_K$. Since $D_M \subseteq \mathbf{C}_G(x_M)$, observe that D_M acts on A_{KLM} by $(x, y)^d = (x^d, y^d)$, where $d \in D_M$ and

80

$(x, y) \in A_{KLM}$. Since each orbit has p-power size and $a_{KLM} \neq 0$, it follows that D_M centralizes some element of K (and some element of L). Then $D_M \subseteq_G D_K$, and the claim is proved.

Now, if K is a class with defect $D_K \subseteq_G D$ and L is any other class, we see that $\hat{K}\hat{L}$ is sum of classes $\hat{M}$ such that $D_M \subseteq_G D_K \subseteq_G D$. This proves that $\mathbf{Z}_D(FG)$ is an ideal of $\mathbf{Z}(FG)$, as desired. ∎

(4.2) LEMMA (Rosenberg). *Suppose that $e_B \in I + J$, where I and J are ideals of $\mathbf{Z}(FG)$. Then $e_B \in I$ or $e_B \in J$.*

Proof. We know that the algebra $\mathbf{Z}(B) = \mathbf{Z}(FG)e_B$ is local by Corollary (3.12). Therefore, it has a unique maximal right ideal $\mathbf{J}(\mathbf{Z}(B))$. Now, if $Ie_B < \mathbf{Z}(B)$ and $Je_B < \mathbf{Z}(B)$, then, since these are ideals of $\mathbf{Z}(B)$, it follows that both are contained in $\mathbf{J}(\mathbf{Z}(B))$. Write $e_B = x + y$, where $x \in I$ and $y \in J$. Then

$$e_B = e_B e_B = xe_B + ye_B \in Ie_B + Je_B \subseteq \mathbf{J}(\mathbf{Z}(B)).$$

This is not possible since $0 \neq e_B = (e_B)^2$ and $\mathbf{J}(\mathbf{Z}(B))$ is a nilpotent ideal. Hence, $Ie_B = \mathbf{Z}(B)$ or $Je_B = \mathbf{Z}(B)$. If, for instance, $Ie_B = \mathbf{Z}(B)$, then $e_B \in Ie_B \subseteq I$, as desired. ∎

The next theorem associates to each p-block of G a uniquely determined conjugacy class of p-subgroups of G.

(4.3) THEOREM. *If B is a block, then there is a unique G-conjugacy class of p-subgroups D of G with $e_B \in \mathbf{Z}_D(FG)$ and such that if $e_B \in \mathbf{Z}_{D_0}(FG)$, where D_0 is a p-subgroup of G, then $D \subseteq_G D_0$.*

Proof. Notice that if $P \in \mathrm{Syl}_p(G)$, then $\mathbf{Z}_P(FG) = \mathbf{Z}(FG)$, and thus $e_B \in \mathbf{Z}_P(FG)$. We claim that

$$\mathbf{Z}_{D_1}(FG) \cap \mathbf{Z}_{D_2}(FG) = \sum_{x,y \in G} \mathbf{Z}_{D_1^x \cap D_2^y}(FG).$$

If $D_1 \subseteq_G D_2$, it is clear that $\mathbf{Z}_{D_1}(FG) \subseteq \mathbf{Z}_{D_2}(FG)$. Hence, it suffices to show that

$$\mathbf{Z}_{D_1}(FG) \cap \mathbf{Z}_{D_2}(FG) \subseteq \sum_{x,y \in G} \mathbf{Z}_{D_1^x \cap D_2^y}(FG).$$

Suppose that

$$z = \sum_{K \in \mathrm{cl}(G)} z(\hat{K})\hat{K} \in \mathbf{Z}_{D_1}(FG) \cap \mathbf{Z}_{D_2}(FG).$$

Then z may be written as a linear combination of $\{\hat{K} \mid D_K \subseteq_G D_1\}$ and of $\{\hat{K} \mid D_K \subseteq_G D_2\}$. Since $\{\hat{K} \mid K \in \mathrm{cl}(G)\}$ is linearly independent, we conclude that if $z(\hat{K}) \neq 0$, then $D_K \subseteq_G D_1$ and $D_K \subseteq_G D_2$. Therefore, there exist $x, y \in G$ such that $D_K \subseteq D_1^x \cap D_2^y$. Hence, $\hat{K} \in \mathbf{Z}_{D_1^x \cap D_2^y}(FG)$ and thus $z \in \sum_{x,y \in G} \mathbf{Z}_{D_1^x \cap D_2^y}(FG)$. This proves the claim.

Now, suppose that D_1 and D_2 are minimal such that $e_B \in \mathbf{Z}_{D_i}(FG)$ for $i = 1, 2$. Since each $\mathbf{Z}_{D_1^x \cap D_2^y}(FG)$ is an ideal of $\mathbf{Z}(FG)$ for $x, y \in G$, by Rosenberg's lemma and the claim, it follows that there exist x and $y \in G$ such that $e_B \in \mathbf{Z}_{D_1^x \cap D_2^y}(FG)$. Since D_1^x and D_2^y are also minimal with $e_B \in \mathbf{Z}_{D_1^x}(FG) = \mathbf{Z}_{D_1}(FG)$ and $e_B \in \mathbf{Z}_{D_2^y}(FG) = \mathbf{Z}_{D_2}(FG)$, it follows that D_1 and D_2 are G-conjugate, as desired. ∎

If B is a p-block of G, then the **defect groups** of B are the p-subgroups of G uniquely determined by B by applying Theorem (4.3). This is a G-conjugacy class of p-subgroups of G which is denoted by $\delta(B)$. Sometimes, we will fix a $D_B \in \delta(B)$. (Some authors use $\delta(B)$ as a fixed defect group of B.)

Our next objective is to give a second definition of the defect groups of a block. Write

$$e_B = \sum_{K \in \mathrm{cl}(G)} a_B(\hat{K})\hat{K}.$$

Since

$$1 = \lambda_B(e_B) = \lambda_B \left(\sum_{K \in \mathrm{cl}(G)} a_B(\hat{K})\hat{K} \right) = \sum_{K \in \mathrm{cl}(G)} a_B(\hat{K})\lambda_B(\hat{K}),$$

there is at least one conjugacy class $K \in \mathrm{cl}(G)$ such that

$$\lambda_B(\hat{K}) \neq 0 \neq a_B(\hat{K}).$$

We call such a class a **defect class** for B. We will show that the defect groups of the defect classes are the defect groups of the blocks.

(4.4) THEOREM (Min-Max). *Let B be a block of G and let $K, L \in \mathrm{cl}(G)$. If $\lambda_B(\hat{K}) \neq 0$, then $D_B \subseteq_G D_K$. If $a_B(\hat{L}) \neq 0$, then $D_L \subseteq_G D_B$.*

Proof. Since $\lambda_B(\hat{K}) \neq 0$, it follows that $\hat{K} \in \mathbf{Z}_{D_K}(FG)$ is not contained in $\ker(\lambda_B)$. Since $\ker(\lambda_B)$ has codimension 1, we see that

$$\mathbf{Z}(FG) = \ker(\lambda_B) + \mathbf{Z}_{D_K}(FG).$$

By Rosenberg's lemma, we have that $e_B \in \mathbf{Z}_{D_K}(FG)$ because $\lambda_B(e_B) = 1 \neq 0$. By Theorem (4.3), $D_B \subseteq_G D_K$, since D_B is minimal satisfying $e_B \in \mathbf{Z}_{D_B}(FG)$. Now, if $a_B(\hat{L}) \neq 0$, then it follows that $D_L \subseteq_G D_B$ since $e_B \in \mathbf{Z}_{D_B}(FG)$ and, by definition, $e_B = \sum_{K \in \mathrm{cl}(G)} a_B(\hat{K})\hat{K}$. ∎

(4.5) COROLLARY. *If K is a defect class for B, then $\delta(K) = \delta(B)$. In particular, if L is another defect class for B, then $\delta(K) = \delta(L)$.*

Proof. If K is a defect class for B, then $\lambda_B(\hat{K}) \neq 0 \neq a_B(\hat{K})$. Now, apply Theorem (4.4). ∎

Next, we calculate the order of a defect group of B.

We recall a few facts from Chapter 3. Since $f_B \in \mathbf{Z}(SG)$ (Corollary (3.8)) and $e_B = (f_B)^*$, notice that

$$a_B(\hat{K}) = f_B(\hat{K})^* .$$

Also, $a_B(\hat{K}) = 0$ if K does not consist of p-regular elements (Corollary (3.8)), and for every p-regular element x_K, we have that

$$f_B(\hat{K}) = \frac{1}{|G|} \sum_{\chi \in \mathrm{Irr}(B)} \chi(1)\chi(x_K^{-1}) = \frac{1}{|G|} \sum_{\varphi \in \mathrm{IBr}(B)} \Phi_\varphi(1)\varphi(x_K^{-1}),$$

(setting $\mathcal{A} = \mathrm{Irr}(B)$ in Theorem (3.6)).

(4.6) THEOREM. *If D is a defect group of B, then $|D| = p^{\mathrm{d}(B)}$.*

Proof. Write $a = \nu(|G|)$. Let K be a defect class for B and write $|D| = p^f$. Since $D \in \delta(K)$ by Corollary (4.5), we have that $\nu(|K|) = a - f$. Let $\chi \in \mathrm{Irr}(B)$. Since

$$0 \neq \lambda_B(\hat{K}) = \left(\frac{\chi(x_K)|K|}{\chi(1)}\right)^* ,$$

it follows that

$$\omega_\chi(\hat{K}) = \frac{\chi(x_K)|K|}{\chi(1)} \in S - \mathcal{P} .$$

Now, $\chi(x_K) \in \mathbf{R}$, and by Lemma (3.21) on "valuations", it follows that $\nu\left(\frac{|K|}{\chi(1)}\right) \leq 0$ (since otherwise $\frac{|K|}{\chi(1)} \in \mathcal{P}$ and $\omega_\chi(\hat{K}) \in \mathcal{P}$). Hence, $\nu(|K|) = a - f \leq \nu(\chi(1)) = a - \mathrm{d}(B) + h$, where h is the height of χ. Now, if we choose (as we can) χ of height zero, we get $\mathrm{d}(B) \leq f$.

Now, we use the fact that $a_B(\hat{K}) \neq 0$. First of all, recall that x_K is p-regular. Write

$$a_B(\hat{K}) = \left(\frac{1}{|G|} \sum_{\varphi \in \mathrm{IBr}(B)} \Phi_\varphi(1)\overline{\varphi(x_K)}\right)^* .$$

By Dickson's theorem, $\frac{\Phi_\varphi(1)}{|G|} \in S$, and thus

$$0 \neq a_B(\hat{K}) = \sum_{\varphi \in \mathrm{IBr}(B)} \left(\frac{\Phi_\varphi(1)}{|G|}\right)^* \overline{\varphi(x_K)}^* .$$

Therefore, there is a $\varphi \in \mathrm{IBr}(B)$ such that $\overline{\varphi(x_K)} \notin \mathcal{P}$. Since φ is a $\mathbb{Z}$-linear combination of $\{\chi^0 \mid \chi \in \mathrm{Irr}(B)\}$ (Lemma (3.16)), it follows that there is a $\chi \in \mathrm{Irr}(B)$ such that $\overline{\chi(x_K)} \notin \mathcal{P}$. Now, since

$$\frac{\chi(x_K)|K|}{\chi(1)} = \overline{\omega_\chi(\hat{K})} \in \mathbf{R},$$

we get

$$\frac{\chi(1)}{|K|} \notin \mathcal{P}.$$

Therefore, by Lemma (3.21) on "valuations", we have that $\nu(\frac{\chi(1)}{|K|}) \leq 0$. Since $\nu(\chi(1)) = a - \mathrm{d}(B) + h$, it follows that $f \leq \mathrm{d}(B) - h \leq \mathrm{d}(B)$, as required. ∎

Before going into the Brauer correspondence and the first main theorem, we prove that the defect groups of the blocks of G contain $\mathbf{O}_p(G)$. We will use the fact that the irreducible F-representations of G contain $\mathbf{O}_p(G)$ in their kernels (Lemma (2.32).)

(4.7) LEMMA. *Let K be a conjugacy class of G. If $K \cap \mathbf{C}_G(\mathbf{O}_p(G)) = \emptyset$, then $\hat{K} \in \mathbf{J}(\mathbf{Z}(FG))$. In particular, $\hat{K}$ is nilpotent.*

Proof. We have that $P = \mathbf{O}_p(G)$ acts on K by conjugation and, since $K \cap \mathbf{C}_G(P) = \emptyset$, each orbit has length divisible by p. It suffices to show that if $\mathcal{X}$ is any irreducible representation of G, then $\mathcal{X}(\hat{K}) = 0$. Let $x \in K$ and let Θ be the P-orbit of x. Now, notice that if $y \in P$, then $x^y = x[x, y] \in xP$. By Lemma (2.32), P is contained in the kernel of $\mathcal{X}$ and therefore, we have that $\mathcal{X}$ is constant on cosets modulo P. It follows that

$$\sum_{z \in \Theta} \mathcal{X}(z) = |\Theta| \mathcal{X}(x) = 0.$$

Therefore, $\mathcal{X}(\hat{K}) = 0$, as required. ∎

(4.8) THEOREM. *If B is a block of G, then $\mathbf{O}_p(G)$ is contained in every defect group of B.*

Proof. If K is any defect class for B, then $\lambda_B(\hat{K}) \neq 0$. Therefore, $\hat{K}$ is not nilpotent and we have that $K \subseteq \mathbf{C}_G(\mathbf{O}_p(G))$ by Lemma (4.7). Hence, $\mathbf{O}_p(G) \subseteq D_K$, as required. ∎

By Theorem (4.8), we already know a condition for a p-subgroup D of G to be a defect group of a block of G. We will soon see more restrictive conditions: the defect groups of the blocks are always the intersection of two Sylow p-subgroups of G, and they are p-**radical**; that is, $P = \mathbf{O}_p(\mathbf{N}_G(P))$.

Now, we introduce the **Brauer homomorphism**. Suppose that P is a p-subgroup of G and let $\mathbf{C}_G(P) \subseteq H \subseteq \mathbf{N}_G(P)$. We define a map $\mathrm{Br}_P : \mathbf{Z}(FG) \to \mathbf{Z}(FH)$ by defining Br_P on $\hat{K}$ for $K \in \mathrm{cl}(G)$ and extending linearly. We set

$$\mathrm{Br}_P(\hat{K}) = \sum_{x \in K \cap \mathbf{C}_G(P)} x \, ,$$

where it is understood that if $K \cap \mathbf{C}_G(P) = \emptyset$, then $\mathrm{Br}_P(\hat{K}) = 0$. Since $\mathbf{C}_G(P) \subseteq H \subseteq \mathbf{N}_G(P)$, we do have that $\mathrm{Br}_P(\hat{K}) \in \mathbf{Z}(FH)$. Some authors say that Br_P is the Brauer homomorphism with respect to (G, H, P).

(4.9) THEOREM. *The Brauer map* Br_P *defines an algebra homomorphism* $\mathbf{Z}(FG) \to \mathbf{Z}(FH)$.

Proof. It suffices to check that

$$\mathrm{Br}_P(\hat{K})\mathrm{Br}_P(\hat{L}) = \mathrm{Br}_P(\hat{K}\hat{L}) \, .$$

As usual, write

$$\hat{K}\hat{L} = \sum_{M \in \mathrm{cl}(G)} a_{KLM}\hat{M} \, ,$$

where $a_{KLM} = |A_{KLM}|^*$, $A_{KLM} = \{(x,y) \in K \times L \mid xy = x_M\}$, and x_M is an arbitrary (but fixed) element of M. Write $C = \mathbf{C}_G(P)$ and let $c \in C$. Since the sets $M \cap C$ are disjoint for different classes M, the coefficient of c in

$$\mathrm{Br}_P(\hat{K}\hat{L}) = \sum_{M \in \mathrm{cl}(G)} a_{KLM}\mathrm{Br}_P(\hat{M})$$

is a_{KLM_0}, where M_0 is the G-class of $c = x_{M_0}$. Also, the coefficient of c in $\mathrm{Br}_P(\hat{K})\mathrm{Br}_P(\hat{L})$ is

$$|\{(x,y) \in (K \cap C) \times (L \cap C) \mid xy = c\}| \, .$$

Now, P acts on A_{KLM_0} by $(x,y)^z = (x^z, y^z)$ for $(x,y) \in A_{KLM_0}$ and $z \in P$. Since the set of fixed points is

$$\{(x,y) \in (K \cap C) \times (L \cap C) \mid xy = c\} \, ,$$

it follows that

$$|A_{KLM_0}| \equiv |\{(x,y) \in (K \cap C) \times (L \cap C) \mid xy = c\}| \bmod p \, ,$$

and the proof of the theorem is complete. $\blacksquare$

Next, we analyse the kernel of the Brauer homomorphism. (Observe that this ideal of $\mathbf{Z}(FG)$ has nothing to do with the subgroup H.)

(4.10) THEOREM. *If P is any p-subgroup of G, then*

$$\ker(\mathrm{Br}_P) = \sum_{P \not\subseteq_G D_K} F\hat{K} .$$

Proof. Write $C = \mathbf{C}_G(P)$. Notice that if $K \in \mathrm{cl}(G)$, then $K \cap C \neq \emptyset$ if and only if $P \subseteq_G D_K$. Now, if

$$z = \sum_{K \in \mathrm{cl}(G)} z(\hat{K})\hat{K} \in \mathbf{Z}(FG) ,$$

it follows that $z \in \ker(\mathrm{Br}_P)$ if and only if $z(\hat{K}) = 0$ whenever $K \cap C \neq \emptyset$ since for different $K \in \mathrm{cl}(G)$ the subsets $K \cap C$ are disjoint. This proves the theorem. ∎

We give another characterization of the defect groups of the blocks.

(4.11) THEOREM. *Suppose that B is a block with defect group D and let P be a p-subgroup of G. Then $\mathrm{Br}_P(e_B) \neq 0$ if and only if $P \subseteq_G D$. Therefore, D is the unique maximal p-subgroup of G (up to G-conjugacy) such that $\mathrm{Br}_D(e_B) \neq 0$.*

Proof. Write $e_B = \sum_{K \in \mathrm{cl}(G)} a_B(\hat{K})\hat{K}$. By Theorem (4.10), we have that $\mathrm{Br}_P(e_B) \neq 0$ if and only if $e_B \notin \sum_{P \not\subseteq_G D_K} F\hat{K}$. This happens if and only if there is a class K with $a_B(\hat{K}) \neq 0$ such that $P \subseteq_G D_K$.

Now, suppose that $P \subseteq_G D$ and let K be a defect class of B. Since $a_B(\hat{K}) \neq 0$ and D and D_K are G-conjugate, it follows that $\mathrm{Br}_P(e_B) \neq 0$, by the first paragraph.

Conversely, if $\mathrm{Br}_P(e_B) \neq 0$ and K is a class such that $a_B(\hat{K}) \neq 0$ with $P \subseteq_G D_K$, then $D_K \subseteq_G D$ by the min-max theorem (4.4), and therefore $P \subseteq_G D$, as required. ∎

Our main objective in this chapter is to prove the first main theorem of Brauer. If P is any p-subgroup of G, let us denote by $\mathrm{Bl}(G \,|\, P)$ the set of blocks of G with defect group P.

(4.12) THEOREM (First Main Theorem). *There exists a natural bijection $f : \mathrm{Bl}(G \,|\, P) \to \mathrm{Bl}(\mathbf{N}_G(P) \,|\, P)$. In fact,*

$$\mathrm{Br}_P(e_B) = e_{f(B)}$$

for $B \in \mathrm{Bl}(G \,|\, P)$.

We will give a reformulation of the first main theorem which goes the other way around, however. For that, we need the **Brauer correspondence**.

If H is a subgroup of G and $b \in \mathrm{Bl}(H)$, we extend the algebra homomorphism $\lambda_b : \mathbf{Z}(FH) \to F$ to a linear map $\lambda_b^G : \mathbf{Z}(FG) \to F$ by setting

$$\lambda_b^G(\hat{K}) = \lambda_b \left(\sum_{x \in K \cap H} x \right) ,$$

where it is understood that the sum $\sum_{x \in K \cap H} x$ is zero if $K \cap H$ is empty.

It may happen that λ_b^G is an algebra homomorphism. If this is the case, then, by Theorem (3.11), there exists a unique block $b^G \in \mathrm{Bl}(G)$ such that

$$\lambda_b^G = \lambda_{b^G} .$$

We will say in this case that b^G is **defined**. The block b^G is called the **induced block**.

(There are several nonequivalent ways of defining induced blocks. The one we use is due to R. Brauer. Another definition which is often used is due to J. Alperin and D. Burry. See [Alperin, Book]. We should say that these definitions coincide in the "important" cases. We refer the reader to [Wheeler] for a discussion on this subject.)

(4.13) LEMMA. *Let b be a block of $H \subseteq G$. If b^G is defined, then every defect group of b is contained in some defect group of b^G.*

Proof. If K is a defect class of b^G, we have that

$$\lambda_b^G(\hat{K}) = \lambda_b(\sum_{x \in K \cap H} x) \neq 0 .$$

In particular, $K \cap H \neq \emptyset$ and there is $L \in \mathrm{cl}(H)$ contained in $K \cap H$ such that $\lambda_b(\hat{L}) \neq 0$. By the min-max theorem (4.4), it follows that $D_b \subseteq_H D_L$, where $D_b \in \delta(b)$ and $D_L \in \delta(L)$. Since D_L is a p-subgroup of $\mathbf{C}_H(x_L)$, there is a Sylow p-subgroup D of $\mathbf{C}_G(x_L)$ containing D_L. Since $x_L \in L \subseteq K$ and K is a defect class of b^G, it follows that D is a defect group of b^G and the proof of the lemma is complete. ∎

(4.14) THEOREM. *Suppose that P is a p-subgroup of G and let H be a subgroup of G satisfying $P\mathbf{C}_G(P) \subseteq H \subseteq \mathbf{N}_G(P)$. If $b \in \mathrm{Bl}(H)$, then b^G is defined and $\lambda_b^G = \mathrm{Br}_P\lambda_b$. Moreover, if B is any block of G, then $B = b^G$ for some block b of H if and only if P is contained in some defect group of B. In this case,*

$$\mathrm{Br}_P(e_B) = \sum_{b^G = B} e_b .$$

Proof. Since λ_b and Br_P are algebra homomorphisms, to prove that b^G is defined it suffices to show that $\lambda_b^G = \mathrm{Br}_P\lambda_b$.

Write $C = \mathbf{C}_G(P)$. If $K \in \mathrm{cl}(G)$, we have to prove that

$$\lambda_b\left(\sum_{x \in K \cap H} x \right) = \lambda_b\left(\sum_{x \in K \cap C} x \right).$$

We may write

$$\sum_{x \in K \cap H} x = \sum_{x \in K \cap C} x + \sum_{x \in K \cap H - K \cap C} x .$$

Since $K \cap C$ is normalized by H, we have that $K \cap H - K \cap C$ is a union of conjugacy classes of H. If $L \in \mathrm{cl}(H)$ is a class with $C \cap L = \emptyset$, then, since $P \subseteq \mathbf{O}_p(H)$, it follows that $\mathbf{C}_H(\mathbf{O}_p(H)) \subseteq C$ and thus $\mathbf{C}_H(\mathbf{O}_p(H)) \cap L = \emptyset$. By Lemma (4.7), $\hat{L}$ is nilpotent and therefore $\lambda_b(\hat{L}) = 0$. The first part of the theorem follows.

Suppose now that B is a block of G. If $B = b^G$ for some $b \in \mathrm{Bl}(H)$, then D_b is contained in some defect group of B by Lemma (4.13). By Theorem (4.8), $P \subseteq \mathbf{O}_p(H) \subseteq D_b$ and hence $P \subseteq_G D_B$, as desired.

Conversely, suppose that P is contained in some defect group of B. By Theorem (4.11), we know that $\mathrm{Br}_P(e_B) \neq 0$. Since Br_P is an algebra homomorphism, we have that $\mathrm{Br}_P(e_B)$ is an idempotent of the algebra $\mathbf{Z}(FH)$. By Theorem (3.11), we may write

$$\mathrm{Br}_P(e_B) = e_{b_1} + \ldots + e_{b_s}$$

for some distinct blocks $b_i \in \mathrm{Bl}(H)$. Now, by the first part, $b^G = B$ for some block $b \in \mathrm{Bl}(H)$ if and only if $\lambda_b(\mathrm{Br}_P(e_B)) = 1$. Since $\mathrm{Br}_P(e_B) = e_{b_1} + \ldots + e_{b_s}$, this happens if and only if $b = b_i$ for some $i = 1, \ldots, s$. $\blacksquare$

The following lemma is interesting in its own right.

(4.15) LEMMA. If P is a normal p-subgroup of G and $B \in \mathrm{Bl}(G\,|\,P)$, then

$$e_B = \sum_{\delta(K)=\{P\}} a_B(\hat{K})\hat{K} .$$

Proof. By the min-max theorem (4.4), if $a_B(\hat{K}) \neq 0$, then $D_K \subseteq P$. On the other hand, since $B = B^G$, it follows that $\mathrm{Br}_P(e_B) = e_B$, by Theorem (4.14) applied to $H = G$. In particular, if $K \cap C = \emptyset$, then $a_B(\hat{K}) = 0$. Therefore, if $a_B(\hat{K}) \neq 0$, then $K \cap C \neq \emptyset$ and thus $P \subseteq D_K$. Hence, $P = D_K$, as desired. $\blacksquare$

If P is a p-subgroup of G, we denote the set of conjugacy classes of G with defect group P by $\mathrm{cl}(G\,|\,P)$.

(4.16) LEMMA. *If P is a p-subgroup of G, $N = \mathbf{N}_G(P)$ and $C = \mathbf{C}_G(P)$, then the map $K \mapsto K \cap C$ is a bijection $\mathrm{cl}(G \mid P) \to \mathrm{cl}(N \mid P)$.*

Proof. Let $K \in \mathrm{cl}(G \mid P)$ and let $x \in K$ be such that $P \in \mathrm{Syl}_p(\mathbf{C}_G(x))$. In particular, $x \in K \cap C \neq \emptyset$. If $u \in G$ and $x^u \in K \cap C$, then $P \subseteq \mathbf{C}_G(x^u)$ and thus $P^{u^{-1}}$ is a Sylow p-subgroup of $\mathbf{C}_G(x)$. Hence, $P^{vu} = P$ for some $v \in \mathbf{C}_G(x)$ and thus $x^u = x^{vu}$ is N-conjugate with x. This proves that $K \cap C$ is a conjugacy class of N. Since distinct conjugacy classes are disjoint, clearly our map is injective.

To finish the proof, it suffices to check that if $L \in \mathrm{cl}(N \mid P)$, then the unique class K of G containing L has defect group P. Let $x \in L$. Since $P \triangleleft N$, we have that $P \in \mathrm{Syl}_p(\mathbf{C}_N(x))$. Suppose that $P < D \in \mathrm{Syl}_p(\mathbf{C}_G(x))$. In this case, we have that $P < N \cap D \subseteq \mathbf{C}_N(x)$. This is a contradiction which proves the lemma. ∎

The following includes Theorem (4.12).

(4.17) THEOREM (First Main Theorem). *The map $b \mapsto b^G$ defines a bijection $\mathrm{Bl}(\mathbf{N}_G(P) \mid P) \to \mathrm{Bl}(G \mid P)$. Furthermore, $\mathrm{Br}_P(e_{b^G}) = e_b$.*

Proof. Write $N = \mathbf{N}_G(P)$ and $C = \mathbf{C}_G(P)$. By Theorem (4.14), we know that b^G is defined for every $b \in \mathrm{Bl}(N)$. Also, $\lambda_b^G = \mathrm{Br}_P \lambda_b$

If $b \in \mathrm{Bl}(N \mid P)$, we first prove that $b^G \in \mathrm{Bl}(G \mid P)$. Let L be a defect class of b, so that $L \in \mathrm{cl}(N \mid P)$ by Corollary (4.5). By Lemma (4.16), the class K of G which contains L lies in $\mathrm{cl}(G \mid P)$ and is such that $K \cap C = L$. Hence,

$$\lambda_{b^G}(\hat{K}) = \lambda_b(\mathrm{Br}_P(\hat{K})) = \lambda_b(\widehat{K \cap C}) = \lambda_b(\hat{L}) \neq 0.$$

Then, by the min-max theorem (4.4), some defect group of b^G is contained in P. Since P is the defect group of b, by Lemma (4.13) we have that P is contained in some defect group of b^G. This proves that b^G has defect group P.

Now, we prove that the map $b \mapsto b^G$ is surjective. If $B \in \mathrm{Bl}(G \mid P)$, by Theorem (4.14) there are blocks $b_1, \ldots, b_s$ of N such that $B = b_i^G$. In fact,

$$\mathrm{Br}_P(e_B) = e_{b_1} + \ldots + e_{b_s}.$$

Since $P \triangleleft N$, it follows that P is contained in every defect group of b_i (by Theorem (4.8)). By Lemma (4.13), each defect group of b_i is contained in some defect group of $b_i^G = B$. Since B has defect group P, we conclude that P is a defect group for every b_i. This shows that the map $b \mapsto b^G$ is surjective. (Moreover, if we show that our map $b \mapsto b^G$ is injective, this will prove the second part of the theorem.)

Assume that $b_1^G = b_2^G$ for blocks $b_1, b_2 \in \mathrm{Bl}(N \mid P)$. Then

$$\mathrm{Br}_P \lambda_{b_1} = \mathrm{Br}_P \lambda_{b_2}$$

by Theorem (4.14). Thus, $\lambda_{b_1}(\widehat{K \cap C}) = \lambda_{b_2}(\widehat{K \cap C})$ for every class K of G. Therefore, by Lemma (4.16), we have that λ_{b_1} and λ_{b_2} coincide in the conjugacy classes $\hat{L}$ of N with defect group P. Now, by Lemma (4.15) (or Corollary (4.5), if we wish), we have that

$$1 = \lambda_{b_1}(e_{b_1}) = \sum_{\delta(L)=\{P\}} a_{b_1}(\hat{L})\lambda_{b_1}(\hat{L}) = \sum_{\delta(L)=\{P\}} a_{b_1}(\hat{L})\lambda_{b_2}(\hat{L}) = \lambda_{b_2}(e_{b_1}).$$

Therefore, $b_1 = b_2$ and the proof of the theorem is complete. ∎

(4.18) COROLLARY. *If P is a defect group of any block of G, then $P = \mathbf{O}_p(\mathbf{N}_G(P))$.*

Proof. This is clear by the first main theorem and the fact that $\mathbf{O}_p(G)$ is contained in every defect group of every p-block of G. ∎

Much of the research in representation theory these days is devoted to proving the **Alperin weight conjecture.** If P is a p-subgroup of G, a **p-weight** of G is a pair

$$(P, \gamma),$$

where $\gamma \in \mathrm{Irr}(\mathbf{N}_G(P)/P)$ is such that $\gamma(1)_p = |\mathbf{N}_G(P)/P|_p$ (that is, γ has p-defect zero considered as a character of $\mathbf{N}_G(P)/P$). If (P, γ) is a p-weight and $g \in G$, then $(P, \gamma)^g = (P^g, \gamma^g)$ is another p-weight. The Alperin weight conjecture states that the number of p-regular classes of G is the number of G-classes of p-weights.

If B is a p-block of G, a p-weight (P, γ) **belongs** to B if the block $b \in \mathrm{Bl}(\mathbf{N}_G(P))$ of γ induces B. The Alperin weight conjecture in the block form asserts that $l(B)$ is the number of G-classes of p-weights belonging to B. There are many fascinating consequences of this conjecture and the reader is invited to read [Alperin, Weights]. As we see, Alperin's conjecture tells us that important pieces of information on blocks can be computed locally.

(There are even more general conjectures by E. C. Dade and G. Robinson. See [Dade, 1992, 1994] and [Robinson, 1996].)

Another conjecture relating block invariants between Brauer first main correspondents is the **Alperin-McKay** conjecture. J. McKay conjectured that in every finite group G, the number of irreducible characters of p'-degree of G equals the number of irreducible characters of p'-degree of $\mathbf{N}_G(P)$, where P is a Sylow p-subgroup of G. The Alperin-McKay conjecture asserts that if b is the Brauer first main correspondent of B, then $k_0(b) = k_0(B)$, where $k_0(B)$ is the number of height zero irreducible characters in B. It is clear that the Alperin-McKay conjecture implies the McKay conjecture.

Next, we give an important result of G. Robinson concerning the number of blocks with a given defect group. This answered a question of R. Brauer.

We define the **Robinson map** R : $\mathbf{Z}(FG) \to \mathbf{Z}(FG)$ by setting

$$R(x) = \sum_{B \in \mathrm{Bl}(G)} \lambda_B(x) e_B$$

for $x \in \mathbf{Z}(FG)$. If $L \in \mathrm{cl}(G)$, observe that

$$R(\hat{L}) = \sum_{B \in \mathrm{Bl}(G)} \lambda_B(\hat{L}) e_B = \sum_{K \in \mathrm{cl}(G)} \left(\sum_{B \in \mathrm{Bl}(G)} \lambda_B(\hat{L}) a_B(\hat{K}) \right) \hat{K}.$$

Since $a_B(\hat{K}) = 0$ if K does not consist of p-regular elements, notice that $\mathrm{supp}(R(\hat{L})) \subseteq G^0$ (the **support** of an element $z = \sum_{g \in G} z_g g \in RG$, for any ring R, is the set $\mathrm{supp}(z)$ consisting of the elements $g \in G$ such that $z_g \neq 0$).

We try to find a convenient expression for the sums

$$\sum_{B \in \mathrm{Bl}(G)} \lambda_B(\hat{L}) a_B(\hat{K}).$$

Let P be a fixed Sylow p-subgroup of G and write $|P| = p^a$. For $K, L \in \mathrm{cl}(G)$, let

$$\Omega_{K,L} = \{(y, z) \in K \times L \mid Py = Pz\} = \{(y, z) \in K \times L \mid zy^{-1} \in P\}.$$

It is clear that $|\Omega_{K,L}|$ does not depend on the Sylow p-subgroup P. We prove next that

$$\frac{|\Omega_{K,L}|}{|K|} \in S.$$

(An alternative proof is inside that of Lemma (4.19).) By Sylow theory, we choose $x \in K$ such that $D_K = P \cap \mathbf{C}_G(x)$ is a Sylow p-subgroup of $\mathbf{C}_G(x)$. Write $p^{\mathrm{d}(K)} = |D_K|$. Now, notice that P acts on $\Omega_{K,L}$ by conjugation. We claim that $p^{a-\mathrm{d}(K)}$ divides every orbit size. If $(y, z) \in \Omega_{K,L}$, the size of the orbit of (y, z) is the index in P of $\mathbf{C}_P(y) \cap \mathbf{C}_P(z)$. Since $y = x^g$ for some $g \in G$, it follows that $\mathbf{C}_P(y) \subseteq \mathbf{C}_G(x)^g$. Therefore, there exists a $u \in G$ such that $\mathbf{C}_P(y) \subseteq (D_K)^u$. Hence,

$$|\mathbf{C}_P(y) \cap \mathbf{C}_P(z)| \leq |\mathbf{C}_P(y)| \leq p^{\mathrm{d}(K)},$$

and the claim follows. We see that $\frac{|\Omega_{K,L}|}{p^{a-\mathrm{d}(K)}}$ is an integer and thus

$$\frac{|\Omega_{K,L}|}{|K|} \in S.$$

(4.19) LEMMA. *If K is a class of p-regular elements and $L \in \mathrm{cl}(G)$, then*

$$\left(\frac{|\Omega_{K,L}|}{|K|}\right)^* = \sum_{B \in \mathrm{Bl}(G)} \lambda_B(\hat{L}) a_B(\hat{K}).$$

Proof. By Burnside's formula (Problem (3.9) of [Isaacs, *Characters*]), it is well known that, if $z \in G$, then

$$|\{(x,y) \in K \times L \mid xy = z\}|$$

$$= \frac{|G|}{|C_G(x_K)||C_G(x_L)|} \sum_{\chi \in \mathrm{Irr}(G)} \frac{\chi(x_K)\chi(x_L)\chi(z^{-1})}{\chi(1)}.$$

Write $K^{-1} = \{y^{-1} \mid y \in K\} = \mathrm{cl}(x_K^{-1})$. We have that

$$|\Omega_{K,L}| = \sum_{x \in P} |\{(y,z) \in K \times L \mid zy^{-1} = x\}|$$

$$= \sum_{x \in P} |\{(y,z) \in L \times K^{-1} \mid yz = x\}|$$

$$= \sum_{x \in P} \frac{|G|}{|C_G(x_K)||C_G(x_L)|} \sum_{\chi \in \mathrm{Irr}(G)} \frac{\chi(x_K^{-1})\chi(x_L)\chi(x^{-1})}{\chi(1)}$$

$$= \frac{|G|}{|C_G(x_K)||C_G(x_L)|} \sum_{\chi \in \mathrm{Irr}(G)} \left(\sum_{x \in P} \chi(x^{-1})\right) \frac{\chi(x_K^{-1})\chi(x_L)}{\chi(1)}$$

$$= \frac{|G|}{|C_G(x_K)||C_G(x_L)|} \sum_{\chi \in \mathrm{Irr}(G)} |P|[\chi_P, 1_P]\frac{\chi(x_K^{-1})\chi(x_L)}{\chi(1)}$$

and therefore,

$$\frac{|\Omega_{K,L}|}{p^{a-d(K)}} = \frac{1}{|C_G(x_K)|_{p'}} \sum_{\chi \in \mathrm{Irr}(G)} \frac{|G : C_G(x_L)|\chi(x_L)}{\chi(1)} \chi(x_K^{-1})[\chi_P, 1_P].$$

(In particular, this proves that $\frac{|\Omega_{K,L}|}{|K|} \in S$.) Now,

$$\left(\frac{|\Omega_{K,L}|}{p^{a-d(K)}}\right)^* = \frac{1}{|C_G(x_K)|_{p'}^*} \sum_{B \in \mathrm{Bl}(G)} \lambda_B(\hat{L}) \left(\sum_{\chi \in \mathrm{Irr}(B)} \chi(x_K^{-1})[\chi_P, 1_P]\right)^*$$

By the weak block orthogonality, we have that

$$\sum_{\chi \in \mathrm{Irr}(B)} \chi(x_K^{-1})[\chi_P, 1_P] = \frac{1}{|P|} \sum_{x \in P} \sum_{\chi \in \mathrm{Irr}(B)} \chi(x_K^{-1})\chi(x)$$

$$= \frac{1}{|P|} \sum_{\chi \in \mathrm{Irr}(B)} \chi(x_K^{-1})\chi(1) = |G|_{p'} f_B(\hat{K}).$$

Therefore

$$\left(\frac{|\Omega_{K,L}|}{|K|}\right)^* = \left(\frac{|\Omega_{K,L}|}{p^{a-\mathrm{d}(K)}}\right)^* \frac{1}{|K|_{p'}^*} = \sum_{B \in \mathrm{Bl}(G)} \lambda_B(\hat{L}) a_B(\hat{K}),$$

as required. ∎

Now, we are ready to count blocks. First of all, notice that by the first main theorem, when counting the blocks of G with a given defect group D, we may assume that D is a normal subgroup of G.

Let us denote the set of p-regular classes of G with defect group D by $\mathrm{cl}(G^0 \,|\, D)$. Write $|D| = p^d$ and $|G|_p = p^a$. We define the matrix

$$\Omega = \left(\frac{|\Omega_{K,L}|}{p^{a-d}} \bmod p\right)_{K,L \in \mathrm{cl}(G^0 \,|\, D)},$$

which is a matrix in $\mathbb{Z}_p$.

(4.20) THEOREM (Robinson). *Suppose that D is a normal p-subgroup of G. Then the number of blocks of G with defect group D is the rank of the matrix Ω.*

Proof. If U is the F-span of $\{\hat{L} \,|\, L \in \mathrm{cl}(G^0 \,|\, D)\}$, we claim that $\mathrm{R}(U) \subseteq U$. If $L \in \mathrm{cl}(G^0 \,|\, D)$ and $\lambda_B(\hat{L}) \neq 0$, then, by the min-max theorem (4.4), we have that the defect groups of B are contained in D. Since $D \subseteq \mathbf{O}_p(G)$ is contained in every defect group of every block of G, we conclude that the defect group of B is D. Hence, for $L \in \mathrm{cl}(G^0 \,|\, D)$, we have

$$\mathrm{R}(\hat{L}) = \sum_{B \in \mathrm{Bl}(G)} \lambda_B(\hat{L}) e_B = \sum_{B \in \mathrm{Bl}(G \,|\, D)} \lambda_B(\hat{L}) e_B.$$

Now, since $a_B(\hat{K}) = 0$ if K does not consist of p-regular elements, by Lemma (4.15) we have that

$$e_B = \sum_{K \in \mathrm{cl}(G^0 \,|\, D)} a_B(\hat{K})\hat{K}$$

for $B \in \text{Bl}(G \,|\, D)$. Therefore,

$$\text{R}(\hat{L}) = \sum_{K \in \text{cl}(G^0 \,|\, D)} \left(\sum_{B \in \text{Bl}(G \,|\, D)} \lambda_B(\hat{L}) a_B(\hat{K}) \right) \hat{K}.$$

This proves that $\text{R}(U) \subseteq U$, as claimed.

Now, we calculate the rank of the linear map $\text{R}_U : U \to U$. We claim that this rank is $|\text{Bl}(G \,|\, D)|$. To see this, by elementary linear algebra it suffices to show that $\{e_B \,|\, B \in \text{Bl}(G \,|\, D)\}$ is a basis of $\text{R}(U)$. Since the e_B's are linearly independent (recall that they are orthogonal) and $\text{R}(\hat{L}) = \sum_{B \in \text{Bl}(G \,|\, D)} \lambda_B(\hat{L}) e_B$ for every $L \in \text{cl}(G^0 \,|\, D)$, it suffices to show that each e_B actually lies in $\text{R}(U)$. Since $\text{R}(e_B) = e_B$, this is clear.

The matrix of R_U with respect to the basis $\{\hat{L} \,|\, L \in \text{cl}(G^0 \,|\, D)\}$ is

$$\left(\sum_{B \in \text{Bl}(G \,|\, D)} \lambda_B(\hat{L}) a_B(\hat{K})) \right)_{L, K \in \text{cl}(G^0 \,|\, D)} .$$

Since $\lambda_B(\hat{L}) = 0$ if $B \notin \text{Bl}(G \,|\, D)$ by the first paragraph of the proof, we have that this matrix is

$$\left(\sum_{B \in \text{Bl}(G)} \lambda_B(\hat{L}) a_B(\hat{K})) \right)_{L, K \in \text{cl}(G^0 \,|\, D)} = \left((\frac{|\Omega_{K,L}|}{p^{a-d}})^* \frac{1}{|K|_{p'}^*} \right)_{L, K \in \text{cl}(G^0 \,|\, D)}$$

by Lemma (4.19). Since the rank of this matrix equals the rank of

$$\left((\frac{|\Omega_{K,L}|}{p^{a-d}})^* \right)_{L, K \in \text{cl}(G^0 \,|\, D)} ,$$

the proof of the theorem is complete. ∎

The next corollary is a generalization due to G. Robinson of a celebrated result of J. A. Green.

(4.21) COROLLARY (Green). *Suppose that D is a defect group of some block of G and let P be a Sylow p-subgroup of G containing D. Then there is a p-regular $x \in G$ such that $D \in \text{Syl}_p(\mathbf{C}_G(x))$ and $P \cap P^x = D$.*

Proof. First, we prove the theorem when $D \triangleleft G$. By Robinson's theorem (4.20), we know that there are conjugacy classes K and L of G with defect group D and consisting of p-regular elements, such that

$$\frac{|\Omega_{K,L}|}{p^{a-d}} \not\equiv 0 \bmod p ,$$

where $p^a = |G|_p$ and $p^d = |D|$. Let $\mathcal{X} = \{Pg \mid g \in G\}$. Now,

$$\Omega_{K,L} = \{(x,y) \in K \times L \mid Px = Py\} = \bigcup_{Pg \in \mathcal{X}} (Pg \cap K) \times (Pg \cap L) ,$$

and thus

$$|\Omega_{K,L}| = \sum_{Pg \in \mathcal{X}} |Pg \cap K||Pg \cap L| .$$

Now, P acts on $\mathcal{X}$ by right multiplication. If $\mathcal{O}(Pg) = \{Pgu \mid u \in P\}$ is the P-orbit of Pg, we have that $|\mathcal{O}(Pg)| = |P : P \cap P^g|$ since $Pgu = Pg$ if and only if $u \in P \cap P^g$.

Notice that, for $u \in P$, we have that $|Pg \cap K| = |Pgu \cap K|$ and $|Pg \cap L| = |Pgu \cap L|$, since $(Pg \cap K)^u = (Pg)^u \cap K^u = Pgu \cap K$, for instance. Observe then that $P \cap P^g$ acts by conjugation on $Pg \cap K$ and $Pg \cap L$. If Δ is a complete set of representatives of orbits of the right multiplication action of P on $\mathcal{X}$, it follows that

$$|\Omega_{K,L}| = \sum_{Pg \in \Delta} |P : P \cap P^g||Pg \cap K||Pg \cap L| .$$

Now, as we said, $P \cap P^g$ acts by conjugation on $Pg \cap K$ and the orbit size of $x \in Pg \cap K$ is $|P \cap P^g : \mathbf{C}_{P \cap P^g}(x)|$. Since $D \in \mathrm{Syl}_p(\mathbf{C}_G(x))$ and $D \triangleleft G$, we see that $D = \mathbf{C}_{P \cap P^g}(x)$. Therefore,

$$|P \cap P^g : D|^2 \text{ divides } |Pg \cap K||Pg \cap L|$$

and thus

$$|P \cap P^g : D| \text{ divides } \frac{|P : P \cap P^g||Pg \cap K||Pg \cap L|}{p^{a-d}} .$$

Since

$$\frac{|\Omega_{K,L}|}{p^{a-d}} = \sum_{Pg \in \Delta} \frac{|P : P \cap P^g||Pg \cap K||Pg \cap L|}{p^{a-d}} \not\equiv 0 \bmod p ,$$

we conclude that there exists a g such that $|P \cap P^g : D| = 1$ with $|Pg \cap K| \neq 0$. If $x \in Pg \cap K$, then $P \cap P^x = P \cap P^g = D$, $D \in \mathrm{Syl}_p(\mathbf{C}_G(x))$ and x is p-regular, as desired.

Now, we prove the general case. Let $D \subseteq P \cap N \subseteq S$ be a Sylow p-subgroup of $N = \mathbf{N}_G(D)$. By the first main theorem and the case proved above, there is some p-regular $x \in N$ with $D \in \mathrm{Syl}_p(\mathbf{C}_N(x))$ such that $S \cap S^x = D$. We know that the class in G of x has defect group D by Lemma (4.16). Now, if $P \cap P^x > D$, then $P \cap P^x \cap N = (P \cap N) \cap (P \cap N)^x > D$. Since $S \cap S^x = D$, this is a contradiction which proves the corollary. ∎

Next, we give a group theoretic expression of the Robinson map.

Recall that we denote the set of p-elements of G by G_p. If $X \subseteq G$, then we write $\hat{X} = \sum_{x \in X} x$.

(4.22) LEMMA. *If Q is any p-subgroup of G, then*

$$|\{P \in \mathrm{Syl}_p(G) \mid Q \subseteq P\}| \equiv 1 \bmod p .$$

As a consequence,

$$\widehat{G_p} = \sum_{P \in \mathrm{Syl}_p(G)} \hat{P}$$

in $\mathbf{Z}(FG)$.

Proof. We have that Q acts by conjugation on $\mathrm{Syl}_p(G)$, a set of size $|\mathrm{Syl}_p(G)| \equiv 1 \bmod p$. The orbits of the action of Q are of size 1 or a power of p. Since a Q-orbit has size 1 if and only if $Q \subseteq P$, the first part follows. Now, if x is any p-element of G, notice that the coefficient of x in $\sum_{P \in \mathrm{Syl}_p(G)} \hat{P}$ is the number of Sylow p-subgroups containing x. Since this number is $1 \bmod p$ by the first part, the lemma follows. ∎

Let us write $\pi : \mathbf{Z}(FG) \to \mathbf{Z}(FG)$ for the linear map defined by $\pi(\hat{K}) = \hat{K}$ if K consists of p-regular elements and $\pi(\hat{K}) = 0$ otherwise.

(4.23) THEOREM (Külshammer). *If $z \in \mathbf{Z}(FG)$, then*

$$\mathrm{R}(z) = \pi(\widehat{G_p} z) .$$

Proof. It suffices to show that $\mathrm{R}(\hat{L}) = \pi(\widehat{G_p} \hat{L})$ for every conjugacy class L of G. By Lemma (4.19), we know that if K is a class of p-regular elements, then the coefficient of $\hat{K}$ in $\mathrm{R}(\hat{L})$ is

$$(\frac{|\Omega_{K,L}|}{|K|})^* .$$

Now, the coefficient z_K of $\hat{K}$ in $\pi(\widehat{G_pL}) = \pi(\sum_{P\in\mathrm{Syl}_p(G)} \hat{P}\hat{L})$ is

$$\left(\sum_{P\in\mathrm{Syl}_p(G)} |\{(x,y) \in P \times L \,|\, xy = x_K\}| \right)^* \,,$$

where, as usual, x_K is a fixed element of K. Since this coefficient depends not on x_K but on K, we see that

$$z_K = \left(\frac{1}{|K|} \sum_{P\in\mathrm{Syl}_p(G)} |\{(x,y) \in P \times L \,|\, xy \in K\}| \right)^* .$$

Now, notice that the map

$$\{(x,y) \in P \times L \,|\, xy \in K\} \to \{(x,y) \in K \times L \,|\, Px = Py\} = \Omega_{K,L}$$

given by $(x,y) \mapsto (xy,y)$ is a bijection. Hence,

$$z_K = \left(\frac{1}{|K|} \sum_{P\in\mathrm{Syl}_p(G)} |\Omega_{K,L}| \right)^* .$$

Since $|\Omega_{K,L}|$ does not depend on the Sylow p-subgroup P and $\frac{|\Omega_{K,L}|}{|K|} \in S$, it follows that

$$z_K = (|\mathrm{Syl}_p(G)| \frac{|\Omega_{K,L}|}{|K|})^* = (\frac{|\Omega_{K,L}|}{|K|})^* .$$

The result follows.   ∎

PROBLEMS

(4.1) Suppose that $H \subseteq G$ and let b be a block of H such that b^G is defined. Find an example in which there exists $H \subseteq K \subseteq G$ such that b^K is not defined.

(4.2) Suppose that $H \subseteq K \subseteq G$ and let $b \in \mathrm{Bl}(H)$ with b^K defined. Show that b^G is defined iff $(b^K)^G$ is defined, and that, in this case, $(b^G) = (b^K)^G$.

(4.3) Suppose that $\mathbf{N}_G(P) \subseteq H \subseteq G$, where p is a p-subgroup of G. Show that the map $\mathrm{Bl}(H \,|\, P) \to \mathrm{Bl}(G \,|\, P)$ given by $b \mapsto b^G$ is a bijection.

(4.4) Suppose that $\sigma : G \to H$ is an isomorphism of groups. If B is a p-block with defect group D, prove that $B^\sigma = \{\chi^\sigma \mid \chi \in B\}$ is a p-block of H with defect group D^σ. (See Problem (2.1).) If U is a subgroup of G and $b \in \mathrm{Bl}(U)$ is such that b^G is defined, show that $(b^\sigma)^H$ is defined and equals $(b^G)^\sigma$. Also, if $\tau \in \mathrm{Aut}(G)$ fixes some $\chi \in \mathrm{Irr}(B)$, prove that D and D^τ are G-conjugate.

(4.5) If B_1 and B_2 are blocks of G with defect group D, prove that $B_1 = B_2$ if and only if $\lambda_{B_1}(\hat{K}) = \lambda_{B_2}(\hat{K})$ for every conjugacy class $K \in \mathrm{cl}(G^0)$ of defect group D.

(4.6) (Tsushima) Suppose that D is a normal p-subgroup of G. If there are m distinct conjugacy classes of p-regular elements of defect group D inside $\mathbf{O}_{p'}(G)$, show that there are at least m blocks of G with defect group D.

(4.7) (Robinson) If $D \lhd G$, show that the number of p-blocks of defect group D is the number of p-blocks of $G/\Phi(D)$ of defect group $D/\Phi(D)$, and, therefore, if we wish to count p-blocks of a given defect group D, it is no loss to assume that D is elementary abelian.

(4.8) Let $P \in \mathrm{Syl}_p(G)$. Prove that the number of blocks of G of maximal defect is equal to the number of p-regular classes of $\mathbf{N}_G(P)$ contained in $\mathbf{C}_G(P)$.

(4.9) Let G be p-solvable with $\mathbf{O}_{p'}(G) = 1$. Show that G has a unique p-block.

(HINT: Use the fact that $\mathbf{C}_G(\mathbf{O}_p(G)) \subseteq \mathbf{O}_p(G)$.)

(4.10) Let $\mathrm{Id}(\mathbf{Z}(FG))$ be the span of the block idempotents of G. Prove that

$$\mathbf{Z}(FG) = \mathrm{Id}(\mathbf{Z}(FG)) \oplus \mathbf{J}(\mathbf{Z}(FG)).$$

Show that the Robinson map is the projection of $\mathbf{Z}(FG)$ over $\mathrm{Id}(\mathbf{Z}(FG))$.

5. The Second Main Theorem

One of the main reasons why R. Brauer studied modular characters was to obtain a deeper insight into the nature of the ordinary characters of a finite group. We already know that the value of an irreducible character on a p-regular element can be computed once we know the irreducible Brauer characters and the decomposition numbers. He found it, however, disturbing that he had to restrict himself to p-regular elements. By allowing the decomposition numbers to be not necessarily rational, it is possible to overcome this difficulty.

Our objective in this chapter is to prove Brauer's second main theorem. Without a doubt, this is one of the highlights of modular representation theory. To state it, we need to introduce the generalized decomposition numbers.

(5.1) LEMMA. *Let x be a p-element of G and let $H = \mathbf{C}_G(x)$. For $\chi \in \mathrm{Irr}(G)$ and $\varphi \in \mathrm{IBr}(H)$, there exist unique $d^x_{\chi\varphi} \in \mathbb{C}$ such that*

$$\chi(xy) = \sum_{\varphi \in \mathrm{IBr}(H)} d^x_{\chi\varphi}\varphi(y)$$

for all p-regular $y \in H$.

Proof. Let $\psi \in \mathrm{Irr}(H)$ and let $\mathcal{X}$ be a representation affording ψ. Since $x \in \mathbf{Z}(H)$, we know by Theorem (1.20) that $\mathcal{X}(x)$ is a scalar matrix. By taking traces, it easily follows that, for $y \in H$, we have $\psi(xy) = \lambda_\psi(x)\psi(y)$, where $\psi_{(x)} = \psi(1)\lambda_\psi$ and λ_ψ is a linear character of $\langle x \rangle$. Now, if $y \in H$ is p-regular, then

$$\chi(xy) = \sum_{\psi \in \mathrm{Irr}(H)} [\chi_H, \psi]\psi(xy) = \sum_{\psi \in \mathrm{Irr}(H)} \frac{[\chi_H, \psi]\psi(x)}{\psi(1)}\psi(y)$$

$$= \sum_{\varphi \in \mathrm{IBr}(H)} \left(\sum_{\psi \in \mathrm{Irr}(H)} \frac{[\chi_H, \psi]\psi(x)}{\psi(1)} d_{\psi\varphi} \right) \varphi(y) = \sum_{\varphi \in \mathrm{IBr}(H)} d^x_{\chi\varphi}\varphi(y),$$

where

$$d^x_{\chi\varphi} = \sum_{\psi\in\mathrm{Irr}(H)} \frac{[\chi_H,\psi]\psi(x)}{\psi(1)} d_{\psi\varphi}.$$

The uniqueness follows from the linear independence of the elements of $\mathrm{IBr}(H)$. ∎

The numbers

$$d^x_{\chi\varphi} = \sum_{\psi\in\mathrm{Irr}(H)} \frac{[\chi_H,\psi]\psi(x)}{\psi(1)} d_{\psi\varphi}$$

are called the **generalized decomposition numbers**. Notice that if ξ is a primitive $o(x)$th root of unity, then $d^x_{\chi\varphi} \in \mathbb{Z}[\xi]$. We will see that, in some aspects, they behave like decomposition numbers.

Recall that if x is a p-element of G, $H = \mathbf{C}_G(x)$ and $b \in \mathrm{Bl}(H)$, then we know by Theorem (4.14) that b^G is defined.

(5.2) THEOREM (Second Main Theorem). *Let x be a p-element of G and let $b \in \mathrm{Bl}(\mathbf{C}_G(x))$. If $\chi \in \mathrm{Irr}(G)$ is not in $\mathrm{Irr}(b^G)$, then $d^x_{\chi\varphi} = 0$ for every $\varphi \in \mathrm{IBr}(b)$.*

We need a few lemmas.

(5.3) LEMMA. *Let A be a ring and let R be a subring of $\mathbf{Z}(A)$ (with the identity of A in R). Assume that A is finitely generated as an R-module. Then $\mathbf{J}(R) \subseteq \mathbf{J}(A)$.*

Proof. First of all, recall that A is an R-algebra. Let M be a simple A-module. We must show that $M\mathbf{J}(R) = 0$. Write $A = \sum a_i R$, a finite sum, and let $0 \neq m \in M$. Note that $M = mA = \sum ma_i R$, so that M is finitely generated as an R-module. By Nakayama's lemma (1.7), it follows that $M\mathbf{J}(R) < M$. Since $M\mathbf{J}(R)$ is an A-submodule of M (because $\mathbf{J}(R) \subseteq \mathbf{Z}(A)$), it must be 0, as desired. ∎

If $\theta : T \to R$ is a surjective ring homomorphism, notice that $\theta(\mathbf{J}(T)) \subseteq \mathbf{J}(R)$. This follows from the fact that for every simple R-module V, the multiplication $vt = v\theta(t)$ makes V into a simple T-module. Therefore, $0 = V\mathbf{J}(T) = V\theta(\mathbf{J}(T))$.

We have been using in these notes the ring homomorphism $^* : S \to F$ and its extension $^* : SG \to FG$. Notice that $\ker(^*) = \mathbf{J}(S)G = \mathcal{P}G$.

(5.4) LEMMA. *Let $f \in SG$ be an idempotent (not necessarily central) and suppose that $x \in f(SG)f$ satisfies $x^* = f^*$. Then there exists a unique element $y \in f(SG)f$ such that $xy = f = yx$.*

Proof. Write $A = f(SG)f$. Note that A is a ring with identity f and that we are trying to prove x is invertible in A. By Theorem (1.6), it suffices to show that $f - x \in \mathbf{J}(A)$. Let $R = fS$. Observe that R is a subring of $\mathbf{Z}(A)$ and that R contains the identity of A. Also, A is finitely generated over R (since the finite set fGf generates it). By Lemma (5.3), we conclude that $\mathbf{J}(R) \subseteq \mathbf{J}(A)$.

The map $s \mapsto fs$ is a surjective homomorphism of S onto R, and thus $f\mathbf{J}(S) \subseteq \mathbf{J}(R)$. It follows that $f\mathbf{J}(S) \subseteq \mathbf{J}(A)$. By hypothesis, $f - x \in \mathbf{J}(S)G$. Hence, we may write $f - x = \sum_{g \in G} u_g g$, where $u_g \in \mathbf{J}(S)$ for all $g \in G$. Since $x \in f(SG)f$, we have that

$$f - x = f(f - x)f = \sum_{g \in G} fu_g(fgf) \in f\mathbf{J}(S)(fSGf) \subseteq \mathbf{J}(A)A \subseteq \mathbf{J}(A),$$

as desired. ∎

Notice that if x is an element in any ring (recall that our rings have an identity), then

$$(1 + x + \ldots + x^{n-1})(1 - x) = 1 - x^n = (1 - x)(1 + x + \ldots + x^{n-1}).$$

Therefore, if x is nilpotent, then it follows that $1 - x$ is invertible. We will use this fact in the following lemma.

(5.5) LEMMA. *Suppose that B is a block of G and let $x \in \mathbf{Z}(SG)$. If $\lambda_B(x^*) = 1$, then there exists a $y \in f_B\mathbf{Z}(SG)$ such that $xy = f_B$.*

Proof. Since $\lambda_B((f_Bx)^*) = \lambda_B(e_B)\lambda_B(x^*) = 1$, we can replace x by f_Bx without affecting either the hypothesis or the conclusion of the lemma. We therefore assume that $x \in f_B\mathbf{Z}(SG)$. In particular, $x^*e_B = x^*$. Thus, $\lambda_{B'}(x^*) = \lambda_{B'}(x^*e_B) = 0$ for all blocks $B' \neq B$. Since $\lambda_B(e_B - x^*) = 0$ by hypothesis, we have that $\lambda_{B'}(e_B - x^*) = 0$ for every block B'. Therefore, $e_B - x^* \in \mathbf{J}(\mathbf{Z}(FG))$ and thus, $e_B - x^*$ is nilpotent. Hence, x^* is invertible in the ring $e_B\mathbf{Z}(FG)$ with identity e_B. It follows that there exists a $u \in f_B\mathbf{Z}(SG)$ with $(xu)^* = e_B$. Since $xu \in f_B(SG)f_B$, Lemma (5.4) yields a unique element $v \in f_B(SG)f_B$ with $xuv = f_B$. Since $xu \in \mathbf{Z}(SG)$, the uniqueness implies that v is centralized by G. So we can take $y = uv$. ∎

(5.6) LEMMA. *Let $H \subseteq G$ and let b be a block of H with $b^G = B \in \mathrm{Bl}(G)$. Then there exists a $w \in SG$ such that*
(a) $(1 - f_B)f_b = (1 - f_B)w$,
(b) $wf_b = w$,
(c) H centralizes w, and
(d) w is supported on $G - H$.

Proof. Write $f_B = a - c$, where $\operatorname{supp}(a) \subseteq H$ and $\operatorname{supp}(c) \subseteq G - H$. Note that $a \in \mathbf{Z}(SH)$ and that H centralizes c. Since $b^G = B$, we have $\lambda_b^G = \lambda_B$ and hence

$$1 = \lambda_B(e_B) = \lambda_b(a^*).$$

By Lemma (5.5), there exists a $y \in f_b \mathbf{Z}(SH)$ such that $ay = f_b$.

Now let $w = cy$, and note that conclusions (b), (c) and (d) are immediate. We have $f_B y = (a - c)y = f_b - w$, and hence

$$(1 - f_B)(f_b - w) = (1 - f_B)f_B y = 0,$$

proving (a).   ∎

(5.7) THEOREM. *Let $H \subseteq G$ and suppose that b is a block of H with $b^G = B \in \mathrm{Bl}(G)$. Suppose that $h \in H$ and that $\mathbf{C}_G(h_p) \subseteq H$, where h_p is the p-part of h. If $\chi \in \mathrm{Irr}(G)$ and $\chi \notin B$, then $\chi(f_b h) = 0$.*

Proof (Isaacs). Let M be a $\mathbb{C}G$-module affording χ and let $V = M f_b$. Since $f_b \in \mathbf{Z}(SH)$, it follows that V is a $\mathbb{C}H$-submodule of M. Now, M is the direct sum of the subspaces V and $W = M(1 - f_b)$. Since $f_b h$ acts like h on V and annihilates W, we see that $\chi(f_b h)$ is the sum of the eigenvalues of the action of h on V (counting multiplicities). We may certainly assume that $V \neq 0$ (since otherwise $M f_b h = 0$ and $\chi(f_b h) = 0$).

Let ω be a primitive pth root of unity in $\mathbf{R} \subseteq S$. We will show that if α is any complex number, then α and $\omega\alpha$ have equal multiplicities as eigenvalues for the action of h on V. If this is proven and α is an eigenvalue for the action of h on V, then $\alpha, \omega\alpha, \omega^2\alpha, \ldots, \omega^{p-1}\alpha$ are all eigenvalues with the same multiplicity. Since the sum of these numbers is zero, the result will then follow.

If U_α is the eigenspace of the action of h for the eigenvalue α, then the multiplicity of the eigenvalue α is $\dim_\mathbb{C}(U_\alpha)$. (Recall that V has a basis of eigenvectors for the action of h because V is a completely reducible $\mathbb{C}\langle h \rangle$-module and the irreducible $\mathbb{C}\langle h \rangle$-modules are one dimensional.)

Let w be as in Lemma (5.6). We claim that there exist $w_1, w_2, \ldots, w_p \in f_b SG f_b$ such that $w = \sum_{i=1}^p w_i$ and $(w_i)^h = w_{i+1}$, with the subscripts interpreted modulo p.

Since H centralizes w, we see that H acts on the elements of $\operatorname{supp}(w)$ by conjugation. Also, the coefficient of x in w is the coefficient of x^z for every $z \in H$. Since $\operatorname{supp}(w) \subseteq G - H$ and $\mathbf{C}_G(h_p) \subseteq H$, each orbit of $\langle h \rangle$ on $\operatorname{supp}(w)$ has size divisible by p because, otherwise, there would exist an $x \in G - H$ such that $|\langle h \rangle : \mathbf{C}_{\langle h \rangle}(x)|$ is a p'-number. In this case, $\langle h_p \rangle \subseteq \mathbf{C}_G(x)$ and this is impossible since $\mathbf{C}_G(h_p) \subseteq H$. Therefore, if $x \in \operatorname{supp}(w)$, we have that p divides $|\langle h \rangle : \mathbf{C}_{\langle h \rangle}(x)|$. In particular, $\mathbf{C}_{\langle h \rangle}(x) \subseteq \langle h^p \rangle$. Hence, if

$$\{\mathbf{C}_{\langle h \rangle}(x)y_j\}_{j=1,\ldots,t}$$

is the set of right cosets of $\langle h^p \rangle$ modulo $\mathbf{C}_{\langle h \rangle}(x)$, it follows that

$$\{\mathbf{C}_{\langle h \rangle}(x) y_j h^i\}_{\substack{j=1,\ldots,t \\ i=1,\ldots,p}}$$

is the set of right cosets of $\langle h \rangle$ modulo $\mathbf{C}_{\langle h \rangle}(x)$. Now, if $\mathcal{O}$ is the $\langle h \rangle$-orbit of $x \in \text{supp}(w)$, then we may write

$$\mathcal{O} = \Delta^h \cup \Delta^{h^2} \cup \ldots \cup \Delta^{h^p},$$

where $\Delta = \{x^{y_1}, \ldots, x^{y_t}\}$. Furthermore, since $h^p \in \langle h^p \rangle$, we have that

$$\langle h^p \rangle = \bigcup_{j=1,\ldots,t} \mathbf{C}_{\langle h \rangle}(x) y_j = \bigcup_{j=1,\ldots,t} \mathbf{C}_{\langle h \rangle}(x) y_j h^p.$$

This easily implies that $\Delta^{h^p} = \Delta$. If we do this for each orbit $\mathcal{O}$, it is clear that we may construct elements $w_1, w_2, \ldots, w_p \in SG$ such that $w = \sum_{i=1}^{p} w_i$ and $(w_i)^h = w_{i+1}$, with the subscripts interpreted modulo p. Since $w = f_b w f_b$ (by Lemma (5.6.b,c)) and $f_b^h = f_b$, we can replace w_i by $f_b w_i f_b$ and so assume that $w_i \in f_b(SG)f_b$. This proves the claim. Since $w_i \in f_b(SG)f_b$, notice that $Vw_i \subseteq V$.

Now define $s \in SG$ by setting

$$s = \sum_{i=1}^{p} \omega^{-i} w_i.$$

Observe that $Vs \subseteq V$. Also,

$$s^h = \sum_{i=1}^{p} \omega^{-i} w_{i+1} = \omega \sum_{i=1}^{p} \omega^{-(i+1)} w_{i+1} = \omega s.$$

Now, for $u \in U_\alpha$, we have that

$$(us)h = uhs^h = \alpha\omega(us).$$

So multiplication by s maps U_α into the eigenspace for $U_{\alpha\omega}$. We will show that the map $v \mapsto vs$ is injective on V. In this case, we will have that

$$\dim_{\mathbb{C}}(U_\alpha) = \dim_{\mathbb{C}}(U_\alpha s) \leq \dim_{\mathbb{C}}(U_{\alpha\omega}).$$

By repeating the argument with the eigenvalue $\alpha\omega^i$, we will get

$$\dim_{\mathbb{C}}(U_\alpha) \leq \dim_{\mathbb{C}}(U_{\alpha\omega}) \leq \dim_{\mathbb{C}}(U_{\alpha\omega^2}) \leq \dim_{\mathbb{C}}(U_{\alpha\omega^p}) = \dim_{\mathbb{C}}(U_\alpha),$$

and the proof of the theorem will be complete.

Now, all that remains to show is that multiplication by s is an injection on V. Note that $\omega^* = 1$ since F has no nontrivial elements with p-power order; thus, $s^* = w^*$. By Lemma (5.6.a), this gives

$$((1 - f_B)f_b)^* = ((1 - f_B)w)^* = (1 - f_B)^* w^* = (1 - f_B)^* s^* = ((1 - f_B)s)^*.$$

Write $f = (1 - f_B)f_b$. First, we show that $vf = v$ for all $v \in V$. Since $V = Mf_b$, it is clear that $vf_b = v$. But also, since χ does not lie in B, by Lemma (3.13) we have that $Mf_B = 0$. Thus $v(1 - f_B) = v$ for $v \in V$. (In particular, observe that $f \neq 0$ because $V \neq 0$.) Now, f is an idempotent (since f_b commutes with $1 - f_B$ and both are idempotents) with $f^* = ((1 - f_B)s)^*$. We wish to apply Lemma (5.4), so we need to establish that $(1 - f_B)s \in f(SG)f$. Since $s = \sum_{i=1}^{p} \omega^{-i}w_i$ and $f_b w_i f_b = w_i$, we see that $f_b s f_b = s$. Now, $1 - f_B \in \mathbf{Z}(SG)$ and thus

$$f(1 - f_B)sf = (1 - f_B)f_b(1 - f_B)s(1 - f_B)f_b = (1 - f_B)f_b s f_b = (1 - f_B)s,$$

as required. Since we have the hypotheses of Lemma (5.4), we conclude that there exists a $y \in SG$ with $(1 - f_B)sy = f$. Now, if $v \in V$, since $v = v(1 - f_B)$ and $vf = v$, we have that

$$vsy = v(1 - f_B)sy = vf = v.$$

Hence, if $vs = 0$, we deduce that $v = 0$. This proves that right multiplication by s is injective, as desired. ∎

Proof of the Second Main Theorem. Let x be a p-element of G and write $H = \mathbf{C}_G(x)$. Let b be a block of H and recall that b^G is defined. Write $b^G = B$. If $\varphi \in \mathrm{IBr}(b)$ and $\chi \in \mathrm{Irr}(G)$ with $\chi \notin B$, then we must show that the generalized decomposition number $d^x_{\chi\varphi} = 0$.

If $\psi \in \mathrm{Irr}(H)$ and $z \in H$, by Lemma (3.31) we have that $\psi(f_b z) = \psi(z)$, if $\psi \in \mathrm{Irr}(b)$ while $\psi(f_b z) = 0$ if $\psi \notin \mathrm{Irr}(b)$. By Theorem (5.7), for p-regular $y \in H$, we have that

$$0 = \chi(f_b xy) = \sum_{\psi \in \mathrm{Irr}(H)} [\chi_H, \psi]\psi(f_b xy) = \sum_{\psi \in \mathrm{Irr}(b)} [\chi_H, \psi]\psi(xy)$$

$$= \sum_{\psi \in \mathrm{Irr}(b)} \frac{[\chi_H, \psi]\psi(x)}{\psi(1)}\psi(y) = \sum_{\psi \in \mathrm{Irr}(b)} \sum_{\varphi \in \mathrm{IBr}(b)} d_{\psi\varphi} \frac{[\chi_H, \psi]\psi(x)}{\psi(1)}\varphi(y)$$

$$= \sum_{\varphi \in \mathrm{IBr}(b)} \left(\sum_{\psi \in \mathrm{Irr}(H)} d_{\psi\varphi} \frac{[\chi_H, \psi]\psi(x)}{\psi(1)} \right) \varphi(y)$$

$$= \sum_{\varphi \in \mathrm{IBr}(b)} d^x_{\chi\varphi}\varphi(y).$$

The result follows by the linear independence of the irreducible Brauer characters of H. ∎

(5.8) COROLLARY. *Let x be a p-element of G and let $y \in \mathbf{C}_G(x)$ be a p'-element. Suppose that $\chi \in \mathrm{Irr}(B)$, where $B \in \mathrm{Bl}(G)$. Then*

$$\chi(xy) = \sum_{\mu} d^x_{\chi\mu}\mu(y)\,,$$

where the sum is over $\mu \in \mathrm{IBr}(b)$ for all blocks $b \in \mathrm{Bl}(\mathbf{C}_G(x))$ inducing B.

Proof. This is clear by Lemma (5.1) and the second main theorem. ∎

Next is a key result in representation theory. As a very particular case of it, we obtain Theorem (3.18) on p-defect zero characters.

(5.9) COROLLARY. *Let $\chi \in \mathrm{Irr}(G)$ and let $g \in G$. If g_p is not contained in any defect group of the block of χ, then $\chi(g) = 0$.*

Proof. Let B be the block of χ. Write $x = g_p$ and $y = g_{p'}$ and suppose that $b \in \mathrm{Bl}(\mathbf{C}_G(x))$ is such that $b^G = B$. By Theorem (4.14), we know that $\langle x \rangle$ is contained in some defect group of B. Since this is not possible by hypothesis, we see that such a block b cannot exist. By Corollary (5.8), $\chi(xy) = 0$ and the proof is complete. ∎

Given a p-element x of G, we define the p-**section** of x as $S(x) = \{u \in G \mid u_p \text{ is } G\text{-conjugate to } x\}$. Notice that $S(x)$ is a union of conjugacy classes of G. In fact, if $\{y_1, \ldots, y_l\}$ is a complete set of representatives of conjugacy classes of p-regular elements of $\mathbf{C}_G(x)$, we claim that $\{xy_1, \ldots, xy_l\}$ is a complete set of representatives of the G-classes inside $S(x)$. By the uniqueness of the p-p' decomposition of the elements of G, observe that $(xy_i)^g = xy_j$ if and only if $g \in \mathbf{C}_G(x)$ and $(y_i)^g = y_j$, which happens if and only if $i = j$. On the other hand, if $z \in S(x)$, then $(z_p)^u = x$ for some $u \in G$. In this case, $(z_{p'})^u$ is a p-regular element in $\mathbf{C}_G(x)$. Therefore, there exists a $v \in \mathbf{C}_G(x)$ such that $(z_{p'})^{uv} = y_j$ for some j. Now,

$$z^{uv} = xy_j$$

as desired.

If $\theta \in \mathrm{cf}(G)$ and $B \in \mathrm{Bl}(G)$, recall that we write

$$\theta_B = \sum_{\chi \in \mathrm{Irr}(B)} [\theta, \chi]\chi$$

for the B-part of θ. Next is another important application of the second main theorem.

(5.10) THEOREM. *If $\theta \in \mathrm{cf}(G)$ vanishes on $S(x)$, then so does θ_B for all $B \in \mathrm{Bl}(G)$.*

Proof. Let $H = \mathbf{C}_G(x)$ and let $y \in H^0$. Then

$$\theta_B(xy) = \sum_{\chi \in \mathrm{Irr}(B)} [\theta, \chi]\chi(xy) = \sum_{\chi \in \mathrm{Irr}(B)} [\theta, \chi] \sum_{\varphi \in \mathrm{IBr}(H)} d^x_{\chi\varphi}\varphi(y)$$

$$= \sum_{\varphi \in \mathrm{IBr}(H)} \left(\sum_{\chi \in \mathrm{Irr}(B)} [\theta, \chi]d^x_{\chi\varphi} \right) \varphi(y).$$

Since

$$0 = \theta(xy) = \sum_{B \in \mathrm{Bl}(G)} \theta_B(xy)$$

for every $y \in H^0$, we have that

$$0 = \sum_{\varphi \in \mathrm{IBr}(H)} \left(\sum_{\chi \in \mathrm{Irr}(G)} [\theta, \chi]d^x_{\chi\varphi} \right) \varphi(y).$$

Therefore,

$$\sum_{\chi \in \mathrm{Irr}(G)} [\theta, \chi]d^x_{\chi\varphi} = 0$$

by the linear independence of the elements of $\mathrm{IBr}(H)$. What we wish to show is that

$$\sum_{\chi \in \mathrm{Irr}(B)} [\theta, \chi]d^x_{\chi\varphi} = 0$$

for every $\varphi \in \mathrm{IBr}(H)$. By the second main theorem, this is clear if the block of φ does not induce B. Assume now that the block of φ induces B. By the second main theorem, we have that

$$\sum_{\chi \in \mathrm{Irr}(G) - \mathrm{Irr}(B)} [\theta, \chi]d^x_{\chi\varphi} = 0.$$

Since

$$0 = \sum_{\chi \in \mathrm{Irr}(G)} [\theta, \chi]d^x_{\chi\varphi} = \sum_{\chi \in \mathrm{Irr}(B)} [\theta, \chi]d^x_{\chi\varphi} + \sum_{\chi \in \mathrm{Irr}(G) - \mathrm{Irr}(B)} [\theta, \chi]d^x_{\chi\varphi},$$

the result follows.   ∎

(5.11) COROLLARY (Block Orthogonality). *Let $g, h \in G$ be such that g_p and h_p are not G-conjugate. If B is a block of G, then*

$$\sum_{\chi \in \mathrm{Irr}(B)} \chi(g)\chi(h^{-1}) = 0.$$

Proof. We define a class function θ by

$$\theta = \sum_{\chi \in \text{Irr}(G)} \chi(h^{-1})\chi \, .$$

Let $x = g_p$. If $y \in \mathbf{C}_G(x)$ is p-regular, then xy and h are not G-conjugate. Hence, by the second orthogonality relation, it follows that $\theta(xy) = 0$. Now, θ vanishes on the p-section $S(x)$ and thus, by Theorem (5.10), so does θ_B for every block B of G. Therefore,

$$0 = \theta_B(xg_{p'}) = \theta_B(g) = \sum_{\chi \in \text{Irr}(B)} \chi(g)\chi(h^{-1}) \, ,$$

as required. ∎

Another consequence of the second main theorem which we treat here is the following.

(5.12) THEOREM. *Suppose that* $x_1, \ldots, x_k$ *are representatives of the G-conjugacy classes of p-elements of G. If $B \in \text{Bl}(G)$, then*

$$\text{k}(B) = \sum_{i=1}^{k} \sum_{\substack{b \in \text{Bl}(\mathbf{C}_G(x_i)) \\ b^G = B}} \text{l}(b) \, .$$

We need a lemma which tells us that, in some sense, the generalized decomposition numbers behave like decomposition numbers.

(5.13) LEMMA. *Suppose that x and y are p-elements of G. Let $\mu \in \text{IBr}(\mathbf{C}_G(x))$, let $\varphi \in \text{IBr}(\mathbf{C}_G(y))$ and let B be a block of G.*
  (a) Suppose that x and y are not G-conjugate. Then

$$\sum_{\chi \in \text{Irr}(G)} \overline{d_{\chi\mu}^x} d_{\chi\varphi}^y = 0 \, .$$

In fact,

$$\sum_{\chi \in \text{Irr}(B)} \overline{d_{\chi\mu}^x} d_{\chi\varphi}^y = 0 \, .$$

  (b) Suppose that $x = y$. Then

$$\sum_{\chi \in \text{Irr}(G)} \overline{d_{\chi\mu}^x} d_{\chi\varphi}^x = c_{\mu\varphi} \, .$$

In fact, if φ and μ lie in the same block b and $b^G = B$, then

$$\sum_{\chi \in \mathrm{Irr}(B)} \overline{d^x_{\chi\mu}} d^x_{\chi\varphi} = c_{\mu\varphi}.$$

Otherwise,

$$\sum_{\chi \in \mathrm{Irr}(B)} \overline{d^x_{\chi\mu}} d^x_{\chi\varphi} = 0.$$

Proof. Write $H = \mathbf{C}_G(x)$ and let $\mathcal{R}$ be a complete set of representatives of H-classes of p-regular elements of H, so that

$$S(x) = \bigcup_{y \in \mathcal{R}} \mathrm{cl}(xy)$$

is a disjoint union (by the remark before the statement of Theorem (5.10)).

We define $\Phi^x_\mu \in \mathrm{cf}(G)$ as follows. If $g \in G - S(x)$, we set $\Phi^x_\mu(g) = 0$. If $g \in S(x)$, then $g \in \mathrm{cl}(xy)$ for a unique $y \in \mathcal{R}$ and we set $\Phi^x_\mu(g) = \Phi_\mu(y)$. We claim that

$$\Phi^x_\mu = \sum_{\chi \in \mathrm{Irr}(G)} \overline{d^x_{\chi\mu}} \chi.$$

By using Lemma (5.1), we have that

$$[\Phi^x_\mu, \chi] = \frac{1}{|G|} \sum_{y \in \mathcal{R}} |G : \mathbf{C}_G(xy)| \Phi_\mu(y) \overline{\chi(xy)}$$

$$= \frac{1}{|H|} \sum_{y \in \mathcal{R}} |H : \mathbf{C}_H(y)| \Phi_\mu(y) \sum_{\tau \in \mathrm{IBr}(H)} \overline{d^x_{\chi\tau}} \tau(y^{-1})$$

$$= \sum_{\tau \in \mathrm{IBr}(H)} \overline{d^x_{\chi\tau}} [\Phi_\mu, \tau]^0 = \overline{d^x_{\chi\mu}},$$

where the last equality follows from Theorem (2.13). This proves the claim. Now,

$$\sum_{\chi \in \mathrm{Irr}(G)} \overline{d^x_{\chi\mu}} d^y_{\chi\varphi} = [\Phi^x_\mu, \Phi^y_\varphi]$$

and this is clearly zero if x and y are not G-conjugate (because in this case $S(x) \cap S(y) = \emptyset$). This proves the first part of (a). If $x = y$, then

$$[\Phi^x_\mu, \Phi^x_\varphi] = \frac{1}{|G|} \sum_{y \in \mathcal{R}} |G : \mathbf{C}_G(xy)| \Phi_\mu(y) \overline{\Phi_\varphi(y)}$$

$$= \frac{1}{|H|} \sum_{y \in \mathcal{R}} |H : \mathbf{C}_H(y)| \Phi_\mu(y) \overline{\Phi_\varphi(y)} = [\Phi_\mu, \Phi_\varphi] = c_{\mu\varphi},$$

and the first part of (b) follows.

Now, let us write b_φ and b_μ for the blocks which contain φ and μ, respectively. If B' is any block of G, then, by the second main theorem, we have that

$$\sum_{\chi \in \mathrm{Irr}(B')} \overline{d^x_{\chi\mu}} d^y_{\chi\varphi} = 0 \text{ unless } b^G_\varphi = B' = b^G_\mu. \tag{1}$$

Therefore,

$$\sum_{\chi \in \mathrm{Irr}(G)} \overline{d^x_{\chi\mu}} d^y_{\chi\varphi} = \sum_{\chi \in \mathrm{Irr}(b^G_\mu) \cap \mathrm{Irr}(b^G_\varphi)} \overline{d^x_{\chi\mu}} d^y_{\chi\varphi}. \tag{2}$$

Now, assume that x and y are not G-conjugate. If $b^G_\mu \neq B$ or $b^G_\varphi \neq B$, then, by (1), it follows that

$$\sum_{\chi \in \mathrm{Irr}(B)} \overline{d^x_{\chi\mu}} d^y_{\chi\varphi} = 0.$$

If $b^G_\mu = b^G_\varphi = B$, then

$$\sum_{\chi \in \mathrm{Irr}(B)} \overline{d^x_{\chi\mu}} d^y_{\chi\varphi} = 0$$

by using the first part of (a) and (2). This completes the proof of part (a).

Now, suppose that $x = y$. By (2) and the first part of (b), we have that

$$c_{\mu\varphi} = \sum_{\chi \in \mathrm{Irr}(b^G_\mu) \cap \mathrm{Irr}(b^G_\varphi)} \overline{d^x_{\chi\mu}} d^x_{\chi\varphi}. \tag{3}$$

This proves the second part of (b). To prove the third part, assume that $b_\varphi \neq b_\mu$ or that $b_\varphi = b_\mu = b$ with $b^G \neq B$. We wish to show that

$$\sum_{\chi \in \mathrm{Irr}(B)} \overline{d^x_{\chi\mu}} d^x_{\chi\varphi} = 0.$$

By (1), we may assume that $b^G_\varphi = b^G_\mu = B$. Hence, we necessarily have that $b_\varphi \neq b_\mu$. In this case, $c_{\mu\varphi} = 0$ and the result follows from (3). ∎

Proof of Theorem (5.12). If $\mathcal{R}_i$ is a complete set of representatives of conjugacy classes of p-regular elements of $\mathbf{C}_G(x_i)$, then

$$\{x_i y\}_{1 \le i \le k, y \in \mathcal{R}_i}$$

is a complete set of conjugacy classes of G. In particular, the number of conjugacy classes $\mathrm{k}(G)$ of G is

$$\mathrm{k}(G) = \sum_{i=1}^{k} |\mathrm{IBr}(\mathbf{C}_G(x_i))| \, .$$

Now, we form a $\mathrm{k}(G) \times \mathrm{k}(G)$ matrix

$$J = \left(d^{x_i}_{\chi\mu} \right) \, ,$$

where the rows are indexed by $\chi \in \mathrm{Irr}(G)$ and the columns by $\{(x_i, \mu) \mid \mu \in \mathrm{IBr}(\mathbf{C}_G(x_i)), 1 \le i \le k\}$. By Lemma (5.13), we have that

$$\overline{J}^t J = \mathrm{diag}(C^{(x_1)}, \dots, C^{(x_k)}) \, ,$$

where $C^{(x_i)}$ is the Cartan matrix of $\mathbf{C}_G(x_i)$. In particular, notice that the matrix J is regular.

Now, we reorganize the matrix J in the following way. If $B_1, \dots, B_t$ are the blocks of G, then we write first $\mathrm{Irr}(B_1)$, then $\mathrm{Irr}(B_2)$ and so on. Considering columns, we put first all indices (x_i, μ) such that the block of μ induces B_1, then B_2 and so on. By the second main theorem, we get

$$J = \begin{pmatrix} J_{B_1} & 0 & \cdots & 0 \\ 0 & J_{B_2} & \cdots & 0 \\ \vdots & \vdots & \ddots & \vdots \\ 0 & 0 & \cdots & J_{B_t} \end{pmatrix} \, .$$

Since J is regular, it follows that each J_{B_i} is square (and regular). We conclude that

$$\mathrm{k}(B) = \sum_{i=1}^{k} \sum_{\substack{b \in \mathrm{Bl}(\mathbf{C}_G(x_i)) \\ b^G = B}} 1(b) \, ,$$

as desired. $\blacksquare$

Suppose that B is a p-block of G with defect group D. A natural question is whether or not the characters of B determine if a p-element x of G lies in some G-conjugate of D or not. The answer is yes.

(5.14) THEOREM. *Suppose that B is a block of G with defect group D and let $\chi \in \mathrm{Irr}(B)$ be of height zero. Let x be a p-element of G. Then $x \in D^g$ for some $g \in G$ if and only if χ is not zero in the p-section of x.*

If G is a p-group, then every nonlinear irreducible character of G vanishes on some class and yet that class is in the defect group (which is the whole group) of the block of the character (Problem (3.1)). This shows that the height zero hypothesis is needed to identify p-elements in a defect group.

To prove Theorem (5.14) we need a useful lemma.

(5.15) LEMMA. *Let B be a block with defect group P and let $K = \mathrm{cl}(y)$ be a defect class of B with $P \in \mathrm{Syl}_p(\mathbf{C}_G(y))$.*
   (a) *If $x \in P$ and $\chi \in \mathrm{Irr}(B)$ has height zero, then $\chi(xy)^* \neq 0$.*
   (b) *If $x \in \mathbf{Z}(P)$ and $L = \mathrm{cl}(xy)$, then $\lambda_B(\hat{L}) \neq 0$.*

Proof. Since K is a defect class of B, we have that $y \in G^0$ and $\lambda_B(\hat{K}) \neq 0$. Let $\chi \in \mathrm{Irr}(B)$ be of height zero. Hence, $\frac{\chi(1)}{|K|}$ is a unit of the ring S because $\chi(1)_p = |K|_p$. Since

$$0 \neq \lambda_B(\hat{K}) = \left(\frac{|K|\chi(y)}{\chi(1)}\right)^* \neq 0,$$

we deduce that $\chi(y)^* \neq 0$.

Now, suppose that $x \in P$. By the remark after the proof of Lemma (2.4), we have that

$$\chi(xy)^* = \chi(y)^* \neq 0$$

because y is the p'-part of xy. This proves part (a).

Now, suppose that $x \in \mathbf{Z}(P)$ and write $L = \mathrm{cl}(xy)$. Note that $\mathbf{C}_G(xy) = \mathbf{C}_G(x) \cap \mathbf{C}_G(y)$ by the unique decomposition of the element xy as the product of its p-part and its p'-part. Since $x \in \mathbf{Z}(P)$, we have that $P \subseteq \mathbf{C}_G(xy)$. Therefore, $P \in \mathrm{Syl}_p(\mathbf{C}_G(xy))$ because $P \in \mathrm{Syl}_p(\mathbf{C}_G(y))$. Hence, $|L|_p = |K|_p = \chi(1)_p$. We conclude that $\frac{|L|}{\chi(1)}$ is a unit of S. Thus,

$$\left(\frac{|L|}{\chi(1)}\right)^* \neq 0.$$

Now, by using part (a), it follows that

$$\lambda_B(\hat{L}) = \left(\frac{|L|\chi(xy)}{\chi(1)}\right)^* = \left(\frac{|L|}{\chi(1)}\right)^* \chi(xy)^* \neq 0,$$

as required. ∎

Proof of Theorem (5.14). Assume first that χ is not zero in $S(x)$. Then there is a $b \in \mathrm{Bl}(\mathbf{C}_G(x))$ with $b^G = B$ by Corollary (5.9). Therefore, x lies in some G-conjugate of D by Theorem (4.14). (We use that χ has height zero for the other direction.)

Suppose that x lies in some G-conjugate of D. We want to show that χ is not zero in $S(x)$. So we may replace x by some G-conjugate and assume that $x \in D$. Now, if $K = \mathrm{cl}(y)$ is a defect class for B with $D \in \mathrm{Syl}_p(\mathbf{C}_G(y))$, we have that $\chi(xy) \neq 0$ by Lemma (5.15.a). ∎

Next, we give an example on how modular representation theory can be used to give deeper information on ordinary characters. We know no purely character theoretic proof of the following result.

(5.16) THEOREM (Knörr). *Suppose that $\chi \in \mathrm{Irr}(G)$ is such that $\chi(g) = 0$ for every element $g \in G$ of order p. Then $\chi(1)_p = |G|_p$.*

Although the next result will only be used in its full strength later on, we find it convenient to prove it now.

(5.17) THEOREM. *Let $B \in \mathrm{Bl}(G \,|\, P)$ and suppose that $\lambda_B(\hat{K}) \neq 0$. Then there exists an $x \in K \cap \mathbf{C}_G(P)$ such that $x_p \in \mathbf{Z}(P)$.*

Proof. Write $C = \mathbf{C}_G(P)$. By using the first main theorem, let $b \in \mathrm{Bl}(\mathbf{N}_G(P) \,|\, P)$ be such that $b^G = B$. Also, $\lambda_B = \mathrm{Br}_P \lambda_b$ by Theorem (4.14) and thus
$$0 \neq \lambda_B(\hat{K}) = \lambda_b(\widehat{K \cap C}).$$

Hence, there exists a conjugacy class L of $\mathbf{N}_G(P)$ contained in $K \cap C$ such that $\lambda_b(\hat{L}) \neq 0$. Then, if $\gamma \in \mathrm{Irr}(b)$ and $x \in L$, it follows that $\gamma(x) \neq 0$. Since b has defect group P and $P \lhd \mathbf{N}_G(P)$, by Corollary (5.9) we deduce that $x_p \in P$. Also, since $x \in C$ we have that $x_p \in C$. It follows that $x_p \in \mathbf{Z}(P)$, as desired. ∎

Proof of Theorem (5.16) (Isaacs). Let B be the p-block of G which contains χ and suppose that B has defect group P. We wish to prove that $P = 1$.

By the first main theorem, let $b \in \mathrm{Bl}(\mathbf{N}_G(P) \,|\, P)$ be such that $b^G = B$.

Let $X = \{g \in G \,|\, o(g) = p\}$, so that X is the union of those conjugacy classes of G which consist of elements of order p. If $K \subseteq X$ is such a conjugacy class, by hypothesis
$$\omega_\chi(\hat{K}) = \frac{|K|\chi(x_K)}{\chi(1)} = 0$$

and thus $\lambda_B(\hat{X}) = 0$. Now, if $C = \mathbf{C}_G(P)$, we have that
$$0 = \lambda_B(\hat{X}) = \lambda_b(\mathrm{Br}_P(\hat{X})) = \lambda_b(\widehat{X \cap C}).$$

Now, if L is a conjugacy class of $\mathbf{N}_G(P)$ inside $X \cap C$ such that $\lambda_b(\hat{L}) \neq 0$, then, by Theorem (5.17) (applied in $\mathbf{N}_G(P)$), we deduce that $L \subseteq P$. Hence,

$$0 = \lambda_b(X \widehat{\cap C \cap} P) = \lambda_b(X \widehat{\cap \mathbf{Z}(P)}).$$

Notice that $X \cap \mathbf{Z}(P) = H - \{1\}$, where

$$H = \Omega_1(\mathbf{Z}(P)) = \{x \in \mathbf{Z}(P) \,|\, x^p = 1\}$$

is a subgroup of $\mathbf{Z}(P)$. Now, $0 = \lambda_b(\hat{H} - 1)$ and thus $\lambda_b(\hat{H}) = 1$. Since H is a group, we see that

$$\hat{H}\hat{H} = \sum_{z \in H} |\{(x, y) \in H \times H \,|\, xy = z\}|z = |H|^* \hat{H}$$

in the group ring FH. Therefore,

$$1 = \lambda_b(\hat{H})^2 = \lambda_b(\hat{H}^2) = \lambda_b(|H|^* \hat{H}) = |H|^* \lambda_b(\hat{H}) = |H|^* .$$

We conclude that p does not divide $|H|$. Therefore $P = 1$, as desired. ∎

Now, we obtain a new relationship between the defect groups of the induced and the inducing blocks.

(5.18) THEOREM. *Suppose that $b \in \mathrm{Bl}(H \,|\, D)$ is such that $b^G = B \in \mathrm{Bl}(G \,|\, P)$ is defined. If $x \in \mathbf{Z}(P)$, then x is G-conjugate to some element in $\mathbf{Z}(D)$.*

Proof. Let $K = \mathrm{cl}(y)$ be a defect class of B with $P \in \mathrm{Syl}_p(\mathbf{C}_G(y))$. By Lemma (5.15.b), we have that

$$\lambda_B(\hat{L}) \neq 0 \,,$$

where $L = \mathrm{cl}(xy)$. Now,

$$0 \neq \lambda_B(\hat{L}) = \lambda_b(\widehat{L \cap H})$$

and thus, there exists a conjugacy class M of H inside L such that $\lambda_b(\hat{M}) \neq 0$. By Theorem (5.17), there is a $z \in M$ such that $z_p \in \mathbf{Z}(D)$. Since z_p is G-conjugate to x, the proof of the theorem is complete. ∎

(5.19) COROLLARY. *Suppose that $H \subseteq G$ and that b^G is defined, where $b \in \mathrm{Bl}(H)$. Then b has defect zero if and only if b^G has defect zero.*

Proof. Since the defect groups of b are contained in defect groups of b^G (Lemma (4.13)), it is clear that if b^G has defect zero, then b has defect zero. Suppose now that b has defect zero. If b^G has defect group P, by Theorem (5.18) we have that $\mathbf{Z}(P) = 1$ and the result follows. ∎

To end this chapter, we study some functions defined by Brauer which express the second main theorem in a very natural way.

Suppose that x is a p-element of G and let $b \in \mathrm{Bl}(\mathbf{C}_G(x))$. Then we say that (x, b) is a **subsection** associated to b^G.

If $\chi \in \mathrm{cf}(G)$, we define a class function $\chi^{(x,b)}$ as follows. We define $\chi^{(x,b)}$ to be zero off $S(x)$. If $g \in S(x)$, by the remark before Theorem (5.10), there exists a $u \in G$ such that $g^u = xy$ for some $y \in \mathbb{C}^0$, where $C = \mathbf{C}_G(x)$. In fact, y is determined by g up to C-conjugacy. We define

$$\chi^{(x,b)}(g) = \chi(xy f_b).$$

Observe that if $z = y^c$ for $c \in C$, then

$$\chi(xz f_b) = \chi((xy f_b)^c) = \chi(xy f_b).$$

Also, if $y \in C^0$, as we did in the proof of the second main theorem, it follows that

$$\chi^{(x,b)}(xy) = \chi(xy f_b) = \sum_{\psi \in \mathrm{Irr}(b)} [\chi_C, \psi]\psi(xy)$$

$$= \sum_{\psi \in \mathrm{Irr}(b)} \frac{[\chi_C, \psi]\psi(x)}{\psi(1)}\psi(y) = \sum_{\mu \in \mathrm{IBr}(b)} \left(\sum_{\psi \in \mathrm{Irr}(b)} \frac{[\chi_C, \psi]\psi(x)}{\psi(1)} d_{\psi\mu} \right) \mu(y)$$

$$= \sum_{\mu \in \mathrm{IBr}(b)} \left(\sum_{\psi \in \mathrm{Irr}(C)} \frac{[\chi_C, \psi]\psi(x)}{\psi(1)} d_{\psi\mu} \right) \mu(y)$$

$$= \sum_{\mu \in \mathrm{IBr}(b)} d_{\chi\mu}^x \mu(y).$$

(5.20) THEOREM. Let B be a block of G and let (x, b) be a subsection associated to B. If $\chi \in \mathrm{Irr}(G)$ is not in B, then

$$\chi^{(x,b)} = 0.$$

Proof. This is clear by the second main theorem and the previous discussion. ∎

Our final objective in this chapter is to prove a converse.

(5.21) THEOREM (Broué-Puig). *Suppose that* $\chi \in \mathrm{Irr}(B)$ *has height zero. If* (x, b) *is a subsection associated to* B, *then*

$$\chi^{(x,b)} \neq 0.$$

(If χ does not have height zero, then Theorem (5.21) is false. See the remark after the statement of Theorem (5.14).)

We need a lemma.

(5.22) LEMMA. *Suppose that* $B \in \mathrm{Bl}(G)$. *Let* $\chi \in \mathrm{Irr}(B)$ *and let*

$$v = \sum_{g \in G^0} \chi(g^{-1})^* g.$$

Then $v \in \mathbf{Z}(FG)e_B$. *Furthermore,* $v \notin \mathbf{J}(\mathbf{Z}(FG))e_B$ *if and only if* χ *has height zero.*

Proof. Let

$$u = \sum_{g \in G^0} \chi(g^{-1})g.$$

By Lemma (3.32), we know that

$$u = |G|_{p'} \sum_{\psi \in \mathrm{Irr}(B)} \frac{[\tilde{\chi}, \psi]}{\psi(1)} e_\psi.$$

In particular, $uf_B = u$, and thus $u^* e_B = u^* = v \in \mathbf{Z}(FG)e_B$. Now,

$$\omega_\chi(u) = |G|_{p'} \frac{[\tilde{\chi}, \chi]}{\chi(1)}$$

and therefore

$$\lambda_B(v) = (|G|_{p'})^* (\frac{[\tilde{\chi}, \chi]}{\chi(1)})^*.$$

(Recall that $\frac{[\tilde{\chi}, \chi]}{\chi(1)} \in S$ by Lemma (3.22.a).) Since $\mathbf{Z}(FG)e_B$ is a local algebra, we have that $v \in \mathbf{J}(\mathbf{Z}(FG))e_B$ if and only if $\lambda_B(v) = 0$. This happens if and only if $\nu([\tilde{\chi}, \chi]) > \nu(\chi(1))$. By Theorem (3.24), this happens if and only if χ has positive height. ∎

Proof of Theorem (5.21). Let

$$v = \sum_{g \in G^0} \chi(g^{-1})^* g.$$

By Lemma (5.22), we know that $v \in \mathbf{Z}(FG)e_B - \mathbf{J}(\mathbf{Z}(FG))e_B$. Let $C = \mathbf{C}_G(x)$ and denote the Brauer homomorphism by

$$\mathrm{Br} = \mathrm{Br}_{\langle x \rangle} : \mathbf{Z}(FG) \to \mathbf{Z}(FC).$$

Then $\mathrm{Br}(\hat{K}) = \widehat{K \cap C}$ for every conjugacy class K of G. Hence,

$$\mathrm{Br}(v) = \sum_{g \in C^0} \chi(g^{-1})^* g$$

since $G^0 \cap C = C^0$. Now, by Corollary (3.12), we have that

$$\mathbf{Z}(FG)e_B = Fe_B \oplus \mathbf{J}(\mathbf{Z}(FG))e_B.$$

Since $v \notin \mathbf{J}(\mathbf{Z}(FG))e_B$, we may write

$$v = ce_B + j,$$

where $0 \neq c \in F$ and $j \in \mathbf{J}(\mathbf{Z}(FG))e_B$. Now, by applying the Brauer homomorphism, we get

$$\sum_{g \in C^0} \chi(g^{-1})^* g = c\mathrm{Br}(e_B) + \mathrm{Br}(j).$$

By Theorem (4.14), we have that $\mathrm{Br}(e_B)e_b = e_b$, and then

$$\left(\sum_{g \in C^0} \chi(g^{-1})^* g \right)e_b = ce_b + \mathrm{Br}(j)e_b. \qquad (1)$$

Now, $\mathbf{J}(\mathbf{Z}(FG))$ is a nilpotent ideal and thus $j^n = 0$ for some integer n. Therefore $\mathrm{Br}(j)^n = 0$ and we have that $\mathrm{Br}(j) \in \mathbf{J}(\mathbf{Z}(FC))$ by Problem (1.2), for instance. Thus, the decomposition (1) is the decomposition associated to the direct sum $\mathbf{Z}(FC)e_b = Fe_b \oplus \mathbf{J}(\mathbf{Z}(FC))e_b$. In particular, since $c \neq 0$, we deduce that

$$\left(\sum_{g \in C^0} \chi(g^{-1})^* g \right)e_b \neq 0.$$

Now, if, as usual, $(\chi_C)_b$ denotes the b-part of χ_C, by Lemma (3.31) and Lemma (3.32) applied in C, we have that

$$\left(\sum_{g \in C^0} \chi(g^{-1}) g \right) f_b = \sum_{g \in C^0} (\chi_C)_b (g^{-1}) g = \sum_{g \in C^0} \chi(g^{-1} f_b) g \, .$$

Hence,

$$\left(\sum_{g \in C^0} \chi(g^{-1})^* g \right) e_b = \sum_{g \in C^0} \chi(g^{-1} f_b)^* g \neq 0 \, .$$

Therefore, there exists a p-regular $y \in \mathbf{C}_G(x)$ such that

$$\chi(y f_b)^* \neq 0 \, .$$

Now, since for every $\psi \in \mathrm{Irr}(C)$ we have that $\psi(xy)^* = \psi(y)^*$ (see the remark after the proof of Lemma (2.4)), it follows that

$$\chi(xy f_b)^* = \sum_{\psi \in \mathrm{Irr}(b)} [\chi_C, \psi]^* \psi(xy)^*$$

$$= \sum_{\psi \in \mathrm{Irr}(b)} [\chi_C, \psi]^* \psi(y)^* = \chi(y f_b)^* \neq 0 \, .$$

Hence, $\chi^{(x,b)}(xy) = \chi(xy f_b) \neq 0$, as required. ∎

In fact, we have shown in Theorem (5.21) that $(\chi^{(x,b)})^* \neq 0$.

PROBLEMS

(5.1) If $b^G = B$, prove that $e_b e_B \neq 0$.

(5.2) If $g \in G$ and (x, b) is a subsection of G, let us write $(x, b)^g = (x^g, b^g)$, where b^g denotes the block of $\mathbf{C}_G(x)^g$ naturally associated to b via the isomorphism $y \mapsto y^g$. (See Problem (4.4).) This is another subsection of G. Prove that $\chi^{(x,b)} = \chi^{(x^g, b^g)}$.

(5.3) Suppose that $\chi, \psi \in \mathrm{Irr}(G)$. Prove the following:

(a) $[\chi^{(x_1, b_1)}, \psi^{(x_2, b_2)}] = 0$ if $S(x_1) \neq S(x_2)$;

(b) if S is a complete set of representatives of G-classes of subsections of G, then

$$\chi = \sum_{(x,b) \in S} \chi^{(x,b)} \, ;$$

(c) $[\chi^{(x,b)}, \psi^{(x,e)}] = 0$ if $b \neq e$;

(d)

$$[\chi^{(x,b)}, \psi^{(x,b)}] = \sum_{\varphi, \mu \in \mathrm{IBr}(b)} d^x_{\chi\varphi} \overline{d^x_{\psi\mu}} \gamma_{\varphi\mu} \, ,$$

where $(\gamma_{\varphi\mu})_{\varphi, \mu \in \mathrm{IBr}(b)}$ is the inverse of the Cartan matrix C_b.

(5.4) If U is a p-section of G, show that $\hat{U} f_B$ is a linear combination of $\hat{L}$ for conjugacy classes $L \subseteq U$.

(HINT: Use block orthogonality.)

(5.5) If U is a p-section of G and B_0 is the principal block of G, show that $\hat{U} \in e_{B_0} \mathbf{Z}(FG)$.

(HINT: Use Problem (5.4).)

(5.6) Suppose that $B \in \mathrm{Bl}(G \mid D)$. If $x \in \mathbf{Z}(D)$ and $\chi \in \mathrm{Irr}(B)$, prove that χ is not zero in $S(x)$.

(HINT: Use Lemma (5.15.b).)

(5.7) Let B be a block of G. If $\{(x_1, b_1), \ldots, (x_n, b_n)\}$ is a complete set of representatives of the G-conjugacy classes of subsections, prove that

$$ k(B) = \sum_{i=1}^{n} l(b_i) . $$

(HINT: Use Theorem (5.12).)

(5.8) Let $B \in \mathrm{Bl}(G \mid D)$ and suppose that $\sigma \in \mathrm{Aut}(G)$ fixes every $\chi \in \mathrm{Irr}(B)$. If $d \in D$, show that d and d^σ are G-conjugate.

(HINT: Use block orthogonality.)

6. The Third Main Theorem

Let B_0 be the principal block of G. Since $1_G \in \mathrm{Irr}(B_0)$, it is clear that $d(B_0) = \nu(|G|_p)$. Therefore, the defect groups of the principal block are the Sylow p-subgroups of G. Our objective in this chapter is to analyse when a block induces the principal block.

If χ is a (not necessarily irreducible) character of G, we also define $\omega_\chi : \mathbf{Z}(\mathbb{C}G) \to \mathbb{C}$ by setting

$$\omega_\chi(\hat{K}) = \frac{|K|\chi(x_K)}{\chi(1)} ,$$

where x_K is a fixed element of $K \in \mathrm{cl}(G)$. Observe that

$$\chi(1)\omega_\chi = \sum_{\psi \in \mathrm{Irr}(G)} [\chi, \psi]\psi(1)\omega_\psi .$$

Of course, if χ is not irreducible, it is not necessarily true that $\omega_\chi(\hat{K}) \in \mathbf{R}$ for $K \in \mathrm{cl}(G)$. As a consequence of our first lemma, we will see that $\omega_\chi(\hat{K}) \in \mathbf{R}$ if χ is induced from an irreducible character of a subgroup of G.

If $H \subseteq G$ is a subgroup of G and $\lambda : \mathbf{Z}(RH) \to R$ is a linear map (R is any ring), we define an R-linear map $\lambda^G : \mathbf{Z}(RG) \to R$ by setting

$$\lambda^G(\hat{K}) = \lambda(\widehat{K \cap H}),$$

where it is understood that $\lambda^G(\hat{K}) = 0$ if $K \cap H = \emptyset$.

(6.1) LEMMA. If $H \subseteq G$ and ξ is a character of H, then

$$(\omega_\xi)^G = \omega_{\xi^G} .$$

Proof. Since

$$\xi^G(1)\omega_{\xi^G} = \sum_{\chi\in\mathrm{Irr}(G)} [\xi^G,\chi]\chi(1)\omega_\chi,$$

it suffices to prove that

$$\xi^G(1)(\omega_\xi)^G = \sum_{\chi\in\mathrm{Irr}(G)} [\xi^G,\chi]\chi(1)\omega_\chi.$$

Let K be a conjugacy class of G and let $x_K \in K$. If $K \cap H$ is not empty, then write

$$K \cap H = L_1 \cup \ldots \cup L_m,$$

where $L_i = \mathrm{cl}_H(x_i)$ is the conjugacy class of x_i in H. By the ordinary induction formula (see page 64 of [Isaacs, *Characters*]), we have that

$$\xi^G(x_K) = |\mathbf{C}_G(x_K)| \sum_{i=1}^{m} \frac{\xi(x_i)}{|\mathbf{C}_H(x_i)|},$$

where it is understood that $\xi^G(x_K) = 0$ if $K \cap H$ is empty. Now,

$$\sum_{\chi\in\mathrm{Irr}(G)} [\xi^G,\chi]\chi(1)\omega_\chi(\hat{K}) = \sum_{\chi\in\mathrm{Irr}(G)} [\xi^G,\chi]\chi(1)\frac{|K|\chi(x_K)}{\chi(1)}$$

$$= |K|\xi^G(x_K) = \frac{|G|}{|\mathbf{C}_G(x_K)|}\left(|\mathbf{C}_G(x_K)| \sum_{i=1}^{m} \frac{\xi(x_i)}{|\mathbf{C}_H(x_i)|}\right)$$

$$= |G:H| \sum_{i=1}^{m} |L_i|\xi(x_i) = \xi^G(1) \sum_{i=1}^{m} \omega_\xi(\hat{L}_i)$$

$$= \xi^G(1)\omega_\xi(\widehat{K\cap H}) = \xi^G(1)(\omega_\xi)^G(\hat{K}),$$

as required. ∎

If b is a block of $H \subseteq G$, then, for every $\xi \in \mathrm{Irr}(b)$, we have

$$\lambda_b^G(\hat{K}) = \lambda_b(\widehat{K\cap H}) = \omega_\xi(\widehat{K\cap H})^*$$

$$= (\omega_\xi)^G(\hat{K})^* = \omega_{\xi^G}(\hat{K})^*.$$

In particular, if $z \in \mathbf{Z}(SG)$, then

$$\lambda_b^G(z^*) = \omega_{\xi^G}(z)^*.$$

We will use this fact without further reference.

We mention one interesting consequence of Lemma (6.1). Recall that for $\chi \in \mathrm{Irr}(G)$, we write $\lambda_\chi : \mathbf{Z}(FG) \to F$ for the algebra homomorphism defined by $\lambda_\chi(\hat{K}) = \omega_\chi(\hat{K})^*$.

(6.2) COROLLARY. *Suppose that $\xi \in \mathrm{Irr}(b)$, where b is a block of $H \subseteq G$. If $\xi^G \in \mathrm{Irr}(G)$, then the block b^G is defined and contains ξ^G.*

Proof. By the discussion before this corollary, we have that

$$\lambda_b^G(\hat{K}) = \omega_{\xi^G}(\hat{K})^* = \lambda_{\xi^G}(\hat{K})$$

for every conjugacy class K of G. Since λ_{ξ^G} is an algebra homomorphism, the proof is complete. ∎

(6.3) LEMMA. *Suppose that $H \subseteq G$ is a subgroup of G and let $b \in \mathrm{Bl}(H)$. Let $\xi \in \mathrm{Irr}(b)$ and let B be a block of G.*
   (a) *If $K \in \mathrm{cl}(G)$, then*

$$\frac{|K|(\xi^G)_B(x_K)}{\xi^G(1)} \in S.$$

   (b) *Suppose that b^G is defined. If $b^G = B$ and $\chi \in \mathrm{Irr}(B)$, then*

$$\frac{|K|(\xi^G)_B(x_K)}{\xi^G(1)} \equiv \frac{|K|\chi(x_K)}{\chi(1)} \bmod \mathcal{P}$$

for $K \in \mathrm{cl}(G)$.
   (c) *If b^G is defined and $b^G \neq B$, then*

$$\frac{|K|(\xi^G)_B(x_K)}{\xi^G(1)} \equiv 0 \bmod \mathcal{P}$$

for $K \in \mathrm{cl}(G)$.

Proof. To prove part (a), we show something slightly more general. If χ is a character of G such that $\omega_\chi(\hat{K}) \in S$ for every $K \in \mathrm{cl}(G)$, then we prove that

$$\frac{|K|\chi_B(x_K)}{\chi(1)} \in S,$$

for every $K \in \mathrm{cl}(G)$. If this is proven, part (a) is obtained by setting $\chi = \xi^G$ and applying Lemma (6.1).
   As usual, write $f_B = \sum\limits_{\psi \in \mathrm{Irr}(B)} e_\psi \in \mathbf{Z}(SG)$. Now, we have

$$\chi(1)\omega_\chi(f_B\hat{K}) = \sum_{\psi \in \mathrm{Irr}(G)} [\chi, \psi]\psi(1)\omega_\psi(f_B\hat{K})$$

$$= \sum_{\psi \in \mathrm{Irr}(G)} [\chi, \psi]\psi(1)\omega_\psi(f_B)\omega_\psi(\hat{K}) = \sum_{\psi \in \mathrm{Irr}(B)} [\chi, \psi]|K|\psi(x_K)$$

$$= |K|\chi_B(x_K).$$

Thus,

$$\omega_\chi(f_B\hat{K}) = \frac{|K|\chi_B(x_K)}{\chi(1)}.$$

Since by hypothesis $\omega_\chi(\hat{L}) \in S$ for every $L \in \mathrm{cl}(G)$ and $f_B\hat{K} \in \mathbf{Z}(SG)$ for $K \in \mathrm{cl}(G)$, it follows that

$$\omega_\chi(f_B\hat{K}) \in S,$$

as desired.

Suppose now that $\xi \in \mathrm{Irr}(b)$ and that b^G is defined. By the first paragraph, we have that

$$\omega_{\xi^G}(f_B\hat{K}) = \frac{|K|(\xi^G)_B(x_K)}{\xi^G(1)}.$$

Hence, by the discussion before Corollary (6.2), we get

$$\lambda_{b^G}(e_B\hat{K}) = \left(\frac{|K|(\xi^G)_B(x_K)}{\xi^G(1)}\right)^*.$$

Therefore, if $B \neq b^G$, we have that

$$0 = \lambda_{b^G}(e_B)\lambda_{b^G}(\hat{K}) = \lambda_{b^G}(e_B\hat{K}) = \left(\frac{|K|(\xi^G)_B(x_K)}{\xi^G(1)}\right)^*.$$

If $b^G = B$ and $\chi \in \mathrm{Irr}(G)$, then

$$\left(\frac{|K|\chi(x_K)}{\chi(1)}\right)^* = \lambda_B(\hat{K}) = \lambda_{b^G}(e_B\hat{K}) = \left(\frac{|K|(\xi^G)_B(x_K)}{\xi^G(1)}\right)^*,$$

as desired. ∎

There is a useful corollary.

(6.4) COROLLARY. *Suppose that b^G is defined, where b is a block of $H \subseteq G$. Let $\xi \in \mathrm{Irr}(b)$. Then $\nu((\xi^G)_B(1)) > \nu(\xi^G(1))$ if $B \neq b^G$, and $\nu((\xi^G)_B(1)) = \nu(\xi^G(1))$ if $B = b^G$. In particular, there exists some $\chi \in \mathrm{Irr}(b^G)$ over ξ.*

Proof. We put $x_K = 1$ and use Lemma (6.3). By part (a), we know that

$$\frac{(\xi^G)_B(1)}{\xi^G(1)} \in \mathbb{Q} \cap S.$$

By parts (b) and (c),

$$\frac{(\xi^G)_B(1)}{\xi^G(1)} \equiv 1 \bmod \mathcal{P}$$

if $b^G = B$ and

$$\frac{(\xi^G)_B(1)}{\xi^G(1)} \equiv 0 \bmod \mathcal{P}$$

if $b^G \neq B$. Now, by Lemma (3.21) on "valuations", the proof of the first part is complete. Since $(\xi^G)_{b^G} \neq 0$, the second part is clear. ∎

Next is the key lemma for the third main theorem.

(6.5) LEMMA. *Let H be a subgroup of G. Suppose that b^G is defined, where $b \in \mathrm{Bl}(H)$, and let $\chi \in \mathrm{Irr}(G)$ and $\psi \in \mathrm{Irr}(b)$.*
   (a) *If $\chi \notin \mathrm{Irr}(b^G)$, then*

$$\frac{[\widetilde{\chi_H}, \psi]}{\psi(1)} \equiv 0 \bmod \mathcal{P}.$$

   (b) *If $\chi \in \mathrm{Irr}(b^G)$ has height zero, then*

$$\frac{[\widetilde{\chi_H}, \psi]}{\psi(1)} \not\equiv 0 \bmod \mathcal{P}.$$

Proof. First of all, notice that

$$\widetilde{\chi_H} = \frac{1}{|G:H|_p}\tilde{\chi}_H$$

by the definition of the ˜ functions in Chapter 2. Also, recall that $\tilde{\chi}$ is a generalized character of G (Lemma (2.15)).

Now, suppose that $\chi \in \mathrm{Irr}(B)$ for some block B. Since $\tilde{\chi}$ only involves characters in B (by Lemma (3.20)), it follows that

$$[\tilde{\chi}, \theta] = [\tilde{\chi}, \theta_B]$$

for every class function θ. Now, if as usual $\mathrm{cl}(G^0)$ denotes the set of conjugacy classes of G of p-regular elements, then, using Frobenius reciprocity we get

$$\frac{[\widetilde{\chi_H}, \psi]}{\psi(1)} = \frac{1}{|G:H|_p}\frac{[\tilde{\chi}_H, \psi]}{\psi(1)} = \frac{1}{|G:H|_p}\frac{[\tilde{\chi}, \psi^G]}{\psi(1)}$$

$$= \frac{1}{|G:H|_p} \frac{[\tilde{\chi},(\psi^G)_B]}{\psi(1)} = \frac{|G:H|}{|G:H|_p} \frac{[\tilde{\chi},(\psi^G)_B]}{\psi^G(1)}$$

$$= |G:H|_{p'} \frac{[(\psi^G)_B,\tilde{\chi}]}{\psi^G(1)} = |G:H|_{p'} \frac{|G|_p}{|G|} \sum_{K\in\mathrm{cl}(G^0)} \frac{|K|(\psi^G)_B(x_K)}{\psi^G(1)} \chi(x_K^{-1})$$

$$= \frac{1}{|H|_{p'}} \sum_{K\in\mathrm{cl}(G^0)} \frac{|K|(\psi^G)_B(x_K)}{\psi^G(1)} \chi(x_K^{-1}).$$

By Lemma (6.3.c), this is zero modulo $\mathcal{P}$ whenever $b^G \neq B$. If, on the contrary, $b^G = B$, then, by Lemma (6.3.b), we have that

$$\frac{[\widetilde{\chi_H},\psi]}{\psi(1)} \equiv \frac{1}{|H|_{p'}} \sum_{K\in\mathrm{cl}(G^0)} \frac{|K|\chi(x_K)}{\chi(1)} \chi(x_K^{-1}) \equiv |G:H|_{p'} \frac{[\tilde{\chi},\chi]}{\chi(1)} \bmod \mathcal{P}.$$

If $\chi \in \mathrm{Irr}(B)$ has height zero, by Theorem (3.24) we know that

$$\frac{[\tilde{\chi},\chi]}{\chi(1)} \not\equiv 0 \bmod \mathcal{P}.$$

This finishes the proof of the lemma. ∎

(6.6) THEOREM (Okuyama). *Let $H \subseteq G$ be a subgroup of G and suppose that $\chi \in \mathrm{Irr}(G)$ is such that $\chi_H \in \mathrm{Irr}(H)$. Assume further that χ and χ_H have height zero in their respective blocks B and b. If $e \in \mathrm{Bl}(H)$ is such that e^G is defined, then $e = b$ if and only if $e^G = B$.*

Proof. Write $\psi = \chi_H$. Suppose first that $e = b$, so that b^G is defined. We want to show that $b^G = B$. It suffices to show that $\chi \in b^G$. Since $\psi \in \mathrm{Irr}(b)$ has height zero, it follows by Theorem (3.24) that

$$\frac{[\widetilde{\chi_H},\psi]}{\psi(1)} = \frac{[\tilde{\psi},\psi]}{\psi(1)} \not\equiv 0 \bmod \mathcal{P}.$$

Then $\chi \in b^G$ by Lemma (6.5.a).

Suppose now that $e^G = B$. We wish to show that $b = e$. Let $\xi \in \mathrm{Irr}(e)$. Since $\chi \in \mathrm{Irr}(e^G)$ has height zero, by Lemma (6.5.b) we have that

$$[\tilde{\psi},\xi] = [\widetilde{\chi_H},\xi] \neq 0.$$

Since ψ lies in b, necessarily $\xi \in \mathrm{Irr}(b)$ by Lemma (3.20). Therefore, $e = b$, as desired. ∎

Now we are ready to prove the third main theorem.

(6.7) THEOREM (Third Main Theorem). *Let b_0 be the principal block of H and let B_0 be the principal block of G, where H is a subgroup of G. If $b \in \mathrm{Bl}(H)$ and b^G is defined, then $b^G = B_0$ if and only if $b = b_0$.*

Proof. This follows from Theorem (6.6) since $(1_G)_H = 1_H$ and 1_G and 1_H have height zero in their respective blocks. ∎

There is a consequence of Lemma (6.5) which is worth mentioning.

(6.8) THEOREM (Blau). *Suppose that b^G is defined, where $b \in \mathrm{Bl}(H)$ and $H \subseteq G$. If $\chi \in \mathrm{Irr}(b^G)$ has height zero, then χ_H contains an irreducible constituent ξ of height zero in b such that*

$$[\chi_H, \xi] \not\equiv 0 \bmod p \, .$$

Proof. Recall that if $\alpha, \beta \in \mathrm{cf}(G)$, then

$$\widetilde{\alpha + \beta} = \tilde{\alpha} + \tilde{\beta} \, .$$

Hence,

$$\widetilde{\chi_H} = \sum_{\xi \in \mathrm{Irr}(H)} [\chi_H, \xi] \tilde{\xi} \, .$$

If $\psi \in \mathrm{Irr}(b)$, then, since χ has height zero in b^G, we have that

$$\frac{[\widetilde{\chi_H}, \psi]}{\psi(1)} = \sum_{\xi \in \mathrm{Irr}(H)} [\chi_H, \xi] \frac{[\tilde{\xi}, \psi]}{\psi(1)} \not\equiv 0 \bmod \mathcal{P} \, ,$$

by Lemma (6.5.b). Since $\frac{[\tilde{\xi}, \psi]}{\psi(1)}$ is zero if ξ is not in b and $\nu(\frac{[\tilde{\xi}, \psi]}{\psi(1)}) > 0$ if ξ does not have height zero (by Lemma (3.22.a) and Theorem (3.24)), the result easily follows. ∎

We now define the kernel of a block.

(6.9) DEFINITION. If B is a p-block of G, we define the **kernel of the block** B as

$$\ker(B) = \bigcap_{\chi \in \mathrm{Irr}(B)} \ker(\chi) \, .$$

(6.10) THEOREM. *If $\chi \in \mathrm{Irr}(B)$, then $\ker(B) = \mathbf{O}_{p'}(\ker(\chi))$. In particular, if B_0 is the principal block of G, then $\ker(B_0) = \mathbf{O}_{p'}(G)$.*

Proof. By elementary character theory, we have that $\ker(B)$ is the kernel of the character $\Xi = \sum_{\chi \in \mathrm{Irr}(B)} \chi(1)\chi$. Since $\Xi(g) = 0$ if g is not p-regular (by the weak block orthogonality), it follows that $\ker(B)$ is a p'-group. Hence, $\ker(B) \subseteq \mathbf{O}_{p'}(\ker(\chi))$. Now, let $N = \mathbf{O}_{p'}(\ker(\chi)) \triangleleft G$ and write

$$u = \frac{1}{|N|} \sum_{n \in N} n \in \mathbf{Z}(SG).$$

If $\psi \in \mathrm{Irr}(G)$, notice that

$$\omega_\psi(u) = \frac{1}{|N|} \sum_{\substack{K \in \mathrm{cl}(G) \\ K \subseteq N}} \omega_\psi(\hat{K}) = \frac{[\psi_N, 1_N]}{\psi(1)}.$$

In particular, $\omega_\chi(u) = 1$. Now, if $\psi \in \mathrm{Irr}(B)$, then

$$(\frac{[\psi_N, 1_N]}{\psi(1)})^* = \omega_\psi(u)^* = \lambda_B(u^*) = \omega_\chi(u)^* = 1 \neq 0.$$

Therefore, $[\psi_N, 1_N] \neq 0$ for $\psi \in \mathrm{Irr}(B)$ and it follows by Clifford's theorem that $N \subseteq \ker(\psi)$ for all $\psi \in \mathrm{Irr}(B)$. This proves the theorem. ∎

It is natural to ask what happens if, in Definition (6.9), we take the intersection of the kernels of the Brauer characters in a p-block. First, we need a lemma.

(6.11) LEMMA. *If $\varphi \in \mathrm{IBr}(G)$ and g is a p-regular element, then $g \in \ker(\varphi)$ if and only if $\varphi(g) = \varphi(1)$.*

Proof. If $g \in \ker(\varphi)$, then it is clear that $\varphi(g) = \varphi(1)$. Now, we prove the converse. Let $\mathcal{X}$ be a representation affording φ. Since $\langle g \rangle$ is a p'-group, we have that the representation $\mathcal{X}_{\langle g \rangle}$ is completely reducible. Hence, the matrix $\mathcal{X}(g)$ is similar to a diagonal matrix $\mathrm{diag}\{(\epsilon_1)^*, \ldots, (\epsilon_n)^*\}$, where $\epsilon_i \in \mathbf{U}$ for all i. By hypothesis, we have that $n = \epsilon_1 + \ldots + \epsilon_n$. Now, since $|\epsilon_i| = 1$ for all i, it follows that $\epsilon_i = 1$ for all i. Hence, $\mathcal{X}(g) = I$ and this proves the lemma. ∎

(6.12) COROLLARY. *If*

$$M = \bigcap_{\varphi \in \mathrm{IBr}(B)} \ker(\varphi),$$

then

$$M/\ker(B) = \mathbf{O}_p(G/\ker(B)).$$

Proof. Since $N = \ker(B)$ consists of p-regular elements by Theorem (6.10) and φ is a linear combination of the elements of $\{\chi^0 \,|\, \chi \in \mathrm{Irr}(B)\}$ (by Lemma (3.16)), it is clear that $N \subseteq M$ by applying Lemma (6.11). Also, if $x \in M$ is p-regular, then $\chi(x) = \chi(1)$ for all $\chi \in \mathrm{Irr}(B)$ since χ^0 is a linear combination of the elements of $\mathrm{IBr}(B)$ and $\varphi(x) = \varphi(1)$ for $\varphi \in \mathrm{IBr}(B)$ (again by Lemma (6.11)). Thus, $x \in N$ and we conclude that M/N is a p-group. On the other hand, since $N \subseteq \ker(\varphi)$, it follows that $\varphi \in \mathrm{IBr}(G/N)$. (See the discussion before Lemma (2.32).) Hence, by Lemma (2.32), $\mathbf{O}_p(G/N) \subseteq \ker(\varphi)/N$ and the result is proved. ∎

We now give a characterization of groups with a normal p-complement.

(6.13) COROLLARY. *Let B_0 be the principal block of a group G. Then G has a normal p-complement if and only if $\mathrm{IBr}(B_0) = \{1_{G^0}\}$. In this case, $\mathrm{Irr}(B_0) = \mathrm{Irr}(G/\mathbf{O}_{p'}(G))$ and the Cartan matrix of B_0 is $(|G|_p)$.*

Proof. By Corollary (6.12) and Theorem (6.10), we have that

$$\bigcap_{\varphi \in \mathrm{IBr}(B_0)} \ker(\varphi) = \mathbf{O}_{p'p}(G) \,.$$

Now, it is clear that G has a normal p-complement if and only if $\mathrm{IBr}(B_0) = \{1_{G^0}\}$.

Suppose now that G has a normal p-complement. Then $G^0 = \mathbf{O}_{p'}(G)$. Since $\mathrm{IBr}(B_0) = \{1_{\mathbf{O}_{p'}(G)}\}$, it follows that

$$\mathrm{Irr}(B_0) = \{\chi \in \mathrm{Irr}(G) \,|\, \chi^0 = \chi(1)1_{\mathbf{O}_{p'}(G)}\} = \mathrm{Irr}(G/\mathbf{O}_{p'}(G)) \,.$$

In particular, observe that $d_{\chi 1} = \chi(1)$ for $\chi \in \mathrm{Irr}(B_0)$. Then the Cartan matrix of B_0 is

$$\left(\sum_{\chi \in \mathrm{Irr}(G/\mathbf{O}_{p'}(G))} \chi(1)^2 \right) = (|G|_p) \,.$$

∎

We close this chapter by obtaining a formula of B. Külshammer for the principal block idempotent. (As we will see in the Problems, this gives an alternative proof of the third main theorem for blocks of subgroups H containing a p-subgroup Q such that $Q\mathbf{C}_G(Q) \subseteq H \subseteq \mathbf{N}_G(Q)$.)

(6.14) THEOREM (Külshammer). *Let B_0 be the principal block of G. If $e_{B_0} = \sum_{g \in G} e_g g$, then*

$$e_g = \frac{1}{|G^0|^*} |\{(x, y) \in G_p \times G^0 \,|\, xy = g\}|^*$$

for every p-regular element g and $e_g = 0$ otherwise.

Proof. First of all, notice that by Problem (2.3), $|G^0|$ is not divisible by p.

As in Theorem (4.23), we let π be the map $\mathbf{Z}(FG) \to \mathbf{Z}(FG)$ such that $\pi(\hat{K}) = \hat{K}$ for p-regular classes and $\pi(\hat{K}) = 0$ for classes of p-singular elements. We wish to prove that

$$|G^0|^* e_{B_0} = \pi(\widehat{G_p}\widehat{G^0}).$$

In the language of Lemma (3.32), observe that

$$\widehat{G^0} = u_{1_G} = \sum_{g \in G^0} 1_G(g)g.$$

By Lemma (3.32), we have that $\widehat{G^0} f_{B_0} = \widehat{G^0}$. Then $\widehat{G^0} = \widehat{G^0} e_{B_0}$. In particular,

$$\lambda_B(\widehat{G^0}) = 0$$

for blocks $B \neq B_0$. Now, since

$$\lambda_{B_0}(\hat{K}) = \omega_{1_G}(\hat{K})^* = |K|^*,$$

it follows that

$$\lambda_{B_0}(\widehat{G^0}) = |G^0|^*.$$

Finally, if R is the Robinson map, by Theorem (4.23) we have that

$$\pi(\widehat{G_p}\widehat{G^0}) = \mathrm{R}(\widehat{G^0}) = \sum_{B \in \mathrm{Bl}(G)} \lambda_B(\widehat{G^0})e_B = \lambda_{B_0}(\widehat{G^0})e_{B_0} = |G^0|^* e_{B_0},$$

as required. ∎

As a consequence of Theorem (6.14), notice that the coefficients of e_{B_0} lie in the prime field $\mathbb{Z}_p$ of F.

PROBLEMS

(6.1) (Külshammer) If Q is a p-subgroup of G and $g \in \mathbf{C}_G(Q)$, then

$$|\{(x, y) \in G_p \times G^0 \mid xy = g\}|$$
$$\equiv |\{(x, y) \in \mathbf{C}_G(Q)_p \times \mathbf{C}_G(Q)^0 \mid xy = g\}| \bmod p.$$

Deduce that if

$$e_{B_0} = \sum_{g \in G} e_g g$$

and

$$e_{b_0} = \sum_{g \in \mathbf{C}_G(Q)} d_g g$$

are the principal block idempotents of G and $\mathbf{C}_G(Q)$, respectively, then $e_g = d_g$ for $g \in \mathbf{C}_G(Q)$. This gives an alternative proof of the third main theorem (for subgroups $Q\mathbf{C}_G(Q) \subseteq H \subseteq \mathbf{N}_G(Q)$).

(6.2) (Külshammer) If $N \lhd G$ and $g \in N$, then

$$|\{(x,y) \in G_p \times G^0 \mid xy = g\}| = |\{(x,y) \in N_p \times N^0 \mid xy = g\}|.$$

Deduce that if

$$e_{B_0} = \sum_{g \in G} e_g g$$

and

$$e_{b_0} = \sum_{g \in N} d_g g$$

are the principal block idempotents of G and N, respectively, then

$$|G^0|^* e_g = |N^0|^* d_g$$

for $g \in N$.

(6.3) (Külshammer) Prove that $\mathbf{O}_{p'}(G) \subseteq \mathrm{supp}(e_{B_0})$.

(HINT: Use Problem (6.2).)

(6.4) Suppose that H is a subgroup of G. Let $\chi \in \mathrm{Irr}(G)$ and let $b \in \mathrm{Bl}(H)$. If $\chi_b = (\chi_H)_b = \sum_{\psi \in \mathrm{Irr}(b)} [\chi_H, \psi] \psi$, prove that

$$\frac{|G : H| \chi_b(1)}{\chi(1)} \in S.$$

(HINT: If T is a complete set of representatives of right cosets of G mod H, prove that $\sum_{t \in T} (f_b)^t \in \mathbf{Z}(SG)$. Apply the algebra homomorphism ω_χ.)

(6.5) Suppose that H is a subgroup of G with p'-index and let $\chi \in \mathrm{Irr}(G)$ be of p-power degree. Show that all the irreducible constituents of χ_H lie in the same block.

(6.6) (Isaacs-Scott) Let H be a subgroup of G and let $b \in \mathrm{Bl}(H)$. Suppose that $\chi \in \mathrm{Irr}(B_0(G))$, where $B_0(G)$ is the principal block of G. Prove

$$\left(\frac{|G : H| \chi_b(1)}{\chi(1)} \right)^* = 0$$

unless p does not divide $|G : H|$ and $b = B_0(H)$, the principal block of H. (See Problem (6.4).) In this case, show

$$\left(\frac{|G : H| \chi_b(1)}{\chi(1)} \right)^* = |G : H|^*.$$

(HINT: Use the fact that $\lambda_{B_0(G)} = \delta_{\mathbf{Z}(FG)}$ and $\lambda_{B_0(H)} = \delta_{\mathbf{Z}(FH)}$, where δ is the augmentation map defined in Problem (1.5).)

(6.7) Let H be a subgroup of G and let $b \in \mathrm{Bl}(H)$ be such that b^G is defined. Assume that b and b^G have the same defect. If $\xi \in \mathrm{Irr}(b)$ has height zero, prove that there exists an irreducible character $\chi \in \mathrm{Irr}(B)$ of height zero such that $[\xi^G, \chi]$ is not divisible by p.

(6.8) (Brauer) Suppose that B_0 is the principal block of G. Prove that

$$\sum_{\chi \in \mathrm{Irr}(B_0)} \chi(1)^2$$

is a power of p if and only if G has a normal p-complement.

(HINT: Use Theorem (3.28) and Corollary (6.13).)

(6.9) Suppose that G has a normal p-complement K. Let B be a block of G of maximal defect. Prove that $\mathrm{IBr}(B) = \{\theta\}$, where $\theta \in \mathrm{Irr}(K)$ is G-invariant. Also, $\mathrm{Irr}(B) = \{\chi\hat{\theta} \,|\, \chi \in \mathrm{Irr}(G/K)\}$, where $\hat{\theta}$ is a character of G extending θ.

(6.10) (Glauberman) Suppose that A is a p-group acting coprimely on a p'-group G. Let $\mathrm{Irr}_A(G) = \{\chi \in \mathrm{Irr}(G) \,|\, \chi^a = \chi \text{ for all } a \in A\}$ and $C = \mathbf{C}_G(A)$. Prove that if $\chi \in \mathrm{Irr}_A(G)$, then there is a unique $\chi' \in \mathrm{Irr}(C)$ such that

$$[\chi_C, \chi'] \not\equiv 0 \bmod p .$$

Furthermore, the map $\chi \mapsto \chi'$ from $\mathrm{Irr}_A(G) \to \mathrm{Irr}(C)$ is a bijection.

(HINT: Use Problem (6.9), Corollary (6.4) and the first main theorem in the semidirect group $\Gamma = GA$.)

(NOTE: This is known as the **Alperin argument**.)

(6.11) (Glauberman) Suppose that a solvable group A acts coprimely on a group G. Let $\mathrm{Irr}_A(G) = \{\chi \in \mathrm{Irr}(G) \,|\, \chi^a = \chi \text{ for all } a \in A\}$ and $C = \mathbf{C}_G(A)$. Prove that there is a canonical bijection from $\mathrm{Irr}_A(G) \to \mathrm{Irr}(C)$.

(HINT: Let p be the smallest prime such that $\mathbf{O}_p(A) > 1$. Use Problem (6.10) and induction on $|A|$.)

(6.12) Let $\varphi \in \mathrm{IBr}(B)$, where B is a p-block of G. Show that $\ker(B) = \mathbf{O}_{p'}(\ker(\varphi))$.

7. The Z^*-Theorem

In this chapter, we finally give a proof of Glauberman's Z^*-theorem. First, we need to prove the Brauer-Suzuki theorem.

(7.1) THEOREM (Brauer-Suzuki). *Suppose that G has a generalized quaternion Sylow 2-subgroup. If $\mathbf{O}_{2'}(G) = 1$, then $|\mathbf{Z}(G)| = 2$.*

If the Sylow 2-subgroup of G is a quaternion group of order greater than 8, the Brauer-Suzuki theorem follows from ordinary character theory without too much effort (see Theorem (7.8) in [Isaacs, *Characters*]). The deep part of the theorem involves the case when the Sylow 2-subgroup is the quaternion group of order 8, and this is what we will focus on here.

G. Glauberman showed in 1974 that the Brauer-Suzuki theorem can be proved without modular representation theory (although certainly there is no known purely group theoretic proof). However, Glauberman's proof is not easy. Since this course is about modular representation theory, we prove the Brauer-Suzuki theorem by using blocks. We will follow R. Brauer's proof, so that the reader can see how blocks are used to study finite groups.

We start with an easier case which will be used later on.

(7.2) THEOREM. *Suppose that $P = \mathbb{Z}_2 \times \mathbb{Z}_2$ is a Sylow 2-subgroup of G. Then G has either one or three conjugacy classes of involutions. In fact, G has three classes of involutions if and only if G has a normal 2-complement. If G has one class of involutions and B_0 is the principal 2-block of G, then*

$$\mathrm{Irr}(B_0) = \{1_G = \chi_0, \chi_1, \chi_2, \chi_3\},$$

and $|\mathrm{IBr}(B_0)| = 3$. Also, for $i = 1, 2, 3$, there are $\epsilon_i = \pm 1$ such that $\chi_i(u) = \epsilon_i$ for every 2-singular element $u \in G$. Furthermore,

$$\chi_i(1) \equiv \epsilon_i \bmod 4$$

and

$$1 + \epsilon_1 \chi_1(s) + \epsilon_2 \chi_2(s) + \epsilon_3 \chi_3(s) = 0$$

for every 2-regular $s \in G$.

Proof. If G has a normal 2-complement, it is clear that G has three classes of involutions.

Suppose that $\mathbf{N}_G(P) = \mathbf{C}_G(P)$. Then G has a normal 2-complement by Burnside's transfer theorem, and therefore G has three classes of involutions. Suppose that $\mathbf{C}_G(P) < \mathbf{N}_G(P)$. Then $\mathbf{N}_G(P)/\mathbf{C}_G(P)$ is a $2'$-subgroup of $\mathrm{Aut}(P) = \mathrm{Sym}(3)$, the symmetric group on three symbols. In this case, $\mathbf{N}_G(P)/\mathbf{C}_G(P)$ has order 3 and every generator of it permutes the three nontrivial elements of P. Hence, we see that G has a unique class of involutions. This proves the first part of the theorem.

From now on assume that G has one class of involutions. Let $t \in P$ be one of them and write $C = \mathbf{C}_G(t)$. Since P is abelian, $P \subseteq C$.

First, notice that C cannot have a unique class of involutions (because t is central in C). By the first part applied to C, we conclude that C has three classes of involutions and, therefore, that C has a normal 2-complement. Thus, if b_0 is the principal block of C, we have that $\mathrm{IBr}(b_0) = \{1_{C^0}\}$ by Corollary (6.13). Now, by Theorem (5.12) and the third main theorem, we have that

$$\mathrm{k}(B_0) - \mathrm{l}(B_0) = \mathrm{l}(b_0) = 1 \,.$$

By Corollary (5.8) and the third main theorem, we have that

$$\chi(ts) = d^t_{\chi 1_{C^0}}$$

for every $\chi \in \mathrm{Irr}(B_0)$ and every $s \in \mathbf{C}_G(t)^0$. To simplify notation, write $d^t_{\chi 1_{C^0}} = d^t_{\chi 1}$.

We claim that $d^t_{\chi 1} \neq 0$ for every $\chi \in \mathrm{Irr}(B_0)$. Otherwise, $\chi(t) = 0$, and therefore $\chi(u) = 0$ for all involutions (since they are G-conjugate). In this case we conclude that B_0 has defect zero by Theorem (3.18). Since the defect groups of the principal block of G are the Sylow 2-subgroups of G, this is a contradiction, which proves the claim.

Now, since C has a normal 2-complement, by Corollary (6.13) the Cartan matrix of b_0 is a 1×1 matrix with the unique entry being the order of a Sylow 2-subgroup of C; that is, $C_{b_0} = (4)$. By Lemma (5.13.b), we get

$$\sum_{\chi \in \mathrm{Irr}(B_0)} |d^t_{\chi 1}|^2 = 4 \,.$$

We know that the generalized decomposition numbers associated with involutions are integers (see the remark after their definition). We also know (by the very definition) that $d^t_{1_G, 1} = 1$. Since we have already proven that they are nonzero, it follows that $\chi(t) = d^t_{\chi 1} = \pm 1$ for all $\chi \in \mathrm{Irr}(B_0)$. Hence,

$$\mathrm{Irr}(B_0) = \{1_G = \chi_0, \chi_1, \chi_2, \chi_3\} \,.$$

Since all the involutions of G are G-conjugate, we have that $\chi_i(t_0) = \chi_i(t)$ for every involution t_0 in G. Write $\chi_i(t_0) = \epsilon_i$ for $i = 0, 1, 2, 3$. Hence, by the above discussion, we see that

$$\chi_i(u) = \epsilon_i$$

for every 2-singular element $u \in G$. Now,

$$[(\chi_i)_P, 1_P] = \frac{1}{4}(\chi_i(1) + 3\epsilon_i)$$

is an integer, and we have that

$$\chi_i(1) \equiv \epsilon_i \bmod 4 .$$

The last equation is just block orthogonality (Corollary (5.11)). ∎

In general, it is not easy to find the set of irreducible Brauer characters of a group G (once the maximal ideal M has been fixed). As we are about to see, for some specific problems, it is not so important to know the irreducible Brauer characters of a block or to have a "basic set".

(7.3) DEFINITION. Suppose that B is a block of G. A **basic set** for B is a basis of the abelian group $\{\sum_{\chi \in \mathrm{Irr}(B)} a_\chi \chi^0 \,|\, a_\chi \in \mathbb{Z}\} \subseteq \mathrm{cf}(G^0)$.

Of course, $\mathrm{IBr}(B)$ (which is contained in $\{\sum_{\chi \in \mathrm{Irr}(B)} a_\chi \chi^0 \,|\, a_\chi \in \mathbb{Z}\}$ by Lemma (3.16)) is a basic set of B. If $\mathcal{B}$ is a basic set for the block B and $\chi \in \mathrm{Irr}(B)$, then

$$\chi^0 = \sum_{\varphi \in \mathcal{B}} d_{\chi\varphi} \varphi$$

for some uniquely defined integers $d_{\chi\varphi}$.

(The reader should be aware of the fact that there is some ambiguity in our notation since any element $\varphi \in \{\sum_{\chi \in \mathrm{Irr}(B)} a_\chi \chi^0 \,|\, a_\chi \in \mathbb{Z}\}$ may lie in different basic sets. However, this notation is simple and, as we will see, there is little risk of confusion.)

These integers $d_{\chi\varphi}$ for $\chi \in \mathrm{Irr}(B)$ and $\varphi \in \mathcal{B}$ can be computed from the matrix equation

$$D_\mathcal{B} = D_B U ,$$

where

$$D_\mathcal{B} = (d_{\chi\varphi})_{\chi \in \mathrm{Irr}(B), \varphi \in \mathcal{B}} ,$$

and

$$U = (u_{\mu\varphi})_{\mu \in \mathrm{IBr}(B), \varphi \in \mathcal{B}} \in \mathrm{GL}(n, \mathbb{Z})$$

is the matrix changing the basis from $\mathrm{IBr}(B)$ to $\mathcal{B}$. Now, we may also define the Cartan invariants

$$c_{\varphi\theta} = \sum_{\chi \in \mathrm{Irr}(B)} d_{\chi\varphi}d_{\chi\theta}$$

for $\varphi, \theta \in \mathcal{B}$. The new Cartan matrix is

$$C_{\mathcal{B}} = (D_B U)^t (D_B U) = U^t C_B U.$$

The basic sets that we are going to use are of the following type. Suppose that we have a block B and a subset $\mathcal{Y} \subseteq \mathrm{Irr}(B)$ such that $|\mathcal{Y}| = |\mathrm{IBr}(B)|$. If for every $\chi \in \mathrm{Irr}(B) - \mathcal{Y}$, χ^0 is an integer linear combination of $\{\eta^0 \mid \eta \in \mathcal{Y}\}$, we claim that $\{\epsilon_\eta \eta^0 \mid \eta \in \mathcal{Y}\}$ is a basic set for B for every choice of $\epsilon_\eta = \pm 1$. It is clear that every $\epsilon_\eta \eta^0$ is a $\mathbb{Z}$-linear combination of $\mathrm{IBr}(B)$. Conversely, since every $\varphi \in \mathrm{IBr}(B)$ is an integer combination of $\{\chi^0 \mid \chi \in \mathrm{Irr}(B)\}$ (by Lemma (3.16)), it follows that every $\varphi \in \mathrm{IBr}(B)$ is an integer linear combination of $\{\epsilon_\eta \eta^0 \mid \eta \in \mathcal{Y}\}$. The claim easily follows.

(7.4) COROLLARY. *With the same hypotheses as in (7.2), assume that G has exactly one class of involutions. Then the notation can be chosen so that $\epsilon_1 = 1$ and $\epsilon_3 = -1$. In this case, $\{1_{G^0}, (\chi_1)^0, \epsilon_2(\chi_2)^0\}$ is a basic set for B_0 (with odd degrees), and the decomposition and Cartan matrices of B_0 with respect to this basic set are*

$$D = \begin{pmatrix} 1 & 0 & 0 \\ 0 & 1 & 0 \\ 0 & 0 & \epsilon_2 \\ 1 & 1 & 1 \end{pmatrix} \quad \text{and} \quad C = \begin{pmatrix} 2 & 1 & 1 \\ 1 & 2 & 1 \\ 1 & 1 & 2 \end{pmatrix}.$$

Proof. Using the fact that

$$1 + \epsilon_1 \chi_1(s) + \epsilon_2 \chi_2(s) + \epsilon_3 \chi_3(s) = 0$$

for every 2-regular element s of G, by setting $s = 1$ we see that there should be two different signs. Choose notation so that $\epsilon_1 = 1$ and $\epsilon_3 = -1$. In this case, notice that

$$(\chi_3)^0 = 1_{G^0} + (\chi_1)^0 + \epsilon_2(\chi_2)^0.$$

Since we know that $\mathrm{l}(B) = 3$, by the remark before this corollary we have that

$$\{1_{G^0}, (\chi_1)^0, \epsilon_2(\chi_2)^0\}$$

is a basic set of B_0. By Theorem (7.2) we have that $\chi_i(1) \equiv \epsilon_i \bmod 4$ and thus the elements of this basic set have odd degree. Now, calculate the matrices. ∎

We may compute generalized decomposition numbers for basic sets in the same way that we compute decomposition numbers and Cartan invariants for Brauer characters.

Let x be a p-element of G. Write $H = \mathbf{C}_G(x)$, and for each block $b \in \mathrm{Bl}(H)$, let $\mathcal{B}_b$ be a basic set for b. Now, if $\chi \in \mathrm{Irr}(G)$ and $z \in \mathbf{C}_G(x)$ is p-regular, we may write

$$\chi(xz) = \sum_{b \in \mathrm{Bl}(H)} \sum_{\psi \in \mathrm{Irr}(b)} [\chi_H, \psi] \frac{\psi(x)}{\psi(1)} \psi(z)$$

$$= \sum_{b \in \mathrm{Bl}(H)} \sum_{\psi \in \mathrm{Irr}(b)} [\chi_H, \psi] \frac{\psi(x)}{\psi(1)} \sum_{\varphi \in \mathcal{B}_b} d_{\psi\varphi} \varphi(z) = \sum_{b \in \mathrm{Bl}(H)} \sum_{\varphi \in \mathcal{B}_b} d_{\chi\varphi}^x \varphi(z),$$

where for every $\varphi \in \mathcal{B}_b$, we set

$$d_{\chi\varphi}^x = \sum_{\psi \in \mathrm{Irr}(b)} [\chi_H, \psi] \frac{\psi(x)}{\psi(1)} d_{\psi\varphi}.$$

Note that these "new" decomposition numbers $d_{\chi\varphi}^x$ are also uniquely determined by $\chi \in \mathrm{Irr}(G)$, x and $\varphi \in \mathcal{B}_b$. Furthermore, notice that if $U_b = (u_{\mu\varphi})_{\mu \in \mathrm{IBr}(b), \varphi \in \mathcal{B}_b}$ is the matrix changing the basis from $\mathrm{IBr}(b)$ to $\mathcal{B}_b$ for $b \in \mathrm{Bl}(\mathbf{C}_G(x))$, by the definition of the generalized decomposition numbers, it follows that the "new" and "old" generalized decomposition numbers are related by the equation

$$d_{\chi\varphi}^x = \sum_{\mu \in \mathrm{IBr}(b)} d_{\chi\mu}^x u_{\mu\varphi} \tag{1}$$

for $\chi \in \mathrm{Irr}(G)$, $b \in \mathrm{Bl}(\mathbf{C}_G(x))$ and $\varphi \in \mathcal{B}$.

With this equation, it is easy to prove the following lemma.

(7.5) LEMMA. *Let x and y be p-elements of G. Suppose that $\mathcal{B}_b$ is a basic set for every $b \in \mathrm{Bl}(\mathbf{C}_G(x))$ and $\mathcal{B}_e$ is a basic set for every $e \in \mathrm{Bl}(\mathbf{C}_G(y))$. Let $B \in \mathrm{Bl}(G)$.*

(a) *If $\chi \in \mathrm{Irr}(B)$, $\varphi \in \mathcal{B}_b$ and $b^G \neq B$, then $d_{\chi\varphi}^x = 0$.*
(b) *If y is not G-conjugate to x, then*

$$\sum_{\chi \in \mathrm{Irr}(B)} \overline{d_{\chi\varphi}^x} d_{\chi\eta}^y = 0$$

for every $\varphi \in \mathcal{B}_b$ and $\eta \in \mathcal{B}_e$.

(c) Let $\varphi \in \mathcal{B}_{b_1}$ and let $\eta \in \mathcal{B}_{b_2}$, where $b_1, b_2 \in \mathrm{Bl}(\mathbf{C}_G(x))$. Then

$$\sum_{\chi \in \mathrm{Irr}(B)} \overline{d^x_{\chi\varphi}} d^x_{\chi\eta} = 0,$$

unless $b_1 = b_2$ and this block induces B. In this case,

$$\sum_{\chi \in \mathrm{Irr}(B)} \overline{d^x_{\chi\varphi}} d^x_{\chi\eta} = c_{\varphi\eta}.$$

(d) If y is not G-conjugate to x and $z \in \mathbf{C}_G(x)^0$, then

$$\sum_{\chi \in \mathrm{Irr}(B)} \overline{\chi(xz)} d^y_{\chi\eta} = 0$$

for every $\eta \in \mathcal{B}_e$.

Proof. Part (a) is clear by the second main theorem and the equation (1). Now, assume that x and y are not G-conjugate. By Lemma (5.13.a), we know that

$$\sum_{\chi \in \mathrm{Irr}(B)} \overline{d^x_{\chi\mu}} d^y_{\chi\tau} = 0$$

for $\mu \in \mathrm{IBr}(\mathbf{C}_G(x))$ and $\tau \in \mathrm{IBr}(\mathbf{C}_G(y))$. By using equation (1) for the blocks b and e, part (b) easily follows. Part (c) follows from Lemma (5.13.b) and (1).

To prove part (d), we use part (b). We have that

$$\sum_{\chi \in \mathrm{Irr}(B)} \overline{\chi(xz)} d^y_{\chi\eta} = \sum_{b \in \mathrm{Bl}(\mathbf{C}_G(x))} \sum_{\varphi \in \mathcal{B}_b} \sum_{\chi \in \mathrm{Irr}(B)} \overline{d^x_{\chi\varphi} \varphi(z)} d^y_{\chi\eta} = 0,$$

by applying part (b). The proof of the lemma is complete. ∎

We will have the hypothesis of Theorem (7.2) in a factor group. So we need a fact about blocks and factor groups. The general results are proved in Chapter 9.

Recall that if $N \triangleleft G$, we identify the characters of $G/N = \bar{G}$ (ordinary and modular) with characters of G containing N in their kernel. (See the discussion before Lemma (2.32).) Some caution is necessary, however, when doing this identification. For instance, if $\bar{\chi} \in \mathrm{Irr}(\bar{G})$ and $\chi \in \mathrm{Irr}(G)$ is the corresponding character of G with $\chi(g) = \bar{\chi}(gN)$ for $g \in G$, then, for $x \in G^0$, we have

$$\chi(x) = \bar{\chi}(xN) = \sum_{\bar{\varphi} \in \mathrm{IBr}(\bar{G})} d_{\bar{\chi}\bar{\varphi}} \bar{\varphi}(xN) = \sum_{\bar{\varphi} \in \mathrm{IBr}(\bar{G})} d_{\bar{\chi}\bar{\varphi}} \varphi(x).$$

Hence, we see that

$$d_{\bar\chi\bar\varphi} = d_{\chi\varphi} \tag{2}$$

for characters χ and φ containing N in their kernels. Furthermore, in this case, we may conclude that $d_{\chi\mu} = 0$ if $N \not\subseteq \ker(\mu)$. Hence, it is clear that the identification above "respects" decomposition numbers. As we are about to see, this may not be true for the Cartan invariants since, given $\varphi \in \mathrm{IBr}(G)$ with $N \subseteq \ker(\varphi)$, there might be ordinary characters χ not containing N in their kernels and with $d_{\chi\varphi} \neq 0$.

To end this discussion, suppose that $\bar B$ is a block of $\bar G$. If $\bar\chi, \bar\psi \in \mathrm{Irr}(\bar B)$ are linked, then, by (2), $\bar\chi$ and $\bar\psi$ are also linked as irreducible characters of G. Hence, we deduce that there is a unique block $B \in \mathrm{Bl}(G)$ such that $\mathrm{Irr}(\bar B) \subseteq \mathrm{Irr}(B)$. Also, since $\bar\varphi \in \mathrm{IBr}(\bar B)$ if and only if there is a $\bar\chi \in \mathrm{Irr}(\bar B)$ with $d_{\bar\chi\bar\varphi} \neq 0$, again by (2) we deduce that $\mathrm{IBr}(\bar B) \subseteq \mathrm{IBr}(B)$. Hence, we see that each block $\bar B$ of $\bar G$ is contained in a unique block $B \in \mathrm{Bl}(G)$.

(The reader should be aware of the fact that, in general, the map $\bar B \mapsto B$ is not injective (or surjective) since a block of G may contain more than one block of $\bar G$.)

(7.6) THEOREM. *Suppose that G has a normal p-subgroup P such that $G/\mathbf{C}_G(P)$ is a p-group. Write $\bar G = G/P$. If $\bar B \in \mathrm{Bl}(\bar G)$ and $B \in \mathrm{Bl}(G)$ is the unique block of G containing $\bar B$, then the map $\bar B \mapsto B$ is a bijection from $\mathrm{Bl}(\bar G)$ to $\mathrm{Bl}(G)$. Also, $\mathrm{IBr}(\bar B) = \mathrm{IBr}(B)$ and the Cartan matrices of $\bar B$ and B are related by $C_B = |P|C_{\bar B}$.*

Proof. We claim that the map $x \mapsto \bar x$ from $G^0 \to \bar G^0$ is a bijection. Given $x \in G$, notice that xP is a p'-element if and only if $x_p \in P$, which happens if and only if $xP = x_{p'}P$. So this map is (always) surjective. Now, suppose that $x, y \in G^0$ are such that $xP = yP$. Then, $x = yz$ for some $z \in P$. Since $G^0 \subseteq \mathbf{C}_G(P)$ by hypothesis, we see that $\mathrm{o}(yz) = \mathrm{o}(y)\mathrm{o}(z)$. Now, $yz = x$ is a p'-element and we conclude that $z = 1$. This proves the claim.

Since $P \subseteq \mathbf{O}_p(G) \subseteq \ker(\varphi)$ for every $\varphi \in \mathrm{IBr}(G)$, we have that the map $\varphi \mapsto \bar\varphi$ is a bijection from $\mathrm{IBr}(G)$ onto $\mathrm{IBr}(\bar G)$, where $\varphi(g) = \bar\varphi(\bar g)$. (See the discussion before Lemma (2.32).)

If $\varphi, \theta \in \mathrm{IBr}(G)$, then

$$[\varphi, \theta]^0 = \frac{1}{|G|} \sum_{g \in G^0} \varphi(g)\theta(g^{-1}) = \frac{1}{|P|}\frac{1}{|\bar G|} \sum_{\bar g \in \bar G^0} \bar\varphi(\bar g)\bar\theta(\bar g^{-1}) = \frac{1}{|P|}[\bar\varphi, \bar\theta]^0 .$$

Now, $\big([\varphi, \theta]^0\big)_{\varphi,\theta\in\mathrm{IBr}(G)}$ is the inverse of the Cartan matrix of G, and we conclude that if $\bar C$ is the Cartan matrix of $\bar G$, then $|P|\bar C = C$.

Finally, suppose that $B \in \mathrm{Bl}(G)$. Since P is contained in $\ker(\varphi)$ for every $\varphi \in \mathrm{IBr}(G)$ and every block of $\bar G$ is contained in a unique block of G, notice that we may write $\mathrm{IBr}(B) = \mathrm{IBr}(\bar{B}_1) \cup \ldots \cup \mathrm{IBr}(\bar{B}_s)$, where $\bar{B}_1, \ldots, \bar{B}_s$ are the blocks of $\bar{G}$ contained in B. If $s > 1$, since $c_{\varphi\theta} = |P| c_{\bar{\varphi}\bar{\theta}}$, we will have that the Cartan matrix C_B of B will have the form

$$\begin{pmatrix} * & 0 \\ 0 & * \end{pmatrix}.$$

This contradicts Problem (3.4). Therefore, each $B \in \mathrm{Bl}(G)$ contains a unique $\bar{B} \in \mathrm{Bl}(\bar{G})$ and $|\mathrm{Bl}(G)| = |\mathrm{Bl}(\bar{G})|$. ∎

If P is a normal p-subgroup of G and $\bar{G} = G/P$, notice that we may identify $\mathrm{cf}(\bar{G}^0)$ with $\mathrm{cf}(G^0)$ by using the fact that $\mathrm{IBr}(\bar{G}) = \mathrm{IBr}(G)$. (In connection with this, see also the proof of Theorem (9.11).) The class functions $\bar{\theta} \in \mathrm{cf}(\bar{G}^0)$ and $\theta \in \mathrm{cf}(G^0)$ are identified if and only if $\bar{\theta}(\bar{g}) = \theta(g)$ for every $g \in G^0$. With the hypothesis of the previous theorem, it easily follows that $\mathcal{B}$ is a basic set for B if and only if $\bar{\mathcal{B}} = \{\bar{\theta} \mid \theta \in \mathcal{B}\}$ is a basic set for $\bar{B}$. The Cartan matrices with respect to basic sets are, again, related by

$$C_{\mathcal{B}} = |P| C_{\bar{\mathcal{B}}}.$$

(See the relationship between the matrices $C_{\mathcal{B}}$ and C_B after Definition (7.3).)

Now we are ready to prove the Brauer-Suzuki theorem. We will use several standard group theoretic facts throughout its proof. We use the fact that the automorphism group of the quaternion group of eight elements is $\mathrm{Sym}(4)$, the symmetric group on four symbols. Also, groups with a cyclic Sylow 2-subgroup have a normal 2-complement. An elementary fact on involutions is needed: if t_1, t_2 are involutions of G, then $\langle t_1, t_2 \rangle$ is dihedral of order $2o(t_1t_2)$. Finally, we use the fact that a group G with Sylow p-subgroup P has a normal p-complement if and only if P controls its G-fusion (that is, if for every conjugacy class of p-elements K of G, then $K \cap P$ is a conjugacy class of P).

For the reader's convenience, we write down the character table of Q_8.

	1	t	y	z	yz
τ_1	1	1	1	1	1
τ_2	1	1	-1	1	-1
τ_3	1	1	1	-1	-1
τ_4	1	1	-1	-1	1
τ_5	2	-2	0	0	0

Proof of the Brauer-Suzuki Theorem. Let P be a quaternion Sylow 2-subgroup of G of order 8 and let t be the involution of P. It suffices to prove that $t \in \mathbf{Z}(G)$. In this case, since $\mathbf{O}_{2'}(G) = 1$, we will have that $\mathbf{Z}(G) \subseteq P$ and thus $\mathbf{Z}(G) = \mathbf{Z}(P) = \{1, t\}$, as desired. Of course, we may assume that $P < G$.

We will prove that $t \in N$ for some proper normal subgroup N of G. Suppose that this happens. Note that $P \cap N$ is cyclic or $P \subseteq N$. If $P \cap N$ is cyclic, then N has a normal 2-complement. Since $\mathbf{O}_{2'}(N) \subseteq \mathbf{O}_{2'}(G) = 1$, it follows that N is a cyclic 2-group. In this case, t is the unique involution of N and thus, $t \in \mathbf{Z}(G)$. If $P \subseteq N$, then, arguing by induction on $|G|$, we have that $\mathbf{Z}(N) = \{1, t\} \triangleleft G$ and, again, we are done.

Our objective is to find a nontrivial character in the principal block of G which contains t in its kernel.

Let us write $\{1, t, y, z, yz\}$ for the representatives of the five conjugacy classes of P. Recall that every element of order 4 is P-conjugate with its inverse. We claim that all the elements of order 4 in P are in fact G-conjugate, so that G has a unique conjugacy class of elements of order 4. (In fact, we will prove that all of them are $\mathbf{C}_G(t)$-conjugate.) Since G cannot have a normal 2-complement, it follows that P cannot control its own G-fusion. Therefore, at least two different conjugacy classes of elements of order 4 in P are G-conjugate. Choose the notation so that $y^g = z$ for some $g \in G$. In this case, $\langle y \rangle^g = \langle z \rangle$ and therefore $\mathbf{N}_G(\langle y \rangle)^g = \mathbf{N}_G(\langle z \rangle)$. Since every subgroup of P is normal in P, it follows that P and P^g are Sylow 2-subgroups of $\mathbf{N}_G(\langle z \rangle)$. Therefore, there exists $x \in \mathbf{N}_G(\langle z \rangle)$ such that $P^{gx} = P$. Write $u = gx \in \mathbf{N}_G(P)$. Now, y and y^u are elements of order 4 in P. Notice that y^u and y are not P-conjugate because if $y^u = y^v$ for some $v \in P$, then

$$\langle z \rangle = \langle z \rangle^x = \langle z^x \rangle = \langle y^u \rangle = \langle y^v \rangle = \langle y \rangle^v = \langle y \rangle \,,$$

a contradiction. Changing notation again (and replacing z by its inverse if necessary), we conclude that $y^g = z$ for some $g \in \mathbf{N}_G(P)$. Now, since $\text{Aut}(P) = \text{Sym}(4)$, we have that $\mathbf{N}_G(P)/\mathbf{C}_G(P) \subseteq \text{Sym}(4)$. If $o(g\mathbf{C}_G(P))$ is a 2-power, then $g\mathbf{C}_G(P) \in P\mathbf{C}_G(P)/\mathbf{C}_G(P)$, which is the unique Sylow 2-subgroup of $\mathbf{N}_G(P)/\mathbf{C}_G(P)$. In this case, y and z are P-conjugate, a contradiction. Hence, the order of $g\mathbf{C}_G(P)$ is 3. Thus, $z^g = (yz)^{\pm 1}$, and the claim has been proven. Notice that since $\{1, t\} = \mathbf{Z}(P) \triangleleft \mathbf{N}_G(P)$, it follows that $g \in \mathbf{C}_G(t)$.

Let us write $\text{Irr}(B_0) = \{1_G = \chi_0, \ldots, \chi_r\}$ for the irreducible characters of the principal block B_0 of G.

Analysis at y

The Sylow 2-subgroups of $\mathbf{C}_G(y)$ cannot have order 8 since the quaternion group does not have a central element of order 4. Hence, $\langle y \rangle$ is a Sylow

2-subgroup of $\mathbf{C}_G(y)$ and thus $\mathbf{C}_G(y)$ has a normal 2-complement. There-
fore, if b_0 is the principal block of $\mathbf{C}_G(y)$, it follows that $\mathrm{IBr}(b_0) = \{1_{\mathbf{C}_G(y)^0}\}$
by Corollary (6.13). By Corollary (5.8) and the third main theorem, we have
that

$$\chi_i(y) = d^y_{\chi_i 1_{\mathbf{C}_G(y)^0}}.$$

Note that since the irreducible characters of P are integer valued, it follows
that $\chi_i(y) \in \mathbb{Z}$. Now, the Cartan matrix of b_0 is (4) (Corollary (6.13)), and
by Lemma (5.13.b) we have that

$$\sum_{i=0}^{r} \chi_i(y)^2 = 4.$$

By block orthogonality, we also have

$$\sum_{i=0}^{r} \chi_i(y)\chi_i(1) = 0.$$

These relations can be written in the following way. If we have two columns

$$a = \begin{pmatrix} a_0 \\ \vdots \\ a_r \end{pmatrix} \quad \text{and} \quad b = \begin{pmatrix} b_0 \\ \vdots \\ b_r \end{pmatrix},$$

we write

$$(a, b) = \sum_{i=0}^{r} \overline{a_i} b_i.$$

Thus, if we write

$$\chi(y) = D_0^y = \begin{pmatrix} \chi_0(y) \\ \vdots \\ \chi_r(y) \end{pmatrix} \quad \text{and} \quad \chi(1) = \begin{pmatrix} \chi_0(1) \\ \vdots \\ \chi_r(1) \end{pmatrix},$$

we have

$$(\chi(y), \chi(y)) = 4 \tag{1}$$

and

$$(\chi(y), \chi(1)) = 0. \tag{2}$$

Recall that these are columns of integers having a 1 as the first entry. By
equation (1), we see that the nonzero entries of the integer column $\chi(y)$ are
necessarily four ± 1.

Analysis at t

We have that P is a Sylow 2-subgroup of $\mathbf{C}_G(t)$, and thus $\mathbf{C}_G(t)/\langle t \rangle$ has Sylow 2-subgroups isomorphic to $\mathbb{Z}_2 \times \mathbb{Z}_2$. Also, notice that all the involutions of $\mathbf{C}_G(t)/\langle t \rangle$ are conjugate since the elements of order 4 of P are $\mathbf{C}_G(t)$-conjugate. So the hypotheses of Corollary (7.4) are satisfied. If $\overline{b}_0$ is the principal block of $\mathbf{C}_G(t)/\langle t \rangle$, then there exists a basic set $\{1 = \psi_0, \psi_1, \psi_2\}$ of $\overline{b}_0$ with odd degrees such that the Cartan matrix of $\overline{b}_0$ with respect to this basic set is $c_{ij} = 1 + \delta_{ij}$. Now, notice that $\{\psi_0, \psi_1, \psi_2\}$ is again a basic set for b_0, the principal block of $\mathbf{C}_G(t)$, by the remark after the proof of Theorem (7.6). Also, the Cartan invariants are twice the ones we had before. By Lemma (7.5.a), the third main theorem and the definition of the generalized decomposition numbers (for basic sets), we have that

$$\chi_i(tu) = \sum_{j=0}^{2} d^t_{\chi_i \psi_j} \psi_j(u)$$

for all $i = 0, \ldots, r$ and $u \in \mathbf{C}_G(t)^0$. (To simplify notation, we write $d^t_{\chi_i \psi_j} = d^t_{ij}$ for all $i = 0, \ldots, r$.) This may also be written in the following column form:

$$\chi(tu) = D^t_0 + \psi_1(u) D^t_1 + \psi_2(u) D^t_2 ,$$

where

$$\chi(tu) = \begin{pmatrix} \chi_0(tu) \\ \vdots \\ \chi_r(tu) \end{pmatrix} \quad \text{and} \quad D^t_j = \begin{pmatrix} d^t_{0j} \\ \vdots \\ d^t_{rj} \end{pmatrix} .$$

Notice that $d^t_{00} = 1, d^t_{01} = 0$ and $d^t_{02} = 0$ by the uniqueness of the generalized decomposition numbers. Also, by the definition of the generalized decomposition numbers, the columns D^t_i are columns of integers since t is an involution. By Lemma (7.5.b,c,d), and for $j = 0, 1, 2$, we have that

$$(D^t_i, D^t_j) = 2(1 + \delta_{ij}) , \tag{3}$$

$$(D^y_0, D^t_j) = 0 \tag{4}$$

and

$$(\chi(1), D^t_j) = 0 . \tag{5}$$

Now, if τ is an irreducible character of P, notice that from the character table of P,

$$\tau(t) \equiv \tau(y) \bmod 2 .$$

Hence,

$$\xi(t) \equiv \xi(y) \bmod 2$$

for every character of ξ of G. Therefore, since $\psi_i(1)$ is odd, we have that

$$D_0^t + D_1^t + D_2^t + D_0^y \equiv D_0^t + \psi_1(1)D_1^t + \psi_2(1)D_2^t + D_0^y = \chi(t) + \chi(y) \equiv 0 \bmod 2 .$$

Then we may define three columns of integers by the equations

$$2u_1 = D_0^y + D_0^t - D_1^t - D_2^t ,$$
$$2u_2 = D_0^y + D_0^t - D_1^t + D_2^t ,$$
$$2u_3 = D_0^y + D_0^t + D_1^t - D_2^t .$$

By using (1) to (5), for $j = 1, 2, 3$, we have that

$$(\chi(1), u_j) = 0, \quad (D_0^y, u_j) = 2 \quad \text{and} \quad (u_i, u_j) = 1 + 2\delta_{ij} .$$

Notice that the first entry in each column u_i is 1. Since $(u_i, u_i) = 3$, it follows that exactly three entries of the column u_i are nonzero. Then the nonzero three entries of each u_i are a 1 and two more ± 1. Moreover, since $(\chi(1), u_i) = 0$ (and the first entry of $\chi(1)$ is 1), it follows that the nonzero entries of u_i are $\{1, 1, -1\}$ in some order. Choose notation so that

$$(u_1)_3 = \begin{pmatrix} 1 \\ \delta_1 \\ \delta_2 \end{pmatrix}$$

where $\delta_1 \delta_2 = -1$ and the subscript 3 means that we only write down the first three rows. We claim that

$$(u_j)_3 = \begin{pmatrix} 1 \\ 0 \\ 0 \end{pmatrix}$$

for $j = 2, 3$. Write

$$(u_2)_3 = \begin{pmatrix} 1 \\ \epsilon_1 \\ \epsilon_2 \end{pmatrix} .$$

Since $(u_1, u_2) = 1$, it follows that $0 = \delta_1 \epsilon_1 + \delta_2 \epsilon_2$. Since $\delta_1 \delta_2 = -1$, we have that $\epsilon_1 = \epsilon_2$. Necessarily, $\epsilon_i = 0$ since the nonzero entries of u_2 are $\{1, 1, -1\}$ in some order. The same argument holds for $(u_3)_3$. This proves the claim.

We already know that

$$(D_0^y)_3 = \begin{pmatrix} 1 \\ f_1 \\ f_2 \end{pmatrix} ,$$

where $f_1, f_2 \in \{0, \pm 1\}$. Since $(D_0^y, u_1) = 2$, we have that $1 + f_1\delta_1 + f_2\delta_2 = 1 + \delta_1(f_1 - f_2) = 2$. Therefore,

$$(D_0^y)_3 = \begin{pmatrix} 1 \\ \delta_1 \\ 0 \end{pmatrix} \quad \text{or} \quad (D_0^y)_3 = \begin{pmatrix} 1 \\ 0 \\ \delta_2 \end{pmatrix}.$$

We may choose notation so that

$$(D_0^y)_3 = \begin{pmatrix} 1 \\ \delta_1 \\ 0 \end{pmatrix}.$$

(In the other case, just interchange the second and the third rows.) Now,

$$2(u_3 - u_1) = 2D_1^t, \quad 2(u_2 - u_1) = 2D_2^t,$$

and we deduce that

$$(D_1^t)_3 = \begin{pmatrix} 0 \\ -\delta_1 \\ -\delta_2 \end{pmatrix} = (D_2^t)_3.$$

Also,

$$(D_0^t)_3 = (2u_1 - D_0^y + D_1^t + D_2^t)_3 = \begin{pmatrix} 1 \\ -\delta_1 \\ 0 \end{pmatrix}.$$

Since $\chi(t) = D_0^t + \psi_1(1)D_1^t + \psi_2(1)D_2^t$, we conclude that

$$\chi_1(t) = -\delta_1(1 + \psi_1(1) + \psi_2(1)), \quad \chi_2(t) = -\delta_2(\psi_1(1) + \psi_2(1))$$

and thus

$$1 + \delta_1\chi_1(t) - \delta_2\chi_2(t) = 0. \tag{6}$$

Since $(\chi(1), u_1) = 0$, we also have that

$$1 + \delta_1\chi_1(1) + \delta_2\chi_2(1) = 0. \tag{7}$$

Now, we claim that the product of two involutions is always an element of odd order. Suppose that t_1, t_2 are involutions of G, and suppose that $s = o(t_1t_2)$ is even. Then $\langle t_1, t_2 \rangle = \langle t_1, t_1t_2 \rangle$ is dihedral of order $2o(s)$ and it follows that G has elementary abelian subgroups of order 4. This is not possible because P does not have any subgroup of this type. Therefore, if

K is the class of t and M is the class of a 2-singular element s, it follows that

$$|\{(x,y) \in K \times K \mid xy = s\}| = 0\,.$$

Therefore, by Burnside's formula (Problem (3.9) of [Isaacs, *Characters*]), we have that

$$\sum_{\chi \in \mathrm{Irr}(G)} \frac{\chi(s^{-1})\chi(t)^2}{\chi(1)} = 0$$

for all 2-singular s. Since s is 2-singular if s^{-1} is 2-singular, we have that the class function

$$\sum_{\chi \in \mathrm{Irr}(G)} \frac{\chi(t)^2}{\chi(1)}\chi$$

is zero in the 2-sections of t and y. It follows by Theorem (5.10) that

$$\sum_{\chi \in \mathrm{Irr}(B_0)} \frac{\chi(s^{-1})\chi(t)^2}{\chi(1)} = 0 \tag{8}$$

for all 2-singular s. By using the column notation and writing

$$\chi(s) = \begin{pmatrix} \chi_0(s) \\ \vdots \\ \chi_r(s) \end{pmatrix} \quad \text{and} \quad \frac{\chi(t)^2}{\chi(1)} = \begin{pmatrix} \frac{\chi_0(t)^2}{\chi_0(1)} \\ \vdots \\ \frac{\chi_r(t)^2}{\chi_r(1)} \end{pmatrix},$$

we may write equation (8) as

$$\left(\chi(s), \frac{\chi(t)^2}{\chi(1)}\right) = 0\,. \tag{9}$$

We claim that $u_1 = \frac{1}{2}(D_0^y + D_0^t - D_1^t - D_2^t)$ is a $\mathbb{C}$-linear combination of columns $\chi(s)$ for 2-singular elements s in G. Since $D_0^y = \chi(y)$, it suffices to show that $D_0^t - D_1^t - D_2^t$ is a $\mathbb{C}$-linear combination of columns $\chi(s)$ for 2-singular elements s in G. Recall that if $u \in \mathbf{C}_G(t)$ is 2-regular, then

$$\chi(tu) = D_0^t + \psi_1(u)D_1^t + \psi_2(u)D_2^t\,.$$

Since $\{1 = \psi_0, \psi_1, \psi_2\}$ is a basic set for b_0, the principal block of $\mathbf{C}_G(t)$, by elementary linear algebra we can find 2-regular elements w and v in $\mathbf{C}_G(t)$ such that the matrix

$$\begin{pmatrix} 1 & 1 & 1 \\ \psi_1(1) & \psi_1(w) & \psi_1(v) \\ \psi_2(1) & \psi_2(w) & \psi_2(v) \end{pmatrix}$$

is invertible. Now, if (α, β, γ) is the solution to the linear system

$$\begin{pmatrix} 1 & 1 & 1 \\ \psi_1(1) & \psi_1(w) & \psi_1(v) \\ \psi_2(1) & \psi_2(w) & \psi_2(v) \end{pmatrix} \begin{pmatrix} \alpha \\ \beta \\ \gamma \end{pmatrix} = \begin{pmatrix} 1 \\ -1 \\ -1 \end{pmatrix},$$

it follows that

$$\alpha\chi(t) + \beta\chi(tw) + \gamma\chi(tv) = D_0^t - D_1^t - D_2^t,$$

as claimed. Now, by using (9), we have

$$(u_1, \frac{\chi(t)^2}{\chi(1)}) = 0$$

and therefore

$$1 + \frac{\delta_1\chi_1(t)^2}{\chi_1(1)} + \frac{\delta_2\chi_2(t)^2}{\chi_2(1)} = 0.$$

Hence,

$$\chi_1(1)\chi_2(1) + \delta_1\chi_1(t)^2\chi_2(1) + \delta_2\chi_2(t)^2\chi_1(1) = 0. \tag{10}$$

By using equations (6) and (7), we have that $\chi_2(1) = (-\delta_2)(1 + \delta_1\chi_1(1))$ and $\chi_2(t) = \delta_2(1 + \delta_1\chi_1(t))$. If we substitute these in (10), we obtain

$$(\chi_1(1) - \chi_1(t))^2 = 0.$$

Therefore, $\chi_1(t) = \chi(1)$ and $t \in \ker(\chi_1)$. This completes the proof. ∎

To prove Glauberman's Z^*-theorem, we need one more result on modular representation theory.

(7.7) THEOREM. *Let $u \in G$ be a p-element and let $v \in \mathbf{O}_{p'}(\mathbf{C}_G(u))$. Suppose that B_0 is the principal block of G. If $\chi \in \mathrm{Irr}(B_0)$, then $\chi(uv) = \chi(u)$.*

Proof. If $w \in \mathbf{C}_G(u)^0$, then, by Corollary (5.8) and the third main theorem, we have that

$$\chi(uw) = \sum_{\mu \in \mathrm{IBr}(b_0)} d^u_{\chi\mu}\mu(w),$$

where b_0 is the principal block of $\mathbf{C}_G(u)$. By Theorem (6.10), we have that $\mathbf{O}_{p'}(\mathbf{C}_G(u)) = \ker(b_0)$. Now, since $v \in \mathbf{O}_{p'}(\mathbf{C}_G(u))$, by Lemma (6.11) and Corollary (6.12), we have that $\mu(v) = \mu(1)$ for all $\mu \in \mathrm{IBr}(b_0)$. Then, by using the above formula with $w = v$ and $w = 1$, we get

$$\chi(uv) = \sum_{\mu \in \mathrm{IBr}(b_0)} d^u_{\chi\mu}\mu(1) = \chi(u)$$

and the proof of the theorem is complete. ∎

(7.8) LEMMA. *Let P be a Sylow 2-subgroup of G and suppose that $u \in P$ is an involution. Then the following conditions are equivalent.*

(a) *If $u, u^g \in P$ for some $g \in G$, then $u = u^g$.*

(b) *$[u, g]$ has odd order for every $g \in G$.*

Proof. Suppose that $[u, g]$ has odd order for every $g \in G$. If $u, u^g \in P$ for some $g \in G$, it follows that $[u, g] = uu^g \in P$. Since, by hypothesis, $[u, g]$ is an element of odd order and lies in a 2-group, we see that $u = u^g$. So let us assume (a) and prove (b). Let $g \in G$ and write $v = u^g$. We want to show that $uv = uu^g = [u, g]$ has odd order. If $u = v$, then $uv = 1$. So we assume that $u \neq v$. Notice that if u, u^x lie in some 2-subgroup of G for some $x \in G$, then, by Sylow theory, there is a $y \in G$ such that $u, u^y, u^{xy} \in P$. By hypothesis, $u = u^y = u^{xy}$ and therefore, $u = u^x$. Now, $(uv)^v = vu = (uv)^{-1}$, and we have that $D = \langle u, v \rangle = \langle uv, u \rangle$ is a dihedral group with $\langle uv \rangle$ having index 2 in D (because $u \neq v$). Let P_0 be a Sylow 2-subgroup of D containing u. Since some D-conjugate of v, say v^d, lies in P_0, we have that u and $v^d = u^{gd}$ lie in some 2-subgroup of G. Hence, $u = v^d$. Then $vD' = uD'$, $uv \in D'$ and therefore, $\langle uv \rangle \subseteq D' < D$. Since $\langle uv \rangle$ has index 2 in D, what we have is $\langle uv \rangle = D'$. Now, $|D'| = o(uv)$ is not divisible by 2 because otherwise there would exist a normal subgroup N of D with D/N (abelian) of order 4. ∎

If G is a finite group, we write

$$Z^*(G)/\mathbf{O}_{2'}(G) = \mathbf{Z}(G/\mathbf{O}_{2'}(G)).$$

If G has a Sylow 2-subgroup P having a unique involution u (that is, if P is cyclic or generalized quaternion), then u is the only G-conjugate of u in P. In the case that P is cyclic, G has a normal 2-complement and, of course, $u \in Z^*(G)$. If P is a generalized quaternion group, by the Brauer-Suzuki theorem, we also have that $u \in Z^*(G)$. The general result is Glauberman's Z^*-theorem.

(7.9) Z^*-THEOREM (Glauberman). *Suppose that u is an involution of $P \in \mathrm{Syl}_2(G)$ such that if $u, u^g \in P$ for some $g \in G$, then $u = u^g$. Then $u \in Z^*(G)$.*

Proof. We argue by induction on $|G|$ and assume that $P < G$. We suppose that $u \notin \mathbf{Z}(G)$ and we seek a contradiction.

Step 1. *If $1 < N \vartriangleleft G$ with $u \notin N$, then $uN \in Z^*(G/N)$. In particular, we may assume that $\mathbf{O}_{2'}(G) = 1$.*

By induction, it suffices to check that G/N satisfies the hypothesis of the theorem. Since uN is an involution of G/N, by Lemma (7.8) it suffices to show that for $g \in G$ the order of $[uN, gN] = [u, g]N$ is odd. However, the order of $[u, g]N$ divides the order of $[u, g]$.

Step 2. *If* $u \in H < G$, *then* $u \in Z^*(H)$.

If $h \in H$, by hypothesis (and Lemma (7.8)), we have that the order of $[u, h]$ is odd. Then, by induction, we conclude that $u \in Z^*(H)$.

Step 3. *If* H *is a proper normal subgroup of* G, *then* $u \notin H$.

If $u \in H \triangleleft G$, then by Step 2 note that $u \in \mathbf{Z}(H)$ (because $\mathbf{O}_{2'}(H) = 1$ by Step 1). Also, $\mathbf{Z}(H)$ is a 2-group. Now, $\mathbf{Z}(H) \triangleleft G$ and thus $[u, g] \in \mathbf{Z}(H)$ for $g \in G$. Since $[u, g]$ has odd order for every $g \in G$ by Lemma (7.8), it follows that $[u, g] = 1$ for every $g \in G$. Then $u \in \mathbf{Z}(G)$, a contradiction.

Step 4. $\mathbf{Z}(G) = 1$.

Suppose that $1 < Z = \mathbf{Z}(G)$. Then uZ is an involution of G/Z, a group of order less than $|G|$. Hence, $uZ \in Z^*(G/Z)$ by Step 1. Since $\mathbf{O}_{2'}(G/Z) = 1$ (using that $\mathbf{O}_{2'}(G) = 1$ and the Schur-Zassenhaus theorem), we have that $u \in \mathbf{O}_2(G) < G$. This is a contradiction since u lies in no proper normal subgroup of G by Step 3.

Step 5. *If* u *and* u^g *lie in some 2-subgroup of* G *for some* $g \in G$, *then* $u = u^g$. *In particular,* $u \in \mathbf{Z}(D)$ *for every 2-subgroup* D *which contains* u.

By Sylow theory, we have that $u, u^y, u^{gy} \in P$ for some $y \in G$. By hypothesis, $u = u^y = u^{gy}$ and thus $u = u^g$, as desired.

Since G cannot have a normal 2-complement, we know that P is not cyclic. By the Brauer-Suzuki theorem, P is not generalized quaternion either (by Steps 1 and 4). Since 2-groups with a unique involution are cyclic or generalized quaternion, we conclude that there exists at least one involution v in P different from u (and thus not G-conjugate to u). Since $u \in \mathbf{Z}(P)$, notice that $uv = vu$ is an involution of P. Now, we work with the fixed involution v.

Step 6. *Let* $g \in G$ *and write* $v^g u = zx$, *where* z *is the 2-part of* $v^g u$ *and* x *is the* $2'$-*part of* $v^g u$. *Then* $z \in \mathbf{C}_G(u)$, *zu is* G-*conjugate to* v *and* $x \in \mathbf{O}_{2'}(\mathbf{C}_G(z))$.

Write $w = v^g$. The statement is clear if w and u commute because in this case, $z = wu$, $zu = w$ is G-conjugate to v and $x = 1 \in \mathbf{O}_{2'}(\mathbf{C}_G(z))$. So we may assume that w and u do not commute. Write $D = \langle w, u \rangle$, a dihedral group. We have that $(wu)^u = uw = (wu)^{-1}$. Since z is the 2-part of wu, we have that $\langle z, u \rangle = P_0$ is a Sylow 2-subgroup of D. Now, $u \in \mathbf{Z}(P_0)$ (because u lies in the center of every 2-group in which it is contained by Step 5). In particular, $z \in \mathbf{C}_G(u)$. However, $z^u = z^{-1}$ (because u inverts wu and z is a power of wu). Therefore, $z^u = z = z^{-1}$, and we have that $P_0 = \{1, z, u, zu\}$ has order 4. Also, notice that $z \in \mathbf{Z}(D)$ because z is the unique involution in $\langle wu \rangle$. Now, since w is an involution in D, some D-conjugate of w lies in P_0. Then it follows that w is D-conjugate to either z, u or zu. Certainly, w is not D-conjugate to u since u and v are not G-conjugate. Also, w is not D-conjugate to z since $z \in \mathbf{Z}(D)$ and w and u do not commute. So the only possibility left is that w is D-conjugate to zu. Hence, zu is G-conjugate to $v = w^{g^{-1}}$. It remains to show that $x \in \mathbf{O}_{2'}(\mathbf{C}_G(z))$. Since $u \in \mathbf{C}_G(z) < G$, by Step 2 we have that $u \in Z^*(\mathbf{C}_G(z))$. In particular, $u \in \mathbf{O}_{2'2}(\mathbf{C}_G(z))$. Since $x \in \mathbf{C}_G(z)$, we have that $[u, x] \in \mathbf{O}_{2'2}(\mathbf{C}_G(z))$. However, $[u, x]$ has odd order (by hypothesis and Lemma (7.8)) and we conclude that $[u, x] \in \mathbf{O}_{2'}(\mathbf{C}_G(z))$. Now, u inverts x^{-1} (because x^{-1} is a power of wu and u inverts wu), and thus $[u, x] = x^2 \in \mathbf{O}_{2'}(\mathbf{C}_G(z))$. Since x has odd order (recall that it is the $2'$-part of wu), it follows that $x \in \mathbf{O}_{2'}(\mathbf{C}_G(z))$, as desired.

Step 7. *With the notation of Step 6, we have that z and vu are G-conjugate.*

Since zu is G-conjugate to v, it follows that there is a $t \in G$ such that $(zu)^t = v \in \mathbf{C}_G(u)$. Hence, $u^{t^{-1}} \in \mathbf{C}_G(zu)$. Also, $u \in \mathbf{C}_G(zu)$ since $z \in \mathbf{C}_G(u)$ (Step 6). Thus, $\langle u, u^{t^{-1}} \rangle \subseteq \mathbf{C}_G(zu)$. Now, by hypothesis, we have that $uu^{t^{-1}} = [u, t^{-1}]$ has odd order. Therefore, the dihedral group $\langle u, u^{t^{-1}} \rangle$ has Sylow 2-subgroups of order 2. Hence, there is an $s \in \langle u, u^{t^{-1}} \rangle \subseteq \mathbf{C}_G(zu)$ such that $u^{t^{-1}} = u^s$. Now,

$$z^{st} = (zuu)^{st} = (zu)^{st} u^{st} = (zu)^t u = vu \,,$$

as desired.

Now, we use representation theory. Let B_0 be the principal block of G.

Step 8. *If $\chi \in \mathrm{Irr}(B_0)$, then $\chi(v^g u^h) = \chi(vu)$ for every $g, h \in G$.*

Certainly, to show this we may assume that $h = 1$ since χ is a class function. By using the notation of the previous steps, we may write $v^g u = zx$, where z is G-conjugate to vu and $x \in \mathbf{O}_{2'}(\mathbf{C}_G(z))$. By Theorem (7.7),

$$\chi(v^g u) = \chi(zx) = \chi(z) = \chi(vu) \,,$$

as desired.

Step 9. *If $1_G \neq \chi \in \mathrm{Irr}(B_0)$, then $\chi(v) = 0$ or $\chi(u) = -\chi(1)$.*

Let $K = \mathrm{cl}(v)$, let $L = \mathrm{cl}(u)$ and, in $\mathbb{C}G$, write

$$\hat{K}\hat{L} = \sum_{M \in \mathrm{cl}(G)} |A_{KLM}| \hat{M} \,,$$

where $A_{KLM} = \{(x, y) \in K \times L \mid xy = x_M\}$. Notice that if $|A_{KLM}| \neq 0$, then $\chi(x_M) = \chi(vu)$ (which does not depend on M) by Step 8. Now, by applying the algebra homomorphism ω_χ, we have that

$$\frac{\chi(v)|K|\chi(u)|L|}{\chi(1)^2} = \sum_{M \in \mathrm{cl}(M)} |A_{KLM}| \frac{\chi(x_M)|M|}{\chi(1)}$$

$$= \frac{\chi(vu)}{\chi(1)} \sum_{M \in \mathrm{cl}(M)} |A_{KLM}||M| = \frac{\chi(vu)}{\chi(1)} |K||L| \,.$$

Therefore,

$$\frac{\chi(u)\chi(v)}{\chi(1)^2} = \frac{\chi(vu)}{\chi(1)} \,.$$

If we write $\alpha(x) = \frac{\chi(x)}{\chi(1)}$, we have shown that $\alpha(u)\alpha(v) = \alpha(uv)$. (Recall that $uv = vu$.)

To arrive here, the only fact that we have used about v is that $v \in P$ is an involution with $v \neq u$. Since $u \in \mathbf{Z}(P)$, we have that $uv \neq u$ is an involution in P. By the same argument applied to the involution uv, we get $\alpha(u)\alpha(uv) = \alpha(v)$. Hence, $\alpha(u)^2\alpha(v) = \alpha(v)$. Therefore, if $\chi(v) \neq 0$, then $\alpha(v) \neq 0$ and we conclude that $\alpha(u)^2 = 1$. Since u is not contained in $\ker(\chi)$, a proper normal subgroup of G, it follows that $\chi(u) = -\chi(1)$ and the Step 9 is shown.

Final Contradiction. By block orthogonality (Corollary (5.11)) applied twice, we have that

$$0 = \sum_{\chi \in \mathrm{Irr}(B_0)} \chi(v)\chi(u) = \sum_{\chi \in \mathrm{Irr}(B_0)} \chi(v)\chi(1) \,.$$

Therefore,

$$0 = \sum_{\chi \in \mathrm{Irr}(B_0)} \chi(v)(\chi(u) + \chi(1)) = 1_G(v)(1_G(u) + 1) = 2 \,,$$

a contradiction. ∎

There is an analogue of the Z^*-theorem for odd primes. Concretely, if x is an element of order p of a finite group G and $[x, g]$ is a p'-element for every $g \in G$, then x is central modulo $\mathbf{O}_{p'}(G)$. See [Guralnick-Robinson]. The proof of this result depends upon the classification of finite simple groups. Contrary to the case where $p = 2$, there is no known block theoretic proof of it.

8. Brauer Characters as Characters

We digress in our study of blocks to analyse in more detail the basic behaviour of Brauer characters. We put special emphasis on the interactions between the Brauer characters of a group and the Brauer characters of its subgroups.

(We suggest that those only interested in blocks read at least as far as Green's theorem (8.11), which we will use later on.)

If $\mathcal{X}$ is an FG-representation affording the Brauer character φ and H is a subgroup of G, then $\mathcal{X}_{FH}$ is an FH-representation affording the Brauer character of H which we denote by φ_H. (See Lemma (2.2.d).) This is the **restriction** of φ to H. Also, if V is an FG-module, then V is an FH-module which is denoted by V_H.

As happens with ordinary characters (although it is more complicated here), we can induce Brauer characters of subgroups to Brauer characters of the group.

If $H \subseteq G$ and W is an FH-module, then the **induced** FG-module W^G is constructed in the following way. Choose a right transversal T of H in G with $1 \in T$ and set

$$W^G = \bigoplus_{t \in T} W \otimes t,$$

where $W \otimes t$ is the vector space over F (naturally isomorphic to W) consisting of the symbols $\{w \otimes t \mid w \in W\}$. Note that

$$\dim_F(W^G) = |G : H| \dim_F(W).$$

If $g \in G$, then $g = ht$ for uniquely determined elements $h \in H$ and $t \in T$, and we write $w \otimes g = wh \otimes t$ for $w \in W$. Now, we define an action of G on W^G by setting

$$(w \otimes t)g = w \otimes tg$$

for $t \in T$, $w \in W$, and $g \in G$ and extending this linearly to W^G. It is straightforward to check that

$$(w \otimes g_1)g_2 = w \otimes (g_1 g_2)$$

for $g_1, g_2 \in G$. With the above action of G, the vector space W^G becomes an FG-module. Note that

$$w \otimes hg = wh \otimes g$$

for $w \in W$, $h \in H$ and $g \in G$.

It is clear that $W \otimes 1$ and W are isomorphic as FH-modules. Hence, $(W^G)_H$ contains an FH-submodule isomorphic to W.

Although it appears that W^G depends on the transversal, it is easy to prove that two different transversals give isomorphic FG-modules. We leave this for the reader to check.

(8.1) DEFINITION. Let H be a subgroup of G and let $\alpha \in \mathrm{cf}(H^0)$. We define $\alpha^G \in \mathrm{cf}(G^0)$ by the formula

$$\alpha^G(x) = \frac{1}{|H|} \sum_{g \in G} \dot{\alpha}(gxg^{-1}),$$

where $x \in G^0$, $\dot{\alpha}(y) = \alpha(y)$ if $y \in H^0$ and $\dot{\alpha}(y) = 0$ if $y \notin H^0$.

Note that if $\mu \in \mathrm{cf}(H)$, then

$$(\mu^0)^G = (\mu^G)^0.$$

As a consequence of this, we may prove all the well-known formulas on inducing ordinary class functions (such as transitivity of induction) by extending $\alpha \in \mathrm{cf}(H^0)$ to any $\delta \in \mathrm{cf}(H)$.

(8.2) THEOREM (Brauer-Nesbitt). *Let $H \subseteq G$ and suppose that W is an FH-module affording the Brauer character α. Then the FG-module W^G affords the Brauer character α^G.*

Proof. Suppose that $\mathcal{X} : FH \to \mathrm{Mat}(n, F)$ is the representation of W in some basis $\mathcal{B}$. Write $\mathcal{X}(h) = (x_{ij}(h))$ for $h \in H$. Let $m = |G : H|$. We construct an FG-representation

$$\mathcal{Y} : FG \to \mathrm{Mat}(nm, F)$$

as follows. Let us write $\mathcal{X}(g) = (x_{ij}(g)) = 0 \in \mathrm{Mat}(n, F)$ for $g \in G - H$. If $T = \{t_1, \ldots, t_m\}$ is a right transversal of H in G (with $t_1 = 1$), we define

$$\mathcal{Y}(g) = \begin{pmatrix} \mathcal{X}(t_1 g t_1^{-1}) & \cdots & \mathcal{X}(t_1 g t_m^{-1}) \\ \vdots & \ddots & \vdots \\ \mathcal{X}(t_m g t_1^{-1}) & \cdots & \mathcal{X}(t_m g t_m^{-1}) \end{pmatrix} \in \mathrm{Mat}(nm, F).$$

We claim that $\mathcal{Y}$ is the representation afforded by the FG-module W^G with respect to the basis $\{v \otimes t \,|\, v \in \mathcal{B}, t \in T\}$ of W^G. Suppose that $\mathcal{B} = \{v_1, \ldots, v_n\}$ and write

$$v_i h = \sum_{1 \le k \le n} x_{ik}(h) v_k$$

for $h \in H$. Now, if $g \in G$ and $t_j \in T$ are fixed, then there are unique $h \in H$ and $t_u \in T$ such that $t_j g = h t_u$. In particular, if $t_j g t_v^{-1} \in H$, then $u = v$. Now,

$$(v_i \otimes t_j) g = v_i \otimes t_j g = v_i h \otimes t_u = \sum_{1 \le k \le n} x_{ik}(t_j g t_u^{-1}) v_k \otimes t_u$$

$$= \sum_{\substack{1 \le k \le n \\ 1 \le v \le m}} x_{ik}(t_j g t_v^{-1}) v_k \otimes t_v \,,$$

which proves the claim. (We ordered the basis of W^G as $\{v_1 \otimes t_1, \ldots, v_n \otimes t_1, \ldots, v_1 \otimes t_m, \ldots, v_n \otimes t_m\}$.)

We now prove that the representation $\mathcal{Y}$ affords the Brauer character α^G. Let $g \in G^0$. We need to find the eigenvalues of the matrix $\mathcal{Y}(g)$. Since g acts on the right cosets of H in G, g induces a permutation on T which can be written as a product of disjoint cycles. The elements in those cycles form the different orbits of the action of $\langle g \rangle$ on T. Suppose that the first cycle in this decomposition is $(s_1 \ldots s_k)$ for some $s_i \in T$. Assume that $k \ge 2$. Then

$$Hs_1 g = Hs_2, \ldots, Hs_i g = Hs_{i+1}, \ldots, Hs_k g = Hs_1.$$

Now, consider the submatrix A of $\mathcal{Y}(g)$ afforded by this first cycle. Notice that

$$A = \begin{pmatrix} 0 & \mathcal{X}(s_1 g s_2^{-1}) & 0 & \cdots & 0 \\ 0 & 0 & \mathcal{X}(s_2 g s_3^{-1}) & \cdots & 0 \\ \vdots & \vdots & \vdots & \ddots & \vdots \\ 0 & 0 & 0 & \cdots & \mathcal{X}(s_{k-1} g s_k^{-1}) \\ \mathcal{X}(s_k g s_1^{-1}) & 0 & 0 & \cdots & 0 \end{pmatrix}.$$

After we have reordered the set T, observe that the matrix $\mathcal{Y}(g)$ is (similar to) a diagonal sum of matrices like A because if $\mathcal{X}(tgs^{-1}) \ne 0$ for some $s, t \in T$, then $tgs^{-1} \in H$, $Htg = Hs$ and, in particular, t and s lie in the same $\langle g \rangle$-orbit of T.

We wish to find the eigenvalues of A. Write $A_i = \mathcal{X}(s_i g(s_{i+1})^{-1})$ for $1 \le i \le k - 1$, set $A_k = \mathcal{X}(s_k g s_1^{-1})$, and consider the matrix

$$B = \begin{pmatrix} I_n & 0 & 0 & \cdots & 0 \\ 0 & A_1 & 0 & \cdots & 0 \\ 0 & 0 & A_1 A_2 & \cdots & 0 \\ \vdots & \vdots & \vdots & \ddots & \vdots \\ 0 & 0 & 0 & \cdots & A_1 \ldots A_{k-1} \end{pmatrix}.$$

It is easy to check that

$$BAB^{-1} = \begin{pmatrix} 0 & I_n & 0 & \cdots & 0 \\ 0 & 0 & I_n & \cdots & 0 \\ \vdots & \vdots & \vdots & \ddots & \vdots \\ 0 & 0 & 0 & \cdots & I_n \\ A_1 \ldots A_k & 0 & 0 & \cdots & 0 \end{pmatrix}.$$

Notice that $A_1 \ldots A_k = \mathcal{X}(h)$, where $h = s_1 g^k s_1^{-1} = (s_1 g s_2^{-1}) \ldots (s_k g s_1^{-1}) \in H^0$.

We claim that the eigenvalues of BAB^{-1} (and therefore of A) can be obtained by taking all the kth roots of the eigenvalues of $\mathcal{X}(h)$. (The reader wishing to use determinants may check at this point that $\det(x I_{nk} - A) = \det(x^k I_n - \mathcal{X}(h))$ and prove the claim.)

Suppose that ϵ is an eigenvalue of $\mathcal{X}(h)$ and let $\{X_1, \ldots, X_{n(\epsilon)}\}$ be a basis of the space of eigenvectors of $\mathcal{X}(h)$ associated with ϵ. (We are writing each X_i as an n-dimensional column vector in F^n.) Since F is algebraically closed, we may find some $\lambda_\epsilon \in F^\times$ such that $\lambda_\epsilon^k = \epsilon$. Now, if we consider the vector

$$\begin{pmatrix} Z_1 \\ \vdots \\ Z_k \end{pmatrix},$$

where $Z_i \in F^n$, note that

$$BAB^{-1} \begin{pmatrix} Z_1 \\ \vdots \\ Z_k \end{pmatrix} = \begin{pmatrix} Z_2 \\ \vdots \\ Z_k \\ \mathcal{X}(h) Z_1 \end{pmatrix}.$$

Hence, using the fact that

$$\lambda_\epsilon^k X_j = \epsilon X_j = \mathcal{X}(h) X_j$$

for $j = 1, \ldots, n(\epsilon)$, we have

$$
BAB^{-1}
\begin{pmatrix}
X_j \\
\lambda_\epsilon X_j \\
\vdots \\
\lambda_\epsilon^{k-1} X_j
\end{pmatrix}
=
\begin{pmatrix}
\lambda_\epsilon X_j \\
\lambda_\epsilon^2 X_j \\
\vdots \\
\lambda_\epsilon^{k-1} X_j \\
\mathcal{X}(h) X_j
\end{pmatrix}
= \lambda_\epsilon
\begin{pmatrix}
X_j \\
\lambda_\epsilon X_j \\
\vdots \\
\lambda_\epsilon^{k-1} X_j
\end{pmatrix} .
$$

Thus, we see that λ_ϵ is an eigenvalue of BAB^{-1} associated to the eigenvector

$$
\begin{pmatrix}
X_j \\
\lambda_\epsilon X_j \\
\vdots \\
\lambda_\epsilon^{k-1} X_j
\end{pmatrix} .
$$

Since the vectors $X_1, \ldots, X_{n(\epsilon)}$ are linearly independent, we also have that the vectors

$$
\begin{pmatrix}
X_1 \\
\lambda_\epsilon X_1 \\
\vdots \\
\lambda_\epsilon^{k-1} X_1
\end{pmatrix}
, \ldots,
\begin{pmatrix}
X_{n(\epsilon)} \\
\lambda_\epsilon X_{n(\epsilon)} \\
\vdots \\
\lambda_\epsilon^{k-1} X_{n(\epsilon)}
\end{pmatrix}
$$

are linearly independent. We conclude that λ_ϵ is an eigenvalue of A whose eigenspace has dimension greater than or equal to $n(\epsilon)$. Now, since p does not divide k (recall that k divides the order of g), we may find a primitive kth root of unity ξ in F. (See Lemma (2.1).) Using the same argument as before, we see that $\lambda_\epsilon, \lambda_\epsilon \xi, \lambda_\epsilon \xi^2, \ldots, \lambda_\epsilon \xi^{k-1}$ are distinct eigenvalues of A whose corresponding eigenspaces have dimension greater than or equal to $n(\epsilon)$. Since h is p-regular, we have that the matrix $\mathcal{X}(h)$ is similar to a diagonal matrix. (Just consider the completely reducible representation $\mathcal{X}_{\langle h \rangle}$.) Hence,

$$
\sum_\epsilon n(\epsilon) = n .
$$

Since A has size kn, we conclude that the eigenspace corresponding to the eigenvalue $\lambda_\epsilon \xi^i$ has dimension exactly $n(\epsilon)$ and the claim is proven.

Now, if we take $\delta_\epsilon, \mu \in \mathbf{U}$ such that $\delta_\epsilon^* = \lambda_\epsilon$ and $\mu^* = \xi$ (see Lemma (2.1)), then

$$
\sum_{j=0}^{k-1} \delta_\epsilon \mu^j = 0 .
$$

(Recall that μ is a root of the polynomial $1 + x + \ldots + x^{k-1}$.) Hence, the contribution of A to the value of the Brauer character afforded by $\mathcal{Y}(g)$ is zero whenever $k > 1$.

We conclude that in the decomposition of the permutation action of g on the right cosets of H in G, we only have to pay attention to those Hs_i such that $Hs_ig = Hs_i$. In this case, $A = \mathcal{X}(s_igs_i^{-1})$ contributes $\alpha(s_igs_i^{-1})$ to the value of the Brauer character afforded by $\mathcal{Y}$. Therefore, the Brauer character afforded by $\mathcal{Y}$ evaluated at g is

$$\sum_{\substack{t \in T \\ tgt^{-1} \in H}} \alpha(tgt^{-1}) = \sum_{t \in T} \dot{\alpha}(tgt^{-1}) = \frac{1}{|H|} \sum_{y \in G} \dot{\alpha}(ygy^{-1}) = \alpha^G(g),$$

as required. ∎

If $H \subseteq G$ and $\alpha \in \mathrm{IBr}(H)$, we write $\mathrm{IBr}(G \mid \alpha)$ to denote the set of irreducible Brauer characters φ of G such that α is an irreducible constituent of φ_H. If $\varphi \in \mathrm{IBr}(G \mid \alpha)$, then it is customary to say that φ **lies over** α or that α **lies under** φ.

(8.3) COROLLARY. *Let $H \subseteq G$ and $\alpha \in \mathrm{IBr}(H)$. Then α is an irreducible constituent of $(\alpha^G)_H$. In particular, $\mathrm{IBr}(G \mid \alpha)$ is nonempty.*

Proof. If W is an FH-module affording α, we already know that W is isomorphic to a simple submodule of $(W^G)_H$. This proves the first part. The second part easily follows from the first. ∎

A big difference between induction and restriction of ordinary and modular characters is that in the latter case Frobenius reciprocity does not necessarily hold. It is easy to find examples of groups G having a subgroup H with irreducible Brauer characters $\alpha \in \mathrm{IBr}(H)$ and $\varphi \in \mathrm{IBr}(G)$ such that φ is an irreducible constituent of the Brauer character α^G and α is not an irreducible constituent of φ_H. ($G = \mathrm{Sym}(3)$ with $p = 3$ already works.) However, it is more complicated to find examples in which α is an irreducible constituent of φ_H and φ is not an irreducible constituent of α^G. The next example is due to W. Willems.

Suppose that $G = A_5$ and let $H = A_4$ be a subgroup of G. Set $p = 2$. In this case, it is clear that H has three irreducible Brauer characters $\{1_{H^o}, \lambda_1, \lambda_2\}$. In fact, these Brauer characters are the restrictions to p-regular elements of the three linear characters of H. The Brauer character table of G was calculated in Chapter 2 and, now, we use the same notation. Let $\varphi_4 \in \mathrm{IBr}(G)$ be the unique irreducible Brauer character of G with degree 4. As the reader may check, we have

$$(\varphi_4)_H = 2(1_{H^o}) + \lambda_1 + \lambda_2.$$

However,

$$(\lambda_1)^G = \varphi_1 + \varphi_2 + \varphi_3 = (\lambda_2)^G.$$

Although, as we have seen, Frobenius reciprocity does not necessarily hold for Brauer characters, in some special cases it is still possible to say something interesting. For this purpose, Nakayama's formulas are very useful. We will only use one of them, leaving the other one to the problems.

(8.4) LEMMA (Nakayama). *Let $H \subseteq G$. Suppose that W is an FH-module and that V is an FG-module. Then*

$$\dim_F(\mathrm{Hom}_{FG}(W^G, V)) = \dim_F(\mathrm{Hom}_{FH}(W, V_H)).$$

Proof. Let T be a right transversal of H in G with $1 \in T$. We define the map $\eta : W \to W^G$ by $\eta(w) = w \otimes 1$. It is clear that η is an FH-homomorphism.

We prove that the linear map

$$\mathrm{Hom}_{FG}(W^G, V) \to \mathrm{Hom}_{FH}(W, V_H)$$

given by $\tau \mapsto \eta\tau$ is a bijection. Notice that if $\eta\tau = 0$, then $\tau(w \otimes 1) = 0$ for all $w \in W$. Since τ is a G-homomorphism, $\tau(w \otimes t) = \tau(w \otimes 1)t = 0$ for all $t \in T$ and thus $\tau = 0$. (Recall that $W^G = \bigoplus_{t \in T} W \otimes t$.)

To finish, we prove that our map is onto. If $\gamma \in \mathrm{Hom}_{FH}(W, V_H)$, we define $\tau \in \mathrm{Hom}_F(W^G, V)$ by setting

$$\tau(\sum_{t \in T} w_t \otimes t) = \sum_{t \in T} \gamma(w_t)t.$$

Let us prove that $\tau \in \mathrm{Hom}_F(W^G, V)$. For that, it suffices to check that

$$\tau((w \otimes t)g) = \tau(w \otimes t)g$$

for $w \in W$, $t \in T$ and $g \in G$. If $t \in T$ and $g \in G$, write $tg = ht_0$, where $h \in H$ and $t_0 \in T$. Then

$$\tau((w \otimes t)g) = \tau(wh \otimes t_0) = \gamma(wh)t_0 = \gamma(w)ht_0 = \gamma(w)tg = \tau(w \otimes t)g,$$

as desired. Since $\eta\tau = \gamma$, we are done. ∎

We now introduce some notation. If U and V are FG-modules, we set

$$\mathbf{I}(U, V) = \dim_F(\mathrm{Hom}_{FG}(U, V)).$$

Also, if

$$\psi = \sum_{\theta \in \mathrm{IBr}(G)} a_\theta \theta \quad \text{and} \quad \tau = \sum_{\theta \in \mathrm{IBr}(G)} b_\theta \theta$$

are Brauer characters of G, we write

$$\mathbf{I}(\psi, \tau) = \sum_{\theta \in \mathrm{IBr}(G)} a_\theta b_\theta = \mathbf{I}(\tau, \psi)$$

so that $\mathbf{I}(\tau, \tau) = 1$ if and only if $\tau \in \mathrm{IBr}(G)$. Also, $\mathbf{I}(\psi, \tau)$ is the multiplicity of τ in ψ whenever $\tau \in \mathrm{IBr}(G)$.

In general, if U and V are FG-modules affording the Brauer characters ψ and τ, it is not necessarily true that $\mathbf{I}(U, V) = \mathbf{I}(\psi, \tau)$. However, we do have that $\mathbf{I}(U, V) = \mathbf{I}(\psi, \tau)$ whenever U and V are completely reducible. This follows from Schur's lemma and the following elementary facts whose proof is left to the problems:

$$\mathbf{I}(U_1 \oplus U_2, V) = \mathbf{I}(U_1, V) + \mathbf{I}(U_2, V)$$

and

$$\mathbf{I}(U, V_1 \oplus V_2) = \mathbf{I}(U, V_1) + \mathbf{I}(U, V_2)$$

for FG-modules U_1, U_2, V_1 and V_2.

When we study the interaction between the Brauer characters of a group and its normal subgroups, the basic tool is Clifford's theorem.

(8.5) THEOREM (Clifford). *Suppose that V is a simple FG-module and let $N \triangleleft G$. If W is a simple submodule of V_N, then $V = \sum_{g \in G} Wg$. In particular, V_N is completely reducible.*

Proof. First of all, we claim that if U is an FN-submodule of V_N and $g \in G$, then $Ug = \{ug \mid u \in U\}$ is an FN-submodule of V. It is clear that Ug is an F-subspace of V. In addition, $ugn = un^{g^{-1}}g \in Ug$ for $u \in U$ and $n \in N$, and this proves the claim. In particular, we deduce that Wg is a simple FN-submodule of V_N (because U is an FN-submodule of Wg if and only if Ug^{-1} is an FN-submodule of W). Now, $\sum_{g \in G} Wg$ is a nonzero G-submodule of V and therefore $V = \sum_{g \in G} Wg$. By Lemma (1.10), we conclude that V_N is completely reducible, as desired. ∎

If θ is a Brauer character of $N \triangleleft G$ afforded by some FN-representation $\mathcal{X}$ and $g \in G$, then $\mathcal{X}^g(n) = \mathcal{X}(n^{g^{-1}})$ defines another representation of N which affords the Brauer character which we denote by θ^g. Note that

$$\theta^g(n) = \theta(n^{g^{-1}}).$$

We say that θ^g is **conjugate** to θ in G. If $\theta^g = \theta$ for all $g \in G$, we say that θ is G-**invariant**. Using the definition, it is easy to check that

$$\mathbf{I}(\mu^g, \theta^g) = \mathbf{I}(\mu, \theta)$$

for Brauer characters θ and μ of N and $g \in G$.

(8.6) THEOREM. *If $N \lhd G$ and W is a simple FN-module, then $(W^G)_N$ is completely reducible.*

Proof. If T is a right transversal of N in G with $1 \in T$, we have defined

$$W^G = \bigoplus_{t \in T} W \otimes t.$$

It suffices to check that $W \otimes t$ is a simple FN-module. (In fact, if W affords the Brauer character θ, then $W \otimes t$ affords the Brauer character θ^t.) Note that if $w \in W$ and $n \in N$, then

$$(w \otimes t)n = w \otimes tn = w \otimes n^{t^{-1}}t = wn^{t^{-1}} \otimes t \in W \otimes t.$$

If U is a subspace of W, it is now clear that $U \otimes t$ is N-stable if and only if U is N-stable. This shows that $W \otimes t$ is simple, as desired. ∎

(8.7) COROLLARY. *Suppose that $N \lhd G$. Let $\theta \in \mathrm{IBr}(N)$ and let $\varphi \in \mathrm{IBr}(G)$. Then φ is an irreducible constituent of θ^G if and only if θ is an irreducible constituent of φ_N. In this case, if $\theta = \theta_1, \ldots, \theta_t$ are the distinct conjugates of θ in G, then*

$$\varphi_N = e \sum_{i=1}^{t} \theta_i.$$

Furthermore, if p does not divide $|G : N|$, then $\mathbf{I}(\theta^G, \varphi) = \mathbf{I}(\theta, \varphi_N)$.

Proof. First, we show that if θ is an irreducible constituent of φ_N, then φ is an irreducible constituent of θ^G. Suppose that V is a simple FG-module affording φ. Since V_N is completely reducible by Clifford's theorem, it follows that the simple constituents of V_N are isomorphic to simple submodules of V_N by the remark after the proof of Theorem (1.19). Therefore, θ is afforded by a simple submodule W of V_N. Then $0 \neq \mathbf{I}(W, V_N)$. (Just consider the inclusion map.) Now, by Nakayama's lemma (8.4), we have that $0 \neq \mathbf{I}(W^G, V)$, and we conclude that V is a simple constituent of W^G.

Conversely, suppose that φ is an irreducible constituent of θ^G. Hence, we may write $\theta^G = \varphi + \delta$, where δ is some Brauer character of G or zero. Using the definition of the function θ^G, we see that

$$(\theta^G)_N = \frac{1}{|N|} \sum_{g \in G} \theta^g.$$

Therefore, since $(\theta^G)_N = \varphi_N + \delta_N$, we have that the Brauer character φ_N should be a sum of G-conjugates of θ. Since φ is a class function of G, we see that the Brauer character φ_N is G-invariant. Hence, $\mathbf{I}(\varphi_N, \theta^g) = \mathbf{I}(\varphi_N, \theta)$ for $g \in G$, and it follows that

$$\varphi_N = e \sum_{i=1}^{t} \theta_i,$$

where $e = \mathbf{I}(\varphi_N, \theta)$ and $\theta_1, \ldots, \theta_t$ are the distinct G-conjugates of θ.

Assume finally that p does not divide $|G : N|$. If W is a simple FN-module, then we have that $(W^G)_N$ is completely reducible by Theorem (8.6). Now, by Problem (1.7), we have that W^G is completely reducible. Therefore, if V is a simple FG-module, we have that the four modules W^G, W, V and V_N are completely reducible. Now, the formula $\mathbf{I}(\theta^G, \varphi) = \mathbf{I}(\theta, \varphi_N)$ follows from Nakayama's lemma (8.4) and the remark before Clifford's theorem. ∎

The Clifford theorem for Brauer characters allows us to prove a Clifford theorem for projective indecomposable characters.

(8.8) COROLLARY. *If $N \lhd G$, $\varphi \in \mathrm{IBr}(G)$ and $\theta \in \mathrm{IBr}(N)$ is an irreducible constituent of φ_N, then*

$$(\Phi_\varphi)_N = e \sum_{i=1}^{t} \Phi_{\theta_i},$$

where $\theta = \theta_1, \ldots, \theta_t$ are the distinct G-conjugates of θ and $e = \mathbf{I}(\theta^G, \varphi)$.

Proof. Since $(\Phi_\varphi)_N$ vanishes off the p-regular elements of N, by Theorem (2.13) we may write

$$(\Phi_\varphi)_N = \sum_{\mu \in \mathrm{IBr}(N)} a_\mu \Phi_\mu$$

for some $a_\mu \in \mathbb{C}$. Now, using Theorem (2.13) twice and the fact that $(\hat{\mu})^0 = \mu$ for $\mu \in \mathrm{IBr}(G)$, we have

$$a_\mu = [(\Phi_\varphi)_N, \mu]^0 = [(\Phi_\varphi)_N, \hat{\mu}] = [\Phi_\varphi, \hat{\mu}^G]$$
$$= [\Phi_\varphi, (\hat{\mu}^G)^0]^0 = [\Phi_\varphi, \mu^G]^0 = \mathbf{I}(\varphi, \mu^G).$$

Hence, if $a_\mu \neq 0$, it follows that φ is an irreducible constituent of μ^G and thus μ is an irreducible constituent of φ_N by Corollary (8.7). Since

$$[(\Phi_\varphi)_N, \theta]^0 = [(\Phi_\varphi)_N, \theta^g]^0$$

for every $g \in G$, the result follows. ∎

As we are about to see, the Clifford correspondence also holds for Brauer characters.

If $\theta \in \mathrm{IBr}(N)$ and $N \lhd G$, the **inertia group** of θ in G is $I_G(\theta) = \{g \in G \mid \theta^g = \theta\}$. Of course, $|G : I_G(\theta)|$ is the number of distinct G-conjugates of θ.

(8.9) THEOREM (Clifford Correspondence). *Suppose that $N \lhd G$ and let $\theta \in \mathrm{IBr}(N)$. If $T = I_G(\theta)$ is the inertia group of θ in G, then the map $\psi \mapsto \psi^G$ is a bijection $\mathrm{IBr}(T \mid \theta) \to \mathrm{IBr}(G \mid \theta)$. Also, $\mathbf{I}((\psi^G)_N, \theta) = \mathbf{I}(\psi_N, \theta)$ and $\mathbf{I}((\psi^G)_T, \psi) = 1$ for $\psi \in \mathrm{IBr}(T \mid \theta)$.*

Proof. Let $\psi \in \mathrm{IBr}(T \mid \theta)$. We claim that we may choose an irreducible constituent φ of ψ^G such that ψ is an irreducible constituent of φ_T. If W is an FT-module affording ψ, this may be proven by choosing a "top" constituent of W^G. If U is a maximal FG-submodule of W^G, then $V = W^G/U$ is a simple FG-module. Clearly, $0 \neq \mathbf{I}(W^G, V)$. By applying Nakayama's formula, we have that $\mathbf{I}(W^G, V) = \mathbf{I}(W, V_T)$. Now, if V affords the irreducible Brauer character φ, it follows that φ is a consituent of ψ^G and ψ is an irreducible constituent of φ_T, as required.

By Clifford's theorem, we may write $\varphi_N = e(\theta_1 + \ldots + \theta_t)$, where $\theta_1, \ldots, \theta_t$ are the distinct G-conjugates of θ and e is some integer. Hence, $t = |G : T|$. Also, by Clifford's theorem, we may write $\psi_N = f\theta$ for some integer f.

Since ψ is an irreducible constituent of φ_T, it follows that $\varphi_T = \psi + \delta$, where δ is a Brauer character of T or zero. Hence, $\varphi_N = \psi_N + \delta_N$ and we conclude that $f \leq e$. In addition, since φ is an irreducible constituent of ψ^G, we see that

$$et\theta(1) = \varphi(1) \leq t\psi(1) = tf\theta(1) \leq te\theta(1).$$

Thus, $f = e$ and $\varphi(1) = t\psi(1)$. Hence, $\varphi(1) = \psi^G(1)$ and we conclude that $\psi^G = \varphi$ is irreducible. Also,

$$\mathbf{I}(\psi_N, \theta) = \mathbf{I}(f\theta, \theta) = f = e = \mathbf{I}(\varphi_N, \theta) = \mathbf{I}((\psi^G)_N, \theta).$$

This clearly implies that
$$\mathbf{I}(\varphi_T, \psi) = 1.$$

Now, if $\psi, \mu \in \mathrm{IBr}(T \mid \theta)$ are distinct and such that $\psi^G = \varphi = \mu^G$, then ψ and μ are different irreducible constituents of φ_T by Corollary (8.3) (or by the first part). Then $\mathbf{I}(\varphi_N, \theta) \geq \mathbf{I}(\psi_N, \theta) + \mathbf{I}(\mu_N, \theta) = 2\mathbf{I}(\varphi_N, \theta)$, by the first part of the proof. This is a contradiction, which proves that our map is injective.

Finally, since θ is an irreducible constituent of φ_N if and only if φ is an irreducible constituent of θ^G by Corollary (8.7), we have that

$$\theta^G = (\theta^T)^G = \sum_{\mu \in \mathrm{IBr}(T \mid \theta)} a_\mu \mu^G$$

proving that our map is onto. ∎

In the situation of the Clifford correspondence theorem, it is customary to say that ψ is the **Clifford correspondent** of ψ^G over θ.

Suppose that $N \lhd G$, and let $\theta \in \mathrm{IBr}(N)$ and $\varphi \in \mathrm{IBr}(G \,|\, \theta)$. If $\psi \in \mathrm{IBr}(T \,|\, \theta)$ induces φ, we know by Theorem (8.9) that $\mathbf{I}(\varphi_N, \theta) = \mathbf{I}(\psi_N, \theta)$. Also, by the last formula in the proof of Theorem (8.9), it is clear that $\mathbf{I}(\theta^G, \varphi) = \mathbf{I}(\theta^T, \psi)$. Therefore, if we wish to study the integers $\mathbf{I}(\varphi_N, \theta)$ and $\mathbf{I}(\theta^G, \varphi)$, it is no loss to assume that θ is G-invariant. Suppose that this is the case and write

$$\theta^G = \sum_{\mu \in \mathrm{IBr}(G \,|\, \theta)} a_\mu \mu$$

and $\mu_N = b_\mu \theta$ for $\mu \in \mathrm{IBr}(G \,|\, \theta)$. Then

$$b_\mu = \frac{\mu(1)}{\theta(1)}.$$

In addition, by Corollary (8.8), we have that

$$a_\mu = \mathbf{I}(\theta^G, \mu) = \frac{\Phi_\mu(1)}{\Phi_\theta(1)}.$$

There is a Clifford correspondence theorem for projective indecomposable characters which easily follows from the next result.

(8.10) THEOREM. *Suppose that $N \lhd G$ and let $\theta \in \mathrm{IBr}(N)$. If $\tau \in \mathrm{IBr}(T \,|\, \theta)$, where $T = I_G(\theta)$, then*

$$(\Phi_\tau)^G = \Phi_{\tau^G}.$$

Proof. If $\varphi \in \mathrm{IBr}(G \,|\, \tau)$, we claim that $\varphi = \tau^G$. Notice that $\varphi \in \mathrm{IBr}(G \,|\, \theta)$ since φ_T contains τ. Hence, $\varphi = \mu^G$ for some $\mu \in \mathrm{IBr}(T \,|\, \theta)$ by the Clifford correspondence for Brauer characters. Now, since $\mathbf{I}(\varphi_N, \theta) = \mathbf{I}(\mu_N, \theta)$ by the Clifford correspondence and τ lies over θ by hypothesis, we necessarily have that $\mu = \tau$ and the claim follows.

Now, $(\Phi_\tau)^G$ vanishes off p-regular elements of G by the definition of the induced character and because Φ_τ vanishes off the p-regular elements of T. Hence, by Theorem (2.13) we may write

$$(\Phi_\tau)^G = \sum_{\varphi \in \mathrm{IBr}(G)} a_\varphi \Phi_\varphi$$

for some $a_\varphi \in \mathbb{C}$. Then

$$a_\varphi = [(\Phi_\tau)^G, \varphi]^0 = [(\Phi_\tau)^G, \hat{\varphi}] = [\Phi_\tau, (\hat{\varphi})_T] = [\Phi_\tau, \varphi_T]^0,$$

which is the multiplicity of τ in φ_T by Theorem (2.13). By the claim above, $a_\varphi = 0$ if $\varphi \neq \tau^G$, and $a_{\tau^G} = 1$ since $\mathbf{I}((\tau^G)_T, \tau) = 1$ by the Clifford correspondence. The proof of the theorem is complete. ∎

Our next objective in this chapter is to analyse some extendibility theorems on Brauer characters. The next result holds in much greater generality (even for finite fields).

(8.11) THEOREM (Green). *Suppose that G/N is a p-group and let $\theta \in$ IBr(N). Then there exists a unique $\varphi \in$ IBr(G) lying over θ. Furthermore, φ_N is the sum of the distinct G-conjugates of θ. In particular, if θ is G-invariant, then $\varphi_N = \theta$.*

Proof. Let $T = I_G(\theta)$. By the Clifford correspondence, we know that

$$|\text{IBr}(T \,|\, \theta)| = |\text{IBr}(G \,|\, \theta)|.$$

Furthermore, if $\psi \in$ IBr$(T \,|\, \theta)$, then

$$\mathbf{I}((\psi^G)_N, \theta) = \mathbf{I}(\psi_N, \theta).$$

Hence, it is clear that we may assume that θ is G-invariant. Let $\varphi \in$ IBr(G) be an irreducible constituent of θ^G. We have to prove that $\varphi_N = \theta$ and that φ is the unique irreducible Brauer character over θ. Since G/N is a p-group, it follows that $N^0 = G^0$. Hence, it clearly suffices to show that $\varphi_N = \theta$.

Write $\varphi_N = e\theta$ (by using Clifford's theorem). By the induction formula, we have that

$$(\theta^G)_N = |G : N|\theta.$$

Hence,

$$e\theta^G(n) = e|G : N|\theta(n) = |G : N|\varphi(n)$$

for $n \in N^0 = G^0$. Therefore,

$$e\theta^G = |G : N|\varphi.$$

By the linear independence of IBr(G), we conclude that θ^G is a multiple of φ and that e is a p-power. If $e > 1$, then $\varphi(g)^* = (e\theta(g))^* = e^*\theta(g)^* = 0$ for every $g \in G^0$. This contradicts the fact that the traces of the irreducible FG-representations are linearly independent. (See Theorem (1.19) and Lemma (2.4).) Hence, $e = 1$ and $\varphi_N = \theta$, as desired. ∎

As we have seen, Green's theorem gives us a criterion for extending an irreducible Brauer character from a normal subgroup N if G/N is a p-group. There is another well-known theorem on extending representations. Its proof is standard and we sketch it for the reader's convenience. It is perhaps remarkable that every known proof of it requires the use of representations (not just characters).

(8.12) THEOREM. *Suppose that* $N \triangleleft G$ *with* G/N *cyclic. If* $\theta \in \mathrm{IBr}(N)$ *is G-invariant, then there is a* $\varphi \in \mathrm{IBr}(G)$ *such that* $\varphi_N = \theta$.

Proof. Write $G = N\langle g \rangle$, and put $a = |G : N| = o(gN)$. Suppose that $\mathcal{Y}$ affords θ. We will prove that there is a representation $\mathcal{X}$ of G extending $\mathcal{Y}$.

Since θ is G-invariant, it follows that the representations $\mathcal{Y}$ and $\mathcal{Y}^{g^{-1}}$ are similar. Therefore, there is a matrix M such that

$$\mathcal{Y}(n^g) = M^{-1}\mathcal{Y}(n)M$$

for all $n \in N$. Also,

$$\mathcal{Y}(n^{g^2}) = \mathcal{Y}((n^g)^g) = M^{-1}\mathcal{Y}(n^g)M = M^{-2}\mathcal{Y}(n)M^2$$

and, in general,

$$\mathcal{Y}(n^{g^j}) = M^{-j}\mathcal{Y}(n)M^j$$

for $j \geq 1$. In particular,

$$M^{-a}\mathcal{Y}(n)M^a = \mathcal{Y}(n^{g^a}) = \mathcal{Y}(g^a)^{-1}\mathcal{Y}(n)\mathcal{Y}(g^a)$$

for all $n \in N$. It then follows that the matrix $\mathcal{Y}(g^a)M^{-a}$ commutes with every matrix $\mathcal{Y}(n)$ for $n \in N$. By Theorem (1.20), there is an $f \in F^\times$ such that

$$\mathcal{Y}(g^a)M^{-a} = fI.$$

Since F is algebraically closed, let $z \in F$ be such that $z^a = f$ and put $A = zM$. Notice that

$$\mathcal{Y}(g^a) = fM^a = A^a$$

and

$$\mathcal{Y}(n^{g^j}) = A^{-j}\mathcal{Y}(n)A^j$$

for every $j \geq 1$. Now, we check that the map $\mathcal{X}(g^i n) = A^i \mathcal{Y}(n)$, where $0 \leq i < a$ and $n \in N$, defines a representation of G extending $\mathcal{Y}$. First of all, we claim that if j is any integer, then

$$\mathcal{X}(g^j n) = A^j \mathcal{Y}(n).$$

To prove the claim, write $j = ad + r$, where d is some integer and $0 \leq r < a$. Then

$$\mathcal{X}(g^j n) = \mathcal{X}(g^r (g^a)^d n) = A^r \mathcal{Y}((g^a)^d n)$$
$$= A^r \mathcal{Y}(g^a)^d \mathcal{Y}(n) = A^r A^{ad} \mathcal{Y}(n) = A^j \mathcal{Y}(n)$$

and the claim is proven. Finally,

$$\mathcal{X}(g^i n g^j m) = \mathcal{X}(g^{i+j} n^{g^j} m) = A^{i+j} \mathcal{Y}(n^{g^j} m) = A^{i+j} \mathcal{Y}(n^{g^j})\mathcal{Y}(m)$$
$$= A^{i+j} A^{-j} \mathcal{Y}(n)A^j \mathcal{Y}(m) = \mathcal{X}(g^i n)\mathcal{X}(g^j m)$$

for $n, m \in N$. ∎

Our next objective is to prove an important extendibility result due to E. C. Dade.

(8.13) THEOREM (Dade). *Suppose that $N \lhd G$ with $(|N|, |G : N|) = 1$. Let $\theta \in \mathrm{IBr}(N)$ be G-invariant. Then there exists a $\mu \in \mathrm{IBr}(G)$ such that $\mu_N = \theta$.*

If $N \subseteq G$ is a normal Hall subgroup of G, it is a fact that every invariant irreducible representation (over any arbitrary field) extends to G. This result is due to M. Isaacs, but it is much harder to prove. In this chapter, we will present two nice proofs of Dade's result. One is due to R. Gow and the other one to D. Passman.

When discussing extendibility of characters, the so called "projective representations" and "factor sets" play a key role. We need to review some of the results of Chapter 11 in [Isaacs, *Characters*].

If A is a (possibly infinite) abelian group, then an A-**factor set** of a group G is a function $\alpha : G \times G \to A$ such that

$$\alpha(xy, z)\alpha(x, y) = \alpha(x, yz)\alpha(y, z)$$

for all $x, y, z \in G$. The set of A-factor sets is denoted by $\mathbf{Z}^2(G, A)$ and it is an abelian group under pointwise multiplication. If $\alpha \in \mathbf{Z}^2(G, A)$, then α^{-1} is the A-factor set defined by $\alpha^{-1}(g, h) = \alpha(g, h)^{-1}$ for $g, h \in G$.

As we are going to see, our interest is focused on $F^\times$-factor sets. If there is no possible confusion, we will refer to them simply as factor sets.

Factor sets are associated to projective representations. We say that $\mathcal{X} : G \to \mathrm{GL}(n, F)$ is a **projective representation** of G if for every $g, h \in G$, there is a scalar $\alpha(g, h) \in F^\times$ such that

$$\mathcal{X}(g)\mathcal{X}(h) = \alpha(g, h)\mathcal{X}(gh).$$

We say that n is the **degree** of $\mathcal{X}$. Note that if $\mathcal{X}$ is a projective representation of G, then

$$\mathcal{X}(x)\mathcal{X}(y)\mathcal{X}(z) = \mathcal{X}(xy)\mathcal{X}(z)\alpha(x, y) = \mathcal{X}(xyz)\alpha(x, y)\alpha(xy, z)$$

and

$$\mathcal{X}(x)\mathcal{X}(y)\mathcal{X}(z) = \mathcal{X}(x)\mathcal{X}(yz)\alpha(y, z) = \mathcal{X}(xyz)\alpha(x, yz)\alpha(y, z)$$

for $x, y, z \in G$. Since $\mathcal{X}(xyz) \neq 0$, we conclude that

$$\alpha(xy, z)\alpha(x, y) = \alpha(x, yz)\alpha(y, z).$$

Hence, $\alpha \in \mathbf{Z}^2(G, F^\times)$.

Observe that if $\mathcal{X}$ is a projective representation, then $\mathcal{X}(1) = \alpha(1, 1)I$. Thus, $\mathcal{X}(1)$ is the identity matrix if and only if $\alpha(1, 1) = 1$.

Projective representations (and factor sets) appear quite naturally in representation theory because of the following elementary result.

If $N \lhd G$ and $\mathcal{Y}$ is an irreducible F-representation of N, we say that $\mathcal{Y}$ is **G-invariant** if the representations $\mathcal{Y}^g$ and $\mathcal{Y}$ are similar for all $g \in G$. Of course, this happens if and only if the Brauer character θ afforded by $\mathcal{Y}$ is G-invariant.

(8.14) THEOREM. *Let $N \lhd G$ and suppose that $\mathcal{Y}$ is an irreducible F-representation of N affording $\theta \in \mathrm{IBr}(N)$. Suppose that θ is G-invariant. Then there exists a projective representation $\mathcal{X}$ of G such that for all $n \in N$ and $g \in G$ we have*
   (a) $\mathcal{X}(n) = \mathcal{Y}(n)$;
   (b) $\mathcal{X}(ng) = \mathcal{X}(n)\mathcal{X}(g)$;
   (c) $\mathcal{X}(gn) = \mathcal{X}(g)\mathcal{X}(n)$.
Furthermore, if $\mathcal{X}_0$ is another projective representation satisfying (a), (b) and (c), then $\mathcal{X}_0(g) = \mathcal{X}(g)\mu(g)$ for some function $\mu : G \to F^\times$ which is constant on cosets of N.

Sketch of Proof. If $F = \mathbb{C}$, then this is Theorem (11.2) of [Isaacs, Characters]. However, in that proof, the only condition needed on the field F is that F is algebraically closed. We find it convenient to review how $\mathcal{X}$ is constructed.

If $g \in G$, then $\mathcal{Y}^g$, defined by

$$\mathcal{Y}^g(n) = \mathcal{Y}(gng^{-1}),$$

is an F-representation of N affording the Brauer character $\theta^g = \theta$. Hence, $\mathcal{Y}$ and $\mathcal{Y}^g$ are similar.

Now, let T be a transversal for N in G with $1 \in T$. Set $P_1 = I$ and for each $1 \neq t \in T$, choose a nonsingular matrix P_t such that

$$P_t \mathcal{Y}(n) P_t^{-1} = \mathcal{Y}^t(n)$$

for all $n \in N$. Define
$$\mathcal{X}(nt) = \mathcal{Y}(n)P_t.$$

It is easy to check that $\mathcal{X}$ satisfies conditions (a), (b) and (c). For the second part of the statement, the key fact used is that if M is any matrix satisfying

$$M\mathcal{Y}(n) = \mathcal{Y}(n)M$$

for all $n \in N$, then M is a scalar matrix (Theorem (1.20)). ∎

(We will see another construction of the projective representation $\mathcal{X}$ in Theorem (8.28).)

We now state a few worthwhile remarks concerning Theorem (8.14). First of all, with the notation above, note that

$$\mathcal{X}(g)\mathcal{Y}(n) = \mathcal{X}(gn) = \mathcal{X}(gng^{-1}g) = \mathcal{Y}(gng^{-1})\mathcal{X}(g)$$

for $g \in G$ and $n \in N$. Also, if α is the factor set associated to $\mathcal{X}$, then

$$\alpha(g,n) = \alpha(n,g) = 1$$

for $n \in N$ and $g \in G$ by parts (b) and (c). Furthermore, for $m,n \in N$ and $g,h \in G$, we have

$$\alpha(gn,hm)\mathcal{X}(gnhm) = \mathcal{X}(gn)\mathcal{X}(hm) = \mathcal{X}(g)\mathcal{X}(nhm)$$

$$= \mathcal{X}(g)\mathcal{X}(hn^h m) = \mathcal{X}(g)\mathcal{X}(h)\mathcal{X}(n^h m)$$

$$= \alpha(g,h)\mathcal{X}(gh)\mathcal{X}(n^h m) = \alpha(g,h)\mathcal{X}(ghn^h m) = \alpha(g,h)\mathcal{X}(gnhm).$$

Thus

$$\alpha(g,h) = \alpha(gn,hm)$$

and therefore α can be seen as a factor set of G/N (see Theorem (8.15) below).

If $\mu : G \to F^\times$ is an arbitrary function, then $\delta(\mu) : G \times G \to F^\times$ defined by

$$\delta(\mu)(g,h) = \mu(g)\mu(h)\mu(gh)^{-1}$$

is also a factor set. In fact, the map

$$\delta : \{\mu : G \to F^\times\} \to \mathbf{Z}^2(G, F^\times)$$

is a homomorphism of groups. The image of δ is denoted by $\mathbf{B}^2(G, F^\times)$ and we will write
$$\mathbf{H}^2(G, F^\times) = \mathbf{Z}^2(G, F^\times)/\mathbf{B}^2(G, F^\times).$$

Note that if $\mathcal{X}$ is a projective representation of G with factor set α and $\mu : G \to F^\times$ is any function, then

$$(\mu\mathcal{X})(g) = \mu(g)\mathcal{X}(g)$$

is another projective representation with factor set $\alpha\delta(\mu)$.

The next result, although not difficult to prove, is of fundamental importance when studying extensions of characters.

(8.15) THEOREM. *Let $N \triangleleft G$ and let $\theta \in \mathrm{IBr}(N)$ be invariant in G. Let $\mathcal{Y}$ be a representation affording θ and let $\mathcal{X}$ be a projective representation of G satisfying conditions (a), (b) and (c) of Theorem (8.14). Let α be the factor set of $\mathcal{X}$. Define $\beta \in \mathbf{Z}^2(G/N, F^\times)$ by $\beta(gN, hN) = \alpha(g, h)$. Then β is well defined and $\beta \mathbf{B}^2(G/N, F^\times) \in \mathbf{H}^2(G/N, F^\times)$ depends only on θ. Also, θ is extendible to G if and only if $\beta \in \mathbf{B}^2(G/N, F^\times)$.*

Proof. This is Theorem (11.7) of [Isaacs, *Characters*] with the field F instead of $\mathbb{C}$. The proof is the same. ∎

   If $N \triangleleft G$, we will view the projective representations of G/N as projective representations $\mathcal{Q}$ of G satisfying $\mathcal{Q}(gn) = \mathcal{Q}(g)$ for all $g \in G$ and $n \in N$.

(8.16) THEOREM (Clifford). *Suppose that $N \triangleleft G$ and let $\mathcal{Y}$ be a G-invariant irreducible representation of N affording $\theta \in \mathrm{IBr}(N)$. Let $\mathcal{X}$ be a projective representation of G associated to $\mathcal{Y}$ satisfying (a), (b) and (c) of Theorem (8.14). Suppose that the factor set of $\mathcal{X}$ is α. If $\mu \in \mathrm{IBr}(G \,|\, \theta)$, then there exist a representation $\mathcal{D}$ of G affording μ and a projective representation $\mathcal{E}$ of G/N with factor set α^{-1} such that for all $g \in G$ we have*

$$\mathcal{D}(g) = \mathcal{E}(g) \otimes \mathcal{X}(g).$$

Furthermore, the order of the element $\beta \mathbf{B}^2(G/N, F^\times) \in \mathbf{H}^2(G/N, F^\times)$ uniquely determined by $\theta \in \mathrm{IBr}(N)$ (by Theorem (8.15)) divides $\frac{\mu(1)}{\theta(1)}$.

Proof. Since $\mu_N = s\theta$, by Clifford's theorem (8.5) we may find an FG-representation $\mathcal{D}$ affording μ such that for $n \in N$, we have

$$\mathcal{D}(n) = \begin{pmatrix} \mathcal{Y}(n) & 0 & \cdots & 0 \\ 0 & \mathcal{Y}(n) & \cdots & 0 \\ \vdots & \vdots & \ddots & \vdots \\ 0 & 0 & \cdots & \mathcal{Y}(n) \end{pmatrix}.$$

For $g \in G$, we write $\mathcal{D}(g)$ in block form:

$$\mathcal{D}(g) = \begin{pmatrix} \mathcal{D}_{11}(g) & \mathcal{D}_{12}(g) & \cdots & \mathcal{D}_{1s}(g) \\ \mathcal{D}_{21}(g) & \mathcal{D}_{22}(g) & \cdots & \mathcal{D}_{2s}(g) \\ \vdots & \vdots & \ddots & \vdots \\ \mathcal{D}_{s1}(g) & \mathcal{D}_{s2}(g) & \cdots & \mathcal{D}_{ss}(g) \end{pmatrix},$$

where $\mathcal{D}_{ij}(g) \in \mathrm{Mat}(\theta(1), F)$. Since $\mathcal{D}(g)\mathcal{D}(n) = \mathcal{D}(gng^{-1})\mathcal{D}(g)$, this easily implies that

$$\mathcal{D}_{ij}(g)\mathcal{Y}(n) = \mathcal{Y}^g(n)\mathcal{D}_{ij}(g)$$

for all $n \in N$, $g \in G$ and $1 \leq i,j \leq s$. Since $\mathcal{X}(g)\mathcal{Y}(n) = \mathcal{Y}^g(n)\mathcal{X}(g)$ (see the remark after the proof of Theorem (8.14)), we have that

$$\mathcal{D}_{ij}(g)\mathcal{Y}(n) = \mathcal{X}(g)\mathcal{Y}(n)\mathcal{X}(g)^{-1}\mathcal{D}_{ij}(g)$$

for all $n \in N$ and $g \in G$. Thus,

$$\mathcal{X}(g)^{-1}\mathcal{D}_{ij}(g)\mathcal{Y}(n) = \mathcal{Y}(n)\mathcal{X}(g)^{-1}\mathcal{D}_{ij}(g)$$

for all $n \in N$ and $g \in G$. By Theorem (1.20), it follows that, for $i,j = 1,\ldots,s$, we have

$$\mathcal{D}_{ij}(g) = e_{ij}(g)\mathcal{X}(g)$$

for some $e_{ij}(g) \in F$. We define the matrix

$$\mathcal{E}(g) = (e_{ij}(g)) \in \mathrm{Mat}(s,F).$$

Note that, for $g \in G$, we have

$$\mathcal{D}(g) = \mathcal{E}(g) \otimes \mathcal{X}(g),$$

the tensor product of matrices. (See the definition and remarks after the statement of Theorem (2.22).) In particular, by elementary linear algebra, we see that $\mathcal{E}(g)$ is regular. We claim that

$$\mathcal{E}(g_1)\mathcal{E}(g_2) = \alpha(g_1,g_2)^{-1}\mathcal{E}(g_1g_2)$$

for $g_1, g_2 \in G$. To see this, note that

$$\mathcal{E}(g_1)\mathcal{E}(g_2) \otimes \mathcal{X}(g_1)\mathcal{X}(g_2) = (\mathcal{E}(g_1) \otimes \mathcal{X}(g_1))(\mathcal{E}(g_2) \otimes \mathcal{X}(g_2))$$

$$= \mathcal{D}(g_1)\mathcal{D}(g_2) = \mathcal{D}(g_1g_2) = \mathcal{E}(g_1g_2) \otimes \mathcal{X}(g_1g_2)$$

$$= \mathcal{E}(g_1g_2) \otimes \alpha(g_1,g_2)^{-1}\mathcal{X}(g_1)\mathcal{X}(g_2)$$

$$= \alpha(g_1,g_2)^{-1}\mathcal{E}(g_1g_2) \otimes \mathcal{X}(g_1)\mathcal{X}(g_2).$$

Since $\mathcal{X}(g_1)\mathcal{X}(g_2) \neq 0$, we conclude that $\mathcal{E}(g_1)\mathcal{E}(g_2) = \alpha(g_1,g_2)^{-1}\mathcal{E}(g_1g_2)$, as desired.

Now,

$$I \otimes \mathcal{Y}(n) = \mathcal{D}(n) = \mathcal{E}(n) \otimes \mathcal{Y}(n)$$

and we conclude that $\mathcal{E}(n) = I$ for $n \in N$. Furthermore, using the fact that $\alpha(n,g) = 1$ (see the remark after Theorem (8.14)), we have

$$\mathcal{E}(ng) = \alpha(n,g)\mathcal{E}(n)\mathcal{E}(g) = \mathcal{E}(g).$$

Thus, we see that $\mathcal{E}$ is a projective representation of G/N with factor set α^{-1}.

Finally, let $\tau : G \to F^{\times}$ be defined by $\tau(g) = \det(\mathcal{E}(g))$. By taking determinants in the expression $\mathcal{E}(g_1)\mathcal{E}(g_2) = \alpha(g_1, g_2)^{-1}\mathcal{E}(g_1 g_2)$, we get that

$$\tau(g_1)\tau(g_2) = (\alpha(g_1, g_2)^{-1})^s \tau(g_1 g_2).$$

Hence, by the definition of δ (before the statement of Theorem (8.15)), we have that

$$(\alpha^{-1})^s = \delta(\tau).$$

Since we may view τ as a function $G/N \to F^{\times}$ because $\tau(gn) = \tau(g)$ for $g \in G$ and $n \in N$, we conclude that

$$(\beta^{-1})^s \mathbf{B}^2(G/N, F^{\times}) = \mathbf{B}^2(G/N, F^{\times}),$$

using the notation in Theorem (8.15). Hence, we see that the order of $\beta \mathbf{B}^2(G/N, F^{\times})$ divides $s = \frac{\mu(1)}{\theta(1)}$, as required. $\blacksquare$

We will prove later that, in fact, all the irreducible F-representations of G over θ are bijectively associated to "irreducible" projective representations of G/N with factor set α^{-1}. But first, we dispose of Dade's theorem.

In Gow's proof of Dade's theorem, we will use the original result of Gallagher for ordinary irreducible characters (Corollary (8.16) of [Isaacs, *Characters*]). We will give the complete proof of that result in Theorem (8.23) below for Brauer characters (and therefore, by choosing a prime p not dividing the order of the group, for ordinary characters). So we suggest the reader who is not familiar with Gallagher's extension theorem first read Theorem (8.23) and its proof, and then come back to this point.

Proof of Dade's Theorem (8.13) (Gow). Let $\hat{\theta}$ be the generalized character of N defined by $\hat{\theta}(n) = \theta(n_{p'})$. (See Lemma (2.15).) Note that

$$(\hat{\theta})^g(n) = \hat{\theta}(n^{g^{-1}}) = \theta((n^{g^{-1}})_{p'}) = \theta((n_{p'})^{g^{-1}}) = \theta(n_{p'}) = \hat{\theta}(n).$$

Hence, $\hat{\theta}$ is G-invariant. Thus, we may write

$$\hat{\theta} = \sum_{\psi \in \mathcal{A}} a_{\psi} \Delta_{\psi} - \sum_{\psi \in \mathcal{B}} b_{\psi} \Delta_{\psi},$$

where $\mathcal{A}, \mathcal{B} \subseteq \mathrm{Irr}(N)$, a_{ψ}, b_{ψ} are positive integers and Δ_{ψ} is the sum of the distinct G-conjugates of ψ for every $\psi \in \mathcal{A} \cup \mathcal{B}$.

By Gallagher's extension theorem, we know that if T_{ψ} is the inertia group of $\psi \in \mathrm{Irr}(N)$ in G, then there exists a $\tau_{\psi} \in \mathrm{Irr}(T_{\psi})$ such that

$(\tau_\psi)_N = \psi$. Also, by the Clifford correspondence for ordinary characters (Theorem (6.11.d) of [Isaacs, *Characters*]), we have that $(\tau_\psi)^G$ is irreducible. Furthermore,

$$((\tau_\psi)^G)_N = \Delta_\psi$$

since $[((\tau_\psi)^G)_N, \psi] = [(\tau_\psi)_N, \psi] = 1$, again by the Clifford correspondence for ordinary characters. Now, consider the characters

$$\Xi = \sum_{\psi \in \mathcal{A}} a_\psi (\tau_\psi)^G$$

and

$$\Psi = \sum_{\psi \in \mathcal{B}} b_\psi (\tau_\psi)^G .$$

Note that $\Xi_N = \sum_{\psi \in \mathcal{A}} a_\psi \Delta_\psi$ and $\Psi_N = \sum_{\psi \in \mathcal{B}} b_\psi \Delta_\psi$. Thus,

$$(\Xi^0)_N - (\Psi^0)_N = (\hat{\theta})^0 = \theta .$$

Now, let

$$\mu = \sum_{\delta \in \mathrm{IBr}(G \,|\, \theta)} \mathbf{I}(\Xi^0, \delta)\delta$$

and

$$\nu = \sum_{\delta \in \mathrm{IBr}(G \,|\, \theta)} \mathbf{I}(\Psi^0, \delta)\delta .$$

Clearly,

$$\mu_N - \nu_N = \theta .$$

Now, if $\delta \in \mathrm{IBr}(G \,|\, \theta)$, we know by Theorem (8.16) that r divides $\frac{\delta(1)}{\theta(1)}$, where r is the order of the element $\beta \mathbf{B}^2(G/N, F^\times)$ which is uniquely determined by θ using Theorem (8.15). It then follows that $\mu_N = ra\theta$ and $\nu_N = rb\theta$ for some integers a and b. Since

$$\theta = \mu_N - \nu_N = r(a - b)\theta ,$$

we conclude that $r = 1$. Hence, θ extends to G by Theorem (8.15), as desired. ∎

We now work toward the promised converse of Theorem (8.16).

(8.17) THEOREM (Jacobson Density Lemma). *Suppose that V is a simple FG-module. Let $\{v_1, \ldots, v_n\}$ be a set of linearly independent vectors in V and let $w_1, \ldots, w_n$ be n vectors of V. Then there exists an $a \in FG$ such that $v_i a = w_i$ for all $i = 1, \ldots, n$.*

Proof. If $a \in FG$, we denote by $a_V \in \operatorname{End}_F(V)$ the map $a_V(v) = va$ for $v \in V$. By Theorem (1.19), we know that the representation $\mathcal{X} : FG \to \operatorname{End}_F(V)$ given by $a \mapsto a_V$ is surjective. If $f \in \operatorname{End}_F(V)$ maps $f(v_i) = w_i$ for all i, we conclude that there exists an $a \in FG$ such that $a_V = f$. This proves the theorem. ∎

The projective representations of a finite group G are naturally associated to representations of certain finite dimensional algebras over F.

Suppose that $\alpha \in \mathbf{Z}^2(G, F^\times)$. Let $F^\alpha G$ be the F-vector space which has the symbols $\{\tilde{g} \,|\, g \in G\}$ as a basis. We define

$$\tilde{g}\tilde{h} = \alpha(g, h)\widetilde{gh}$$

and extend this multiplication linearly to $F^\alpha G$.

By using the fact that $\alpha(xy, z)\alpha(x, y) = \alpha(x, yz)\alpha(y, z)$ for $x, y, z \in G$, it is easy to see that the multiplication in $F^\alpha G$ is associative. If $x \in G$, we have that $\alpha(x^{-1}x, 1)\alpha(x^{-1}, x) = \alpha(x^{-1}, x)\alpha(x, 1)$, $\alpha(1x, x^{-1})\alpha(1, x) = \alpha(1, xx^{-1})\alpha(x, x^{-1})$, and we deduce that

$$\alpha(x, 1) = \alpha(1, x) = \alpha(1, 1)$$

for all $x \in G$. Therefore,

$$\tilde{x}(\alpha(1, 1)^{-1}\tilde{1}) = (\alpha(1, 1)^{-1}\tilde{1})\tilde{x} = \tilde{x}$$

so that $\alpha(1, 1)^{-1}\tilde{1}$ is the identity in $F^\alpha G$. Now, it is straightforward to check that $F^\alpha G$ is a finite dimensional F-algebra with identity $\alpha(1, 1)^{-1}\tilde{1}$. The algebra $F^\alpha G$ is called the **twisted group algebra** with respect to α.

Since $\alpha(xx^{-1}, x)\alpha(x, x^{-1}) = \alpha(x, x^{-1}x)\alpha(x^{-1}, x)$ we have that

$$\alpha(x, x^{-1}) = \alpha(x^{-1}, x).$$

(Of course, as happens with the equality $\alpha(x, 1) = \alpha(1, x) = \alpha(1, 1)$, this equation also holds for every A-factor set α over any abelian group A.)

It is clear that the elements $\{\tilde{g} \,|\, g \in G\}$ are invertible elements of $F^\alpha G$. In fact,

$$(\tilde{g})^{-1} = \alpha(1, 1)^{-1}\alpha(g, g^{-1})^{-1}\widetilde{g^{-1}}.$$

If $\mathcal{X} : F^\alpha G \to \operatorname{Mat}(n, F)$ is a representation of the algebra $F^\alpha G$, then the map $g \mapsto \mathcal{X}(\tilde{g})$ defines a projective representation of G with factor set α. (Note that $\mathcal{X}(\tilde{g}) \in \operatorname{GL}(n, F)$ since $\tilde{g}$ is a unit of $F^\alpha G$.) Conversely, if $\mathcal{X}$ is a projective representation of G with factor set α, then

$$\sum_{g \in G} a_g \tilde{g} \mapsto \sum_{g \in G} a_g \mathcal{X}(g)$$

defines a representation of $F^\alpha G$.

If $\mathcal{X} : G \to \mathrm{GL}(n, F)$ is a projective representation with factor set α and $P \in \mathrm{GL}(n, F)$, then $P\mathcal{X}P^{-1}$ is a projective representation with factor set α. We say that $\mathcal{X}$ and $P\mathcal{X}P^{-1}$ are **similar**. Also, we say that $\mathcal{X}$ is **irreducible** if $\mathcal{X}$ is not similar to a projective representation of the form

$$\begin{pmatrix} * & * \\ 0 & * \end{pmatrix}.$$

By the discussion after the proof of Theorem (1.19) on general A-modules and representations, it follows that $\mathcal{X}$ is irreducible if and only if the associated representation of $F^\alpha G$ is irreducible, which happens if and only if the associated $F^\alpha G$-module is simple.

Observe that if $\mathcal{X}$ is a projective representation with factor set α and $\mathcal{Y}$ is a projective representation with factor set α^{-1}, then the map

$$(\mathcal{X} \otimes \mathcal{Y})(g) = \mathcal{X}(g) \otimes \mathcal{Y}(g)$$

defines an F-representation of G.

(8.18) THEOREM. *Suppose that $N \triangleleft G$ and let $\mathcal{Y}$ be an irreducible G-invariant F-representation of N. Let $\mathcal{X}$ be a projective representation of G associated with $\mathcal{Y}$ satisfying conditions (a), (b) and (c) of Theorem (8.14) and with factor set α. If $\mathcal{E}$ is a projective representation of G/N with factor set α^{-1}, then the FG-representation $\mathcal{E} \otimes \mathcal{X}$ is irreducible if and only if $\mathcal{E}$ is irreducible. Furthermore, if $\mathcal{R}$ and $\mathcal{E}$ are irreducible projective representations of G/N with factor set α^{-1}, then $\mathcal{E} \otimes \mathcal{X}$ and $\mathcal{R} \otimes \mathcal{X}$ are similar if and only if $\mathcal{E}$ and $\mathcal{R}$ are similar.*

Proof. Set $A = F^{\alpha^{-1}}G$ and $B = F^\alpha G$. In order to avoid any possible confusion, we assume that A is the F-algebra which has the symbols $\{\tilde{g} \mid g \in G\}$ as a basis (with multiplication $\tilde{g}\tilde{h} = \alpha(g, h)^{-1}\widetilde{gh}$ for $g, h \in G$) and that B is the F-algebra which has the symbols $\{\bar{g} \mid g \in G\}$ as a basis (with multiplication $\bar{g}\bar{h} = \alpha(g, h)\overline{gh}$ for $g, h \in G$).

Suppose that r is the degree of $\mathcal{X}$ and suppose that s is the degree of $\mathcal{E}$. Let $V = F^r$ and $W = F^s$, so that V and W are B- and A-modules (respectively) via $v\bar{g} = v\mathcal{X}(g)$ and $w\tilde{g} = w\mathcal{E}(g)$. Note that if $n \in N$, then $w\tilde{n} = w$ because $\mathcal{E}(n) = \mathcal{E}(1) = \alpha^{-1}(1, 1)I_s = I_s$. Also, since $\alpha(n, m) = 1$ for $n, m \in N$, it follows that V is a simple FN-module via $vn = v\bar{n}$ for $n \in N$.

Let $\{v_1, \ldots, v_r\}$ be the canonical basis of V and let $\{w_1, \ldots, w_s\}$ be the canonical basis of W, so that $w_i \otimes v_j$ for $1 \le i \le s, 1 \le j \le r$ is an F-basis of the vector space $W \otimes V$. With respect to these bases, we have that

$$w_i\tilde{g} = \sum_{j=1}^{s} e_{ij}(g)w_j \quad \text{and} \quad v_i\bar{g} = \sum_{j=1}^{r} x_{ij}(g)v_j,$$

where
$$\mathcal{E}(g) = (e_{ij}(g)) \quad \text{and} \quad \mathcal{X}(g) = (x_{ij}(g))$$

for $g \in G$.

If we set
$$(w_i \otimes v_j)g = w_i \tilde{g} \otimes v_j \bar{g}$$

for $g \in G$ and extend this by linearity to $W \otimes V$, we have that $W \otimes V$ is an FG-module satisfying
$$(w \otimes v)g = w\tilde{g} \otimes v\bar{g}$$

for $w \in W$ and $v \in V$. It is straightforward to check that the representation afforded by the FG-module $W \otimes V$ with respect to the basis $\{w_i \otimes v_j\}$ is $\mathcal{E} \otimes \mathcal{X}$.

Now, suppose that Y is an A-submodule of W. If $Y \otimes V$ is the F-span of $\{y \otimes v \mid y \in Y, v \in V\}$, then $Y \otimes V$ is an FG-submodule of $W \otimes V$ of dimension $\dim_F(Y)\dim_F(V)$. Therefore, if W is not a simple A-module, it follows that $W \otimes V$ is not a simple FG-module.

So we assume that W is a simple A-module and we prove that $W \otimes V$ is a simple FG-module. Let $X \neq 0$ be an FG-submodule of $W \otimes V$ and let $0 \neq x \in X$. Hence, there are $a_{ij} \in F$ such that

$$x = \sum_{\substack{1 \leq i \leq s \\ 1 \leq j \leq r}} a_{ij}(w_i \otimes v_j) = \sum_{1 \leq j \leq r} \left(\sum_{1 \leq i \leq s} a_{ij}w_i \right) \otimes v_j = \sum_{1 \leq j \leq r} y_j \otimes v_j \, ,$$

where $y_j = \sum_{1 \leq i \leq s} a_{ij}w_i$. Since $x \neq 0$, we may assume that $y_1 \neq 0$ (by reordering the basis $\{v_1, \ldots, v_r\}$).

Now, since V is a simple FN-module, it follows from the Jacobson density lemma that given $v \in V$ there exists an $a = \sum_{n \in N} a_n n \in FN$ such that $v_1 a = v$ and $v_j a = 0$ for $1 < j \leq r$. Now,

$$(y_j \otimes v_j)a = (y_j \otimes v_j)\left(\sum_{n \in N} a_n n \right)$$

$$= \sum_{n \in N} a_n(y_j \tilde{n} \otimes v_j \bar{n}) = \sum_{n \in N} a_n(y_j \otimes v_j n) = y_j \otimes v_j a \, .$$

Thus,

$$xa = \left(\sum_{1 \leq j \leq r} y_j \otimes v_j \right)a = \sum_{1 \leq j \leq r} (y_j \otimes v_j)a = \sum_{1 \leq j \leq r} y_j \otimes v_j a = y_1 \otimes v \, .$$

Therefore, we see that $y_1 \otimes V = \{y_1 \otimes v \mid v \in V\} \subseteq X$. For $g \in G$, note that $V\bar{g} = V$ since $\bar{g}$ is a unit of B. Hence, we have that

$$y_1\tilde{g} \otimes V = y_1\tilde{g} \otimes V\bar{g} = (y_1 \otimes V)g \subseteq Xg = X \, .$$

Finally, let $v \in V$ and let $w \in W$. Since W is a simple A-module and $y_1 \neq 0$, we have that $y_1 A = W$. Therefore, given $w \in W$, we may write $w = \sum_{g \in G} b_g y_1 \tilde{g}$ for some $b_g \in F$. Hence,

$$w \otimes v = \sum_{g \in G} b_g (y_1 \tilde{g} \otimes v) \in X$$

and thus $X = W \otimes V$. This proves the first part.

It is clear that if $\mathcal{R}$ and $\mathcal{E}$ are similar, then $\mathcal{E} \otimes \mathcal{X}$ and $\mathcal{R} \otimes \mathcal{X}$ are similar because if $P \mathcal{E} P^{-1} = \mathcal{R}$, then

$$(P \otimes I_r)(\mathcal{E} \otimes \mathcal{X})(P \otimes I_r)^{-1} = (P \otimes I_r)(\mathcal{E} \otimes \mathcal{X})(P^{-1} \otimes I_r) = \mathcal{R} \otimes \mathcal{X} .$$

Suppose now that the irreducible representations $\mathcal{E} \otimes \mathcal{X}$ and $\mathcal{R} \otimes \mathcal{X}$ are similar. In particular, note that the degree of $\mathcal{R}$ equals the degree of $\mathcal{E}$. As we did for the representation $\mathcal{E} \otimes \mathcal{X}$, we now construct an appropriate FG-module affording $\mathcal{R} \otimes \mathcal{X}$. First of all, we define on $W = F^s$ another A-module structure via

$$w \cdot \tilde{g} = w \mathcal{R}(g)$$

for $w \in W$ and $g \in G$. Note that $w \cdot \tilde{n} = w$ for $n \in N$ for the same reason as $w\tilde{n} = w$. Now, we define on the vector space $W \otimes V$ another FG-module structure by setting

$$(w_i \otimes v_j) \cdot g = w_i \cdot \tilde{g} \otimes v_j \bar{g} = w_i \mathcal{R}(g) \otimes v_j \mathcal{X}(g) ,$$

and extending this by linearity to $W \otimes V$.

Now, since the representations $\mathcal{E} \otimes \mathcal{X}$ and $\mathcal{R} \otimes \mathcal{X}$ are similar, this implies that there exists a bijective linear map

$$f : W \otimes V \to W \otimes V$$

satisfying

$$f((w \otimes v)g) = f(w \otimes v) \cdot g$$

for $w \in W$, $v \in V$ and $g \in G$.

Write

$$f(w_i \otimes v_k) = \sum_{\substack{1 \leq j \leq s \\ 1 \leq l \leq r}} a_{ikjl}(w_j \otimes v_l) = \sum_{1 \leq j \leq s} w_j \otimes f_{ij}(v_k) ,$$

where

$$f_{ij}(v_k) = \sum_{1 \leq l \leq r} a_{ikjl} v_l$$

and $a_{ikjl} \in F$. Hence, we see that for $v \in V$, we may write

$$f(w_i \otimes v) = \sum_{1 \leq j \leq s} w_j \otimes f_{ij}(v)$$

for uniquely defined $f_{ij} \in \text{End}_F(V)$. Now, if $n \in N$, we have that

$$\sum_{1 \leq j \leq s} w_j \otimes f_{ij}(vn) = f(w_i \otimes vn) = f((w_i \otimes v)n)$$

$$= f(w_i \otimes v) \cdot n = \sum_{1 \leq j \leq s} w_j \otimes f_{ij}(v)n \, ,$$

and we deduce that $f_{ij} \in \text{End}_{FN}(V_N)$. Since V_N is simple, by Schur's lemma we conclude that $f_{ij}(v) = \lambda_{ij}v$ for uniquely defined $\lambda_{ij} \in F$. Define $h : W \to W$ by

$$h(w_i) = \sum_{1 \leq j \leq s} \lambda_{ij}w_j$$

and note that

$$f(w_i \otimes v) = h(w_i) \otimes v$$

for all i and all $v \in V$. Hence,

$$f(w \otimes v) = h(w) \otimes v$$

for all $v \in V$ and $w \in W$. Now, choose $v \neq 0$. For $g \in G$, we have that

$$h(w\tilde{g}) \otimes v\tilde{g} = f(w\tilde{g} \otimes v\tilde{g}) = f((w \otimes v)g) = f(w \otimes v) \cdot g = h(w) \cdot \tilde{g} \otimes v\tilde{g} \, .$$

We deduce that $h(w\tilde{g}) = h(w) \cdot \tilde{g}$, proving that h is an A-homomorphism. Finally, if $h(w) = 0$, then $f(w \otimes v) = 0$ and thus, $w \otimes v = 0$. Since $v \neq 0$, we have that $w = 0$. This shows that h is injective and therefore, that h is an A-isomorphism. Hence, the representations $\mathcal{E}$ and $\mathcal{R}$ are similar, as desired. ∎

If $H \subseteq G$, $\alpha \in \text{cf}(G^0)$ and $\beta \in \text{cf}(H^0)$, then it is easy to check that

$$(\alpha_H \beta)^G = \alpha \beta^G \, .$$

(See the remark after Definition (8.1).) We will use this formula in the proof of the next result.

(8.19) COROLLARY. *Suppose that $N \triangleleft G$ and let $\varphi \in \text{IBr}(N)$ be invariant in G. Suppose that $\eta \in \text{IBr}(G)$ is such that $\eta_N\varphi \in \text{IBr}(N)$. Then the map $\beta \mapsto \beta\eta$ is a bijection $\text{IBr}(G \mid \varphi) \to \text{IBr}(G \mid \eta_N\varphi)$.*

Proof. Let $\mathcal{Z}$ be a representation affording η. Let $\mathcal{Y}$ be a representation affording φ and let $\mathcal{X}$ be a projective representation of G associated with $\mathcal{Y}$ satisfying (a), (b) and (c) of Theorem (8.14) and with factor set α. Note that $\mathcal{X} \otimes \mathcal{Z}$ is a projective representation of G with factor set α associated to the invariant irreducible representation $\mathcal{Y} \otimes \mathcal{Z}_N$ satisfying (a), (b) and (c) of Theorem (8.14). (See Theorem (2.23).)

Now, let $\beta \in \mathrm{IBr}(G \,|\, \varphi)$. Using Theorem (8.16), let $\mathcal{D}$ be a representation of G affording β such that $\mathcal{D} = \mathcal{E} \otimes \mathcal{X}$, where $\mathcal{E}$ is an irreducible (by Theorem (8.18)) projective representation of G/N with factor set α^{-1}. By Theorem (8.18), we have that

$$\mathcal{E} \otimes (\mathcal{X} \otimes \mathcal{Z}) = (\mathcal{E} \otimes \mathcal{X}) \otimes \mathcal{Z} = \mathcal{D} \otimes \mathcal{Z}$$

is irreducible. Hence, $\beta\eta \in \mathrm{IBr}(G)$. Now, suppose that $\gamma \in \mathrm{IBr}(G \,|\, \varphi)$ is such that $\beta\eta = \gamma\eta$. Using again Theorems (8.16) and (8.18), we may find an irreducible representation $\mathcal{R}$ affording γ such that $\mathcal{R} = \mathcal{Q} \otimes \mathcal{X}$ for some irreducible projective representation $\mathcal{Q}$ of G/N with factor set α^{-1}. Now, we have that the irreducible representations $\mathcal{D} \otimes \mathcal{Z} = \mathcal{E} \otimes (\mathcal{X} \otimes \mathcal{Z})$ and $\mathcal{R} \otimes \mathcal{Z} = \mathcal{Q} \otimes (\mathcal{X} \otimes \mathcal{Z})$ afford the same Brauer character. By Theorem (8.18), we deduce that $\mathcal{E}$ and $\mathcal{Q}$ are similar. Therefore, $\mathcal{D} = \mathcal{E} \otimes \mathcal{X}$ and $\mathcal{R} = \mathcal{Q} \otimes \mathcal{X}$ are similar. This proves that $\beta = \gamma$.

Finally, notice that if $\beta \in \mathrm{IBr}(G \,|\, \varphi)$, it is clear that $\beta\eta$ lies over $\varphi\eta_N$. Now, by using Corollary (8.7), we have that

$$(\varphi\eta_N)^G = (\varphi^G)\eta = \left(\sum_{\beta \in \mathrm{IBr}(G \,|\, \varphi)} a_\beta \beta \right) \eta = \sum_{\beta \in \mathrm{IBr}(G \,|\, \varphi)} a_\beta \beta\eta \,,$$

and this proves that the map $\beta \mapsto \beta\eta$ is onto. ∎

(8.20) COROLLARY. *Let $N \triangleleft G$ and let $\eta \in \mathrm{IBr}(G)$. If $\eta_N = \theta \in \mathrm{IBr}(N)$, then the characters $\beta\eta$ for $\beta \in \mathrm{IBr}(G/N)$ are irreducible, distinct for distinct β and are all the irreducible constituents of θ^G.*

Proof. This result follows from Corollary (8.19) by setting $\varphi = 1_N$ and identifying $\mathrm{IBr}(G \,|\, \varphi)$ with $\mathrm{IBr}(G/N)$. ∎

Using Corollary (8.20), it is easy to describe the irreducible Brauer characters of direct products of groups.

If G and H are finite groups, $\alpha \in \mathrm{cf}(G)$ and $\beta \in \mathrm{cf}(H)$, then $\alpha \times \beta \in \mathrm{cf}(G \times H)$ is defined by

$$(\alpha \times \beta)(g, h) = \alpha(g)\beta(h) \,.$$

In the same way, if $\theta \in \mathrm{cf}(G^0)$ and $\varphi \in \mathrm{cf}(H^0)$, then $\theta \times \varphi \in \mathrm{cf}((G \times H)^0)$ is defined by

$$(\theta \times \varphi)(g, h) = \theta(g)\varphi(h) \,.$$

(Of course, $(G \times H)^0 = G^0 \times H^0$.)

(8.21) THEOREM. *If G and H are finite groups, then $\mathrm{IBr}(G \times H) = \{\theta \times \varphi \mid \theta \in \mathrm{IBr}(G), \varphi \in \mathrm{IBr}(H)\}$.*

Proof. To simplify notation, we identify G with $G \times 1$ and H with $1 \times H$, so that we view G and H as normal subgroups of $G \times H$. If $\mathcal{X} : G \to \mathrm{GL}(n, F)$ affords $\theta \in \mathrm{IBr}(G)$, it is clear that the irreducible representation $\hat{\mathcal{X}} : G \times H \to \mathrm{GL}(n, F)$ given by $\hat{\mathcal{X}}(g, h) = \mathcal{X}(g)$ affords the Brauer character $\theta \times 1_{H^0}$, which has H in its kernel. By using the same argument with the irreducible Brauer characters φ of H, we apply Corollary (8.20) to conclude that the Brauer characters $(\theta \times 1_{H^0})(1_{G^0} \times \varphi) = \theta \times \varphi$ are irreducible. It is not necessary to appeal to Corollary (8.20) to conclude that they are distinct since

$$(\theta \times \varphi)_G = \varphi(1)\theta$$

and

$$(\theta \times \varphi)_H = \theta(1)\varphi.$$

Finally, since the number of p-regular classes of $G \times H$ is $|\mathrm{IBr}(G)||\mathrm{IBr}(H)|$, the proof of the theorem follows. $\blacksquare$

Using Corollary (8.20), we can easily derive the next result about Brauer character degrees.

(8.22) THEOREM (Swan). *Suppose that $N \vartriangleleft G$ with G/N solvable. Let $\varphi \in \mathrm{IBr}(G)$ and let $\theta \in \mathrm{IBr}(N)$ be an irreducible constituent of φ_N. Then $\frac{\varphi(1)}{\theta(1)}$ divides $|G : N|$.*

Proof. We argue by induction on $|G : N|$. Let $T = I_G(\theta)$ be the inertia group of θ in G and let $\tau \in \mathrm{IBr}(T \mid \theta)$ be such that $\tau^G = \varphi$. Suppose that $T < G$. By induction, we have that $\frac{\tau(1)}{\theta(1)}$ divides $|T : N|$. Since $\varphi(1) = |G : T|\tau(1)$, we deduce in this case that $\frac{\varphi(1)}{\theta(1)}$ divides $|G : N|$. So we may assume that θ is G-invariant.

Now, suppose that $N \vartriangleleft M \vartriangleleft G$ and let $\eta \in \mathrm{IBr}(M)$ be an irreducible constituent of φ_M lying over θ. If $N < M < G$, then, by induction, we will have that $\frac{\varphi(1)}{\eta(1)}$ divides $|G : M|$ and $\frac{\eta(1)}{\theta(1)}$ divides $|M : N|$, yielding the theorem.

Therefore, we may assume that G/N is simple. Hence, G/N is cyclic. Thus, we have that θ extends to G by Theorem (8.12). If $\eta \in \mathrm{IBr}(G)$ extends θ, we have by Corollary (8.20) that $\varphi = \eta\lambda$ for some $\lambda \in \mathrm{IBr}(G/N)$. Since G/N is abelian, we have that $\lambda(1) = 1$. In this case, $\varphi_N = \theta$, and the theorem follows. $\blacksquare$

The hypothesis that G/N is solvable can be replaced by the hypothesis that G/N is p-solvable. We will prove this after we introduce "modular character triples".

Next, we continue to discuss more extendibility theorems.

If $\varphi \in \mathrm{IBr}(G)$, then φ determines up to similarity an irreducible representation $\mathcal{X} : G \rightarrow \mathrm{GL}(n, F)$. Hence, φ uniquely determines an F-representation

$$\det(\mathcal{X}) : G \rightarrow F^\times$$

defined by $\det(\mathcal{X})(g) = \det(\mathcal{X}(g))$. This representation uniquely determines a linear Brauer character of G which we will denote by $\det(\varphi)$. We write $\mathrm{o}(\varphi)$ for the smallest positive integer n such that

$$\det(\varphi)^n = 1 .$$

It is clear that $\mathrm{o}(\varphi)$ is also the smallest positive integer n such that

$$\det(\mathcal{X})^n = 1 .$$

Note that $\mathrm{o}(\varphi)$ is a p'-number since $F^\times$ does not contain nontrivial p-elements. In fact, $\mathrm{o}(\varphi)$ divides $|G : G'|_{p'}$ by Problem (2.7).

Our next objective is to prove the following theorem.

(8.23) THEOREM. *Let $N \triangleleft G$ and let $\theta \in \mathrm{IBr}(N)$ be G-invariant such that $(\mathrm{o}(\theta)\theta(1), |G : N|) = 1$. Then θ extends to G.*

We need one more result.

(8.24) THEOREM. *If G is a finite group, then $\mathbf{H}^2(G, F^\times)$ is finite and each of its elements has order dividing $|G|$.*

Proof. This is Theorem (11.15) of [Isaacs, *Characters*]. ∎

Proof of Theorem (8.23). Let $\mathcal{Y}$ be a representation of N affording θ. First of all, we prove the theorem if θ is linear. Since θ is G-invariant, we have that $\mathcal{Y}$ and $\mathcal{Y}^g$ are similar for $g \in G$. Since $\mathcal{Y}$ has degree 1, this means that $\mathcal{Y}^g = \mathcal{Y}$ for all $g \in G$. Therefore, $\ker(\mathcal{Y}) \triangleleft G$. By working in $G/\ker(\mathcal{Y})$, note that we may assume that $\mathcal{Y}$ is faithful. Since $\mathcal{Y}^g = \mathcal{Y}$ for all $g \in G$, this implies that $N \subseteq \mathbf{Z}(G)$. Also, $N \cong \mathcal{Y}(N) \subseteq F^\times$ and therefore we deduce that N is cyclic. Since $\mathcal{Y}$ is faithful, this easily implies that $\mathrm{o}(\theta) = |N|$. Hence, by hypothesis, we have that N is a central Hall subgroup of G. Using the Schur-Zassenhaus theorem, we may write $G = N \times K$, where $(|N|, |K|) = 1$. In this case, $\theta \times 1_{K^o}$ extends θ.

Now, we prove the general case. Let $\mathcal{X}$ be a projective representation of G associated with $\mathcal{Y}$ satisfying the conditions (a), (b) and (c) of Theorem (8.14). Write $\lambda = \det(\mathcal{Y})$. By the first part, the representation λ extends to a representation $\nu : G \rightarrow F^\times$.

If $g, h \in G$, we know that

$$\mathcal{X}(g)\mathcal{X}(h) = \alpha(g, h)\mathcal{X}(gh). \tag{1}$$

Also, by Theorem (8.15), we know that $\beta(gN, hN) = \alpha(g, h)$ is well defined and that θ extends to G if and only if $\beta \in \mathbf{B}^2(G/N, F^\times)$.

Let $\mu(g) = \det(\mathcal{X}(g))$. Since $\mathcal{X}(n) = \mathcal{Y}(n)$ for $n \in N$, we have that $\mu(n) = \lambda(n)$. Notice that if $r = \theta(1)$, then, by taking determinants in (1), we have

$$\mu(g)\mu(h) = \alpha(g, h)^r \mu(gh).$$

Since $\mathcal{X}(gn) = \mathcal{X}(g)\mathcal{Y}(n)$ for $g \in G$ and $n \in N$, we conclude that

$$\mu(gn) = \mu(g)\lambda(n).$$

Now, let $\omega(g) = \mu(g)\nu(g)^{-1}$. Then

$$\omega(gn) = \mu(gn)\nu(g)^{-1}\nu(n)^{-1} = \mu(g)\lambda(n)\nu(g)^{-1}\lambda(n)^{-1} = \omega(g),$$

so that ω defines a function on G/N. Also,

$$\omega(g)\omega(h)\omega(gh)^{-1} = \mu(g)\mu(h)\mu(gh)^{-1} = \alpha(g, h)^r.$$

Therefore, $\beta^r \in \mathbf{B}^2(G/N, F^\times)$. By Theorem (8.24), we have that the order of $\beta\mathbf{B}^2(G/N, F^\times)$ divides $|G/N|$. Since $(r, |G/N|) = 1$ by hypothesis, it follows that $\beta \in \mathbf{B}^2(G/N, F^\times)$, as desired. $\blacksquare$

In the characteristic zero case, Gallagher's theorem on extendibility of invariant characters from a normal Hall subgroup is usually proved as a consequence of Theorem (8.23). This is not the case in prime characteristic. If N is a normal Hall subgroup of G and $\theta \in \mathrm{IBr}(N)$, it is not necessarily true that $(\theta(1), |G : N|) = 1$ because it is not true, in general, that $\theta(1)$ divides $|N|$, as we know. This is why Dade's theorem has to be proved in a different way.

Now, we discuss modular character triples.

(8.25) DEFINITION. Suppose that $N \lhd G$ and let $\theta \in \mathrm{IBr}(N)$ be G-invariant. We say that (G, N, θ) is a **modular character triple.**

We denote the set of Brauer characters η of G such that η_N is a multiple of θ by $\mathrm{Br}(G \,|\, \theta)$. This is the set of nonnegative integer linear combinations of the elements in $\mathrm{IBr}(G \,|\, \theta)$. Now, suppose that (Γ, M, φ) is another modular character triple and let $\tau : G/N \to \Gamma/M$ be an isomorphism of groups. For $N \subseteq H \subseteq G$, let H^τ be the subgroup of Γ such that

$$H^\tau/M = \tau(H/N).$$

For every such H, suppose that there is a map

$$\sigma_H : \mathrm{Br}(H \,|\, \theta) \to \mathrm{Br}(H^\tau \,|\, \varphi)$$

such that the following conditions hold for $N \subseteq K \subseteq H \subseteq G$ and $\chi, \psi \in \mathrm{Br}(H \,|\, \theta)$:

(a) $\sigma_H : \mathrm{IBr}(H \,|\, \theta) \to \mathrm{IBr}(H^\tau \,|\, \varphi)$ is a bijection;

(b) $\sigma_H(\chi + \psi) = \sigma_H(\chi) + \sigma_H(\psi)$;

(c) $\sigma_K(\chi_K) = \sigma_H(\chi)_{K^\tau}$;

(d) $\sigma_H(\chi\beta) = \sigma_H(\chi)\beta^\tau$ for $\beta \in \mathrm{IBr}(H/N)$, where β^τ is the Brauer character naturally associated with β via the isomorphism τ. (See Problem (2.1).)

We say in this case that

$$(\sigma, \tau) : (G, N, \theta) \to (\Gamma, M, \varphi)$$

is an **isomorphism** from (G, N, θ) to (Γ, M, φ) and that (G, N, θ) and (Γ, M, φ) are **isomorphic** as modular character triples.

(Of course, when p does not divide the orders of the groups G and Γ, an isomorphism of modular character triples is an isomorphism of ordinary character triples (Definition (11.23) of [Isaacs, *Characters*]). We do stress this because the reader will find that this Definition (11.23), apparently, differs from our Definition (8.25) above for p'-groups G and Γ. For Brauer characters, we find it more convenient to work with our definition.)

It is straightforward (and we leave this for the reader to check) that isomorphism defines an equivalence relation on modular character triples.

With the notation in Definition (8.25), we claim that

$$\frac{\chi(1)}{\theta(1)} = \frac{\sigma_H(\chi)(1)}{\varphi(1)}$$

for $N \subseteq H \subseteq G$ and $\chi \in \mathrm{Br}(H \,|\, \theta)$. This is true because

$$\frac{\chi(1)}{\theta(1)}\varphi = \frac{\chi(1)}{\theta(1)}\sigma_N(\theta)$$

$$= \sigma_N\!\left(\frac{\chi(1)}{\theta(1)}\theta\right) = \sigma_N(\chi_N) = \sigma_H(\chi)_M = \frac{\sigma_H(\chi)(1)}{\varphi(1)}\varphi.$$

Suppose that (G, N, θ) is a modular character triple and let $\mu : G \to \Gamma$ be an onto homomorphism with $\ker(\mu) \subseteq \ker(\theta)$. If $M = \mu(N)$ and $\varphi \in \mathrm{IBr}(M)$ is the character corresponding to $\theta \in \mathrm{IBr}(N/\ker(\mu))$, it is easy to show that (G, N, θ) and (Γ, M, φ) are isomorphic modular character triples (with respect to the natural maps). In particular, if $M \triangleleft G$ with $M \subseteq \ker(\theta)$, then (G, N, θ) and $(G/M, N/M, \theta)$ are isomorphic character triples.

More interesting is the next example.

(8.26) LEMMA. *Let (G, N, φ) be a modular character triple and let $\eta \in \mathrm{IBr}(G)$ be such that $\eta_N \varphi = \theta \in \mathrm{IBr}(N)$. For $N \subseteq H \subseteq G$, define $\sigma_H : \mathrm{Br}(H \,|\, \varphi) \to \mathrm{Br}(H \,|\, \theta)$ by $\sigma_H(\psi) = \psi \eta_H$. Let $i : G/N \to G/N$ be the identity map. Then*

$$(\sigma, i) : (G, N, \varphi) \to (G, N, \theta)$$

is an isomorphism of modular character triples.

Proof. If $N \subseteq H \subseteq G$, then, by Corollary (8.19), we have that the map $\psi \mapsto \psi \eta_H$ is a bijection $\mathrm{IBr}(H \,|\, \varphi) \to \mathrm{IBr}(H \,|\, \theta)$. It is clear that conditions (b), (c) and (d) of Definition (8.25) are satisfied. ∎

Note that, given G-invariant $\lambda, \theta \in \mathrm{IBr}(N)$, where $N \triangleleft G$ and $\lambda(1) = 1$, it follows by Lemma (8.26) that (G, N, λ) and (G, N, θ) are isomorphic if $\theta \bar{\lambda}$ is extendible to G.

The next lemma allows us to construct a projective representation associated with a G-invariant representation of a normal subgroup N (satisfying conditions (a), (b) and (c) of Theorem (8.14)) in which we have some control on the values of the factor set. (We will see how to use this in the proof of Theorem (8.28).)

(8.27) LEMMA. *Suppose that $N \triangleleft G$ and let $\mathcal{R}$ be a G-invariant irreducible representation of N over F. Let $x, y \in G$ and let $\mathcal{X}$, $\mathcal{Y}$ and $\mathcal{Z}$ be extensions of $\mathcal{R}$ to $\langle N, x \rangle$, $\langle N, y \rangle$ and $\langle N, xy \rangle$, respectively. Then there is a scalar $\alpha(x, y) \in F^\times$ such that*

$$\mathcal{X}(x)\mathcal{Y}(y) = \alpha(x, y)\mathcal{Z}(xy) \,.$$

Proof. For $n \in N$, we have

$$\mathcal{R}(yny^{-1}) = \mathcal{Y}(yny^{-1}) = \mathcal{Y}(y)\mathcal{Y}(n)\mathcal{Y}(y^{-1}) = \mathcal{Y}(y)\mathcal{R}(n)\mathcal{Y}(y^{-1}) \,.$$

Conjugating by $\mathcal{X}(x)$, we obtain

$$\mathcal{X}(x)\mathcal{Y}(y)\mathcal{R}(n)\mathcal{Y}(y)^{-1}\mathcal{X}(x)^{-1} = \mathcal{X}(x)\mathcal{R}(yny^{-1})\mathcal{X}(x^{-1})$$

$$= \mathcal{X}(x(yny^{-1})x^{-1}) = \mathcal{R}(xyny^{-1}x^{-1}) \,.$$

Also,

$$\mathcal{R}(xyny^{-1}x^{-1}) = \mathcal{Z}(xy)\mathcal{R}(n)\mathcal{Z}(xy)^{-1} \,.$$

Now, the matrix $\mathcal{Z}(xy)^{-1}\mathcal{X}(x)\mathcal{Y}(y)$ commutes with the matrices $\mathcal{R}(n)$ for $n \in N$ and the result follows from Theorem (1.20). ∎

(8.28) THEOREM. *Suppose that (G, N, θ) is a modular character triple. Then there is an isomorphism of modular character triples*

$$(\sigma, \tau) : (G, N, \theta) \to (\Gamma, M, \varphi)$$

where φ is linear and faithful. In particular, M is a central p'-subgroup of Γ.

Proof. Let $\mathcal{Y}$ be an FN-representation affording θ. For each $\bar{g} = gN \in G/N$, we choose a representation $\mathcal{Y}_{\bar{g}}$ of $\langle N, g \rangle$ extending $\mathcal{Y}$ (by applying Theorem (8.12)). We define

$$\mathcal{X}(g) = \mathcal{Y}_{\bar{g}}(g).$$

Note that, by Lemma (8.27), $\mathcal{X}$ is a projective representation of G. Also, $\mathcal{X}(gn) = \mathcal{X}(g)\mathcal{Y}(n)$, $\mathcal{X}(ng) = \mathcal{Y}(n)\mathcal{X}(g)$ and $\mathcal{X}(n) = \mathcal{Y}(n)$ for $n \in N$ and $g \in G$. Therefore, $\mathcal{X}$ is a projective representation associated to $\mathcal{Y}$ satisfying conditions (a), (b) and (c) of Theorem (8.14). In particular, if α is the factor set of $\mathcal{X}$, we have that $\alpha(g, n) = \alpha(n, g) = 1$ for $n \in N$ and $g \in G$. Furthermore, if a is any integer with $g^a \in N$, then

$$\mathcal{X}(g^a) = \mathcal{Y}(g^a) = \mathcal{Y}_{\bar{g}}(g^a) = \mathcal{Y}_{\bar{g}}(g)^a = \mathcal{X}(g)^a.$$

In particular, $\mathcal{X}(g)^{|G|} = I$ and thus $\det(\mathcal{X}(g))^{|G|} = 1$ for all $g \in G$. Since

$$\mathcal{X}(g)\mathcal{X}(h) = \alpha(g, h)\mathcal{X}(gh),$$

we deduce that

$$\alpha(g, h)^{|G|\theta(1)} = 1$$

for all $g, h \in G$. Thus,

$$\alpha(g, h)^{(|G|\theta(1))_{p'}} = 1.$$

Now, let E be the group of $(|G|\theta(1))_{p'}$th roots of unity in $\mathbb{C}$. Since $* : \mathbf{U} \to F^\times$ is an isomorphism of multiplicative groups, it follows that there is a unique $\gamma \in \mathbf{Z}^2(G, \mathbf{U})$ such that $\gamma(g, h)^* = \alpha(g, h)$ for all $g, h \in G$. Let $\tilde{G} = \{(g, \epsilon) \mid g \in G, \epsilon \in E\}$ with multiplication

$$(g_1, \epsilon_1)(g_2, \epsilon_2) = (g_1 g_2, \gamma(g_1, g_2)\epsilon_1\epsilon_2).$$

The fact that γ is in $\mathbf{Z}^2(G, \mathbf{U})$ makes this multiplication associative. Since $\gamma(g, 1) = \gamma(1, g) = \gamma(1, 1) = 1$ and $\gamma(g, g^{-1}) = \gamma(g^{-1}, g)$ for $g \in G$ (see the comments after the proof of Theorem (8.17)), it easily follows that $(1, 1)$ is the identity of $\tilde{G}$ and that $(g^{-1}, \gamma(g, g^{-1})^{-1}\epsilon^{-1})$ is the inverse of (g, ϵ) for $g \in G$ and $\epsilon \in E$. Hence, we have that $\tilde{G}$ is a finite group.

Let $\tilde{N} = \{(n, \epsilon) \mid n \in N, \epsilon \in E\}$. Note that

$$(n_1, \epsilon_1)(n_2, \epsilon_2) = (n_1 n_2, \epsilon_1 \epsilon_2)$$

since $\alpha(n_1, n_2) = 1$ for $n_1, n_2 \in N$. Hence, $\tilde{N} = N \times E$ is a subgroup of $\tilde{G}$. To make the notation easier, we identify E with $1 \times E$ and N with $N \times 1$ whenever we find it necessary. If $n \in N$, $g \in G$ and $\delta \in E$, then, by using the fact that γ is constant on cosets modulo N (because α is), we easily check that

$$(g, \epsilon)(n, \delta)(g, \epsilon)^{-1} = (gng^{-1}, \delta).$$

We see that E is a p'-subgroup of $\mathbf{Z}(\tilde{G})$ and that N and $\tilde{N}$ are normal subgroups of $\tilde{G}$.

We define

$$\tilde{\mathcal{X}}(g, \epsilon) = \epsilon^* \mathcal{X}(g).$$

It is clear that $\tilde{\mathcal{X}}$ is an F-representation of $\tilde{G}$. Suppose that $\tilde{\mathcal{X}}$ affords the Brauer character τ. For $n \in N$, note that

$$\tilde{\mathcal{X}}(n, \epsilon) = \epsilon^* \mathcal{Y}(n).$$

Therefore, if $\{\mu_1, \dots, \mu_{\theta(1)}\}$ are the eigenvalues of $\mathcal{Y}(n)$, we have that

$$\{\epsilon^* \mu_1, \dots, \epsilon^* \mu_{\theta(1)}\}$$

are the eigenvalues of $\tilde{\mathcal{X}}(n, \epsilon)$. Thus, for $n \in N^0$ we have that

$$\tau(n, \epsilon) = \epsilon \theta(n).$$

Now, define $\tilde{\theta} = \theta \times 1_{E^0} \in \mathrm{IBr}(\tilde{N})$ and observe that $\tilde{\theta}_N = \theta = \tau_N \in \mathrm{IBr}(N)$. In particular, $\tau \in \mathrm{IBr}(\tilde{G})$. Also, note that $\tilde{\theta}$ is $\tilde{G}$-invariant because θ is G-invariant.

We have that the map $\tilde{G} \to G$ given by $(g, \epsilon) \mapsto g$ is an onto homomorphism with kernel E. Since $E \subseteq \ker(\tilde{\theta})$, we have that $(\tilde{G}, \tilde{N}, \tilde{\theta})$ and (G, N, θ) are isomorphic modular character triples by the remark preceding the statement of Lemma (8.26).

Now, define $\lambda \in \mathrm{Irr}(\tilde{N})$ by $\lambda(n, \epsilon) = \bar{\epsilon}$, so that λ is a linear character of $\tilde{N}$ with $\ker(\lambda) = N$. Also, $\lambda^0 \in \mathrm{IBr}(\tilde{N})$ and $\ker(\lambda^0) = N$ (since E is a p'-group). Observe that λ, and therefore λ^0, are $\tilde{G}$-invariant. Also,

$$\tau_{\tilde{N}} = \bar{\lambda}^0 \tilde{\theta}.$$

By the remark after the proof of Lemma (8.26), we have that the triples $(\tilde{G}, \tilde{N}, \lambda^0)$ and $(\tilde{G}, \tilde{N}, \tilde{\theta})$ are isomorphic. Also, the triples $(\tilde{G}, \tilde{N}, \lambda^0)$ and $(\tilde{G}/N, \tilde{N}/N, \lambda^0)$ are isomorphic by the remark preceding the statement of Lemma (8.26). Now, write $\Gamma = \tilde{G}/N$, $M = \tilde{N}/N$ and $\varphi = \lambda^0$, and note that φ is faithful. Also, observe that M is a central p'-subgroup of Γ (although this also follows from the fact that φ is linear and faithful). ∎

Next, we can give two interesting results which easily follow by using character triple isomorphisms.

(8.29) THEOREM. *Suppose that $N \lhd G$ and let $\theta \in \mathrm{IBr}(N)$ be G-invariant. For every prime q dividing $|G : N|$, let Q/N be a Sylow q-subgroup of G/N. Then θ extends to G if and only if θ extends to Q for every q.*

Theorem (8.29) appears for ordinary characters as Corollary (11.36) of [Isaacs, *Characters*]. The strategy for proving it there was to reduce to the case where the character θ is linear by using character triple isomorphisms. Afterwards, the case left is elementary and appears as Theorem (6.26) of [Isaacs, *Characters*]. We will use this latter result in our proof of Theorem (8.29).

Proof of Theorem (8.29). We know that if (σ, τ) is a modular character triple isomorphism $(G, N, \theta) \to (\Gamma, M, \varphi)$ and $N \subseteq H \subseteq G$, then

$$\frac{\chi(1)}{\theta(1)} = \frac{\sigma_H(\chi(1))}{\varphi(1)}$$

for $\chi \in \mathrm{IBr}(H \,|\, \theta)$. Hence, note that θ extends to H if and only if φ extends to H^τ. This, together with Theorem (8.28), shows that we may assume that θ is linear. In this case, let $\hat{\theta} : N \to \mathbb{C}$ be defined by $\hat{\theta}(n) = \theta(n_{p'})$. Note that $\hat{\theta}$ is the unique linear character μ of N with $\mathbf{O}^{p'}(N)$ in its kernel such that $\mu^0 = \theta$. Call $\hat{\theta}$ the canonical lift of θ. By definition, we see that $\hat{\theta}$ is G-invariant. Now, if Q/N is a Sylow q-subgroup of G/N, it follows by hypothesis that θ extends to Q. Let δ be some extension of θ to Q and let $\hat{\delta}$ be the canonical lift of δ. Note that $(\hat{\delta})^0 = (\hat{\delta}^0)_N = \delta_N = \theta$. Since $\mathbf{O}^{p'}(G) \subseteq \ker(\hat{\delta})$, we have that $\mathbf{O}^{p'}(N) \subseteq \mathbf{O}^{p'}(G) \cap N \subseteq \ker(\hat{\delta}_N)$. Therefore, $\hat{\delta}_N$ is the canonical lift of θ and by uniqueness, $\hat{\delta}_N = \hat{\theta}$. Hence, by Theorem (6.26) of [Isaacs, *Characters*], it follows that $\hat{\theta}$ extends to G. If $\chi \in \mathrm{Irr}(G)$ is such an extension, then χ^0 is an extension of θ to G. This proves the theorem. ∎

The next result generalizes Swan's theorem (8.22).

(8.30) THEOREM (Dade). *Suppose that $N \lhd G$ and that G/N is p-solvable. Let $\varphi \in \mathrm{IBr}(G)$ and let $\theta \in \mathrm{IBr}(N)$ be an irreducible constituent of φ_N. Then $\frac{\varphi(1)}{\theta(1)}$ divides $|G : N|$.*

Proof. Arguing as in Theorem (8.22), we may assume that θ is G-invariant and that G/N is simple. Since G/N is p-solvable, we have that G/N is a p-group or that G/N is a p'-group. If G/N is a p-group, by Green's theorem we have that $\varphi_N = \theta$ and the result is clear. Suppose that G/N is a p'-group. Using Theorem (8.28), we may assume that N is a central p'-subgroup of G. Therefore, G is a p'-group. Thus, $\mathrm{IBr}(G) = \mathrm{Irr}(G)$ by Theorem (2.12). Now, by Ito's theorem (6.15) of [Isaacs, *Characters*], we have that $\varphi(1)$ divides $|G : N|$, as desired. ∎

To prove Passman's version of Dade's theorem, we need a preliminary result of independent interest.

(8.31) LEMMA. *Let* $C = (c_{ij}) \in \mathrm{Mat}(n, \mathbb{Z})$ *and let* q *be a prime not dividing* $\det(C)$. *Suppose that* Q *is a* q-group acting on the set $I = \{1, \ldots, n\}$. *Assume that for every* $x \in Q$ *we have*

$$c_{ij} = c_{i \cdot x, j \cdot x}$$

for all $i, j \in I$. *If* Q *fixes* $a \in I$, *then there exists* $b \in I$ *fixed by* Q *such that* q *does not divide* c_{ab}.

Proof. We may certainly view Q as a subgroup of $S = \mathrm{Sym}(n)$. (In other words, we may assume that Q acts faithfully on I.) For each $\pi \in S$, write

$$e(\pi) = \prod_{i=1}^{n} c_{i, i \cdot \pi} \, .$$

Thus,

$$\det(C) = \sum_{\pi \in S} \mathrm{sgn}(\pi) e(\pi) \, .$$

For $x \in Q$, we have that

$$e(\pi) = \prod_{i=1}^{n} c_{i \cdot x^{-1}, i \cdot x^{-1} \cdot \pi} = \prod_{i=1}^{n} c_{i, i \cdot x^{-1} \pi x} = e(\pi^x) \, .$$

Also, $\mathrm{sgn}(\pi) = \mathrm{sgn}(\pi^x)$. We see that the functions e and sgn are constant on the orbits Δ of the conjugation action of Q on S. If $\pi \in \Delta$, where Δ is any Q-orbit, write $\mathrm{sgn}(\Delta) = \mathrm{sgn}(\pi)$ and $e(\Delta) = e(\pi)$. Hence, we have that

$$\det(C) = \sum_{\Delta} |\Delta| \mathrm{sgn}(\Delta) e(\Delta)$$

is not divisible by q. Since Q is a q-group, there must exist an orbit $\Delta = \{\pi\}$ such that q does not divide $e(\pi)$. Now, since Q fixes a, it follows that Q fixes $b = a \cdot \pi$. Also, since c_{ab} divides $e(\pi)$, we have that q does not divide c_{ab}, as required. ∎

(8.32) THEOREM (Passman). *Let* $q \neq p$ *and suppose that* Q *is a* q-group acting on $\mathrm{Irr}(G)$ *and* $\mathrm{IBr}(G)$ *such that* $d_{\chi\varphi} = d_{\chi^x \varphi^x}$ *for all* $\chi \in \mathrm{Irr}(G)$, $\varphi \in \mathrm{IBr}(G)$ *and* $x \in Q$. *Suppose that* Q *fixes* $\alpha \in \mathrm{IBr}(G)$. *Then there exists a* Q-fixed $\psi \in \mathrm{Irr}(G)$ *such that* q *does not divide* $d_{\psi\alpha}$.

Proof. Let C be the Cartan matrix of G. If $\varphi, \theta \in \mathrm{IBr}(G)$, we know that

$$c_{\varphi\theta} = \sum_{\chi \in \mathrm{Irr}(G)} d_{\chi\varphi} d_{\chi\theta}.$$

Since $d_{\chi\varphi^x} = d_{\chi^{x-1}\varphi}$ and $d_{\chi\theta^x} = d_{\chi^{x-1}\theta}$, it easily follows that $c_{\varphi\theta} = c_{\varphi^x\theta^x}$ for $x \in Q$. By Corollary (2.18), we have that $\det(C)$ is a power of p and thus is not divisible by q. By Lemma (8.31), we deduce that there exists a Q-fixed $\beta \in \mathrm{IBr}(G)$ such that q does not divide $c_{\alpha\beta}$. Since the decomposition numbers $d_{\psi\alpha}$ and $d_{\psi\beta}$ are constant as ψ runs over a Q-orbit Δ of $\mathrm{Irr}(G)$, we may write

$$c_{\alpha\beta} = \sum_{\Delta} d_{\Delta\alpha} d_{\Delta\beta}.$$

Since $c_{\alpha\beta}$ is not divisible by q, we deduce that there exists a singleton orbit $\{\psi\}$ such that q does not divide $d_{\psi\alpha}$, as desired. ∎

Alternative Proof of Dade's Theorem (Passman). Suppose that $N \lhd G$ is a Hall subgroup of G, and let $\theta \in \mathrm{IBr}(N)$ be G-invariant. We wish to show that θ extends to G. By Theorem (8.29), we may assume that G/N is a q-group for some prime q. If $q = p$, then the result follows by Green's theorem. So we may assume that $q \neq p$.

Let Q be a Sylow q-subgroup of G. By Theorem (8.32), let $\psi \in \mathrm{Irr}(N)$ be Q-invariant such that $d_{\psi\theta}$ is not divisible by q. By Gallagher's extension theorem (Corollary (8.16) of [Isaacs, *Characters*]), let $\chi \in \mathrm{Irr}(G)$ be such that $\chi_N = \psi$. Now,

$$\psi^0 = (\chi_N)^0 = (\chi^0)_N = \sum_{\varphi \in \mathrm{IBr}(G)} d_{\chi\varphi} \varphi_N$$

$$= \sum_{\varphi \in \mathrm{IBr}(G)} \sum_{\eta \in \mathrm{IBr}(N)} d_{\chi\varphi} \sum_{\eta \in \mathrm{IBr}(N)} \mathbf{I}(\varphi_N, \eta) \eta$$

$$= \sum_{\eta \in \mathrm{IBr}(N)} \left(\sum_{\varphi \in \mathrm{IBr}(G)} d_{\chi\varphi} \mathbf{I}(\varphi_N, \eta) \right) \eta.$$

We deduce that

$$d_{\psi\theta} = \sum_{\varphi \in \mathrm{IBr}(G)} d_{\chi\varphi} \mathbf{I}(\varphi_N, \theta).$$

Therefore, there exists a $\varphi \in \mathrm{IBr}(G)$ such that q does not divide $\mathbf{I}(\varphi_N, \theta) = \frac{\varphi(1)}{\theta(1)}$. By Swan's theorem (8.22), we conclude that $\varphi_N = \theta$, as required. ∎

We now use the techniques that we have developed to prove another kind of result which will conclude this chapter.

(8.33) THEOREM. *Suppose that G and H are finite groups and let q be any prime. Assume that FG and FH are isomorphic as F-algebras. Then G has a normal q-complement if and only if H has a normal q-complement.*

We need another result.

(8.34) THEOREM. *Suppose that q is a prime different from p and let $\mathcal{U} = \{\varphi \in \mathrm{IBr}(G) \mid q$ does not divide $\varphi(1)\}$. Then $|G : G'|_q$ divides $\sum_{\varphi \in \mathcal{U}} \Phi_\varphi(1)\varphi(1)$ and G has a normal q-complement if and only if q does not divide*

$$\frac{\sum_{\varphi \in \mathcal{U}} \Phi_\varphi(1)\varphi(1)}{|G : G'|_q} .$$

Proof. Let $N = \mathbf{O}^q(G)$ and write $\mathcal{V} = \{\eta \in \mathrm{IBr}(N) \mid q$ does not divide $\eta(1)$ and η is G-invariant$\}$. If $\eta \in \mathcal{V}$, note that $(o(\eta)\eta(1), q) = 1$ because $o(\eta)$ divides $|N : N'|$ and $\mathbf{O}^q(N) = N$. Therefore, by Theorem (8.23), it follows that there is a $\varphi \in \mathcal{U}$ such that $\varphi_N = \eta$. Furthermore, by Corollary (8.20), all the extensions are of the form $\lambda\varphi$, where $\lambda \in \mathrm{IBr}(G/N)$ is linear. Since p does not divide $|G/N|$, it follows by Theorem (2.12) that there are exactly $|G/N : (G/N)'| = |G : G'|_q$ extensions, all of which lie in $\mathcal{U}$. Conversely, if $\varphi \in \mathcal{U}$, then it follows from Swan's theorem that $\varphi_N \in \mathcal{V}$.

Suppose now that $\varphi \in \mathcal{U}$. By Corollary (8.8), we may write $(\Phi_\varphi)_N = e\Phi_{\varphi_N}$, where $e = \mathbf{I}((\varphi_N)^G, \varphi)$. Since $p \neq q$, this number is $\mathbf{I}(\varphi_N, \varphi_N) = 1$ by the last part of Corollary (8.7). Thus, $(\Phi_\varphi)_N = \Phi_{\varphi_N}$ for $\varphi \in \mathcal{U}$. If we write $\mathcal{U}_\Phi = \{\Phi_\varphi \mid \varphi \in \mathcal{U}\}$ and $\mathcal{V}_\Phi = \{\Phi_\eta \mid \eta \in \mathcal{V}\}$, we see that the map $\mathcal{U}_\Phi \to \mathcal{V}_\Phi$ given by $\Phi_\varphi \mapsto (\Phi_\varphi)_N$ is well defined and surjective. Note that if $\lambda \in \mathrm{IBr}(G)$ is linear and $\varphi \in \mathrm{IBr}(G)$, then

$$\lambda\Phi_\varphi = \Phi_{\lambda\varphi}$$

by Problem (2.13). Also, if $\varphi, \tau \in \mathcal{U}$ are such that $(\Phi_\varphi)_N = (\Phi_\tau)_N$, then $\Phi_{\varphi_N} = \Phi_{\tau_N}$. Hence, $\varphi_N = \tau_N$ and, by Corollary (8.20), we deduce that $\varphi = \lambda\tau$ for some linear $\lambda \in \mathrm{IBr}(G/N)$. Thus, $\Phi_\varphi = \lambda\Phi_\tau$.

By the above discussion, it is clear that

$$\sum_{\varphi \in \mathcal{U}} \Phi_\varphi(1)\varphi(1) = |G : G'|_q \sum_{\eta \in \mathcal{V}} \Phi_\eta(1)\eta(1) .$$

Now, since

$$|N| = \sum_{\eta \in \mathrm{IBr}(N)} \Phi_\eta(1)\eta(1)$$

(by Problem (2.9)) and $(\Phi_\eta)^g = \Phi_{\eta^g}$ for $g \in G$ (Problem (2.1)), we conclude that

$$|N| = \sum_{\eta \in \mathrm{IBr}(N)} \Phi_\eta(1)\eta(1) \equiv \sum_{\eta \in \mathcal{V}} \Phi_\eta(1)\eta(1) \bmod q .$$

Now, G has a normal q-complement if and only if q does not divide $|N|$, which happens if and only if q does not divide

$$\sum_{\eta \in \mathcal{V}} \Phi_\eta(1)\eta(1),$$

which happens if and only if q does not divide

$$\frac{\sum_{\varphi \in \mathcal{U}} \Phi_\varphi(1)\varphi(1)}{|G : G'|_q},$$

as required. ∎

Proof of Theorem (8.33). Suppose that $\alpha : FG \to FH$ is an isomorphism of F-algebras. If $0 < X_1 < \ldots < X_{n-1} < X_n = FG$ is a composition series of FG as a right FG-module, it follows that $0 < \alpha(X_1) < \ldots < \alpha(X_{n-1}) < \alpha(X_n) = FH$ is a composition series of FH as a right FH-module. Also, note that $X_{i+1}/X_i \cong X_{j+1}/X_j$ as FG-modules if and only if $\alpha(X_{i+1})/\alpha(X_i) \cong \alpha(X_{j+1})/\alpha(X_j)$ as FH-modules. Now, observe that if X_{i+1}/X_i affords the irreducible Brauer character φ of G, then $\Phi_\varphi(1)$ is the number of composition factors X_{j+1}/X_j isomorphic to X_{i+1}/X_i because FG affords the Brauer character $(\rho_G)^0 = \sum_{\varphi \in \mathrm{IBr}(G)} \Phi_\varphi(1)\varphi$ (by Problem (2.9)). Therefore, we may conclude that there is a bijection $\tau : \mathrm{IBr}(G) \to \mathrm{IBr}(H)$ such that $\tau(\varphi)(1) = \varphi(1)$ and $\Phi_\varphi(1) = \Phi_{\tau(\varphi)}(1)$ for all $\varphi \in \mathrm{IBr}(G)$.

Now, since $|G : G'|_{p'}$ is the number of linear Brauer characters of G (by Problem (2.7)), it follows that

$$|G : G'|_{p'} = |H : H'|_{p'}.$$

If $q \neq p$, then the theorem follows from the above discussion and Theorem (8.34).

Suppose now that $p = q$. If $\{e_{B_1}, \ldots, e_{B_s}\}$ is the set of all primitive idempotents of $\mathbf{Z}(FG)$, then it is clear that $\{\alpha(e_{B_1}), \ldots, \alpha(e_{B_s})\}$ is the set of all primitive idempotents of $\mathbf{Z}(FH)$ because α isomorphically maps $\mathbf{Z}(FG)$ onto $\mathbf{Z}(FH)$. Since X_{i+1}/X_i is the trivial FG-module if and only if $\alpha(X_{i+1})/\alpha(X_i)$ is the trivial FH-module and $X_{i+1}e_{B_j} \subseteq X_i$ if and only if $\alpha(X_{i+1})\alpha(e_{B_j}) \subseteq \alpha(X_i)$, by Lemma (3.13) we see that the bijection τ in the first paragraph of the proof maps $\mathrm{IBr}(B_0(G))$ onto $\mathrm{IBr}(B_0(H))$, where $B_0(G)$ and $B_0(H)$ are the principal blocks of G and H, respectively. Since G has a normal p-complement if and only if $|\mathrm{IBr}(B_0(G))| = 1$ by Corollary (6.13), the proof of the theorem is complete. ∎

PROBLEMS

(8.1) Prove that $\mathbf{I}(U_1 \oplus U_2, V) = \mathbf{I}(U_1, V) + \mathbf{I}(U_2, V)$ and $\mathbf{I}(U, V_1 \oplus V_2) = \mathbf{I}(U, V_1) + \mathbf{I}(U, V_2)$ for FG-modules U_1, U_2, U, V, V_1 and V_2. Deduce that if U and V are completely reducible FG-modules affording the Brauer characters ψ and τ, respectively, then

$$\mathbf{I}(U, V) = \mathbf{I}(\psi, \tau).$$

(8.2) (Nakayama) Let $H \subseteq G$. Suppose that W is an FH-module and V is an FG-module. Show

$$\dim_F(\operatorname{Hom}_{FG}(V, W^G)) = \dim_F(\operatorname{Hom}_{FH}(V_H, W)).$$

(8.3) Suppose that $N \triangleleft G$ and let $\theta \in \mathrm{IBr}(N)$. If $\theta^G \in \mathrm{IBr}(G)$, show $I_G(\theta) = N$.

(8.4) If φ is Brauer character of G and H is a subgroup of G, prove that $(\varphi_H)^G = \varphi + \delta$, where δ is a Brauer character of G or zero.

(8.5) (Mackey) Let H, K be subgroups of G and let T be a set of double coset representatives so that

$$G = \bigcup_{t \in T} HtK$$

is a disjoint union. Suppose that α is a Brauer character of H. If $g \in G$, we denote by α^g the Brauer character of H^g defined by $\alpha^g(h^g) = \alpha(h)$ for $h \in H$ (see Problem (2.1)). Prove that

$$(\alpha^G)_K = \sum_{t \in T} (\alpha^t_{H^t \cap K})^K.$$

If $G = HK$, deduce that

$$(\alpha^G)_K = (\alpha_{H \cap K})^K.$$

If $(\alpha^G)_K$ is irreducible, deduce that $G = HK$.

(8.6) Let $N \triangleleft G$ and let $\theta \in \mathrm{IBr}(N)$ be G-invariant. Show

$$\sum_{\mu \in \mathrm{IBr}(G \mid \theta)} \frac{\Phi_\mu(1)}{\Phi_\theta(1)} \frac{\mu(1)}{\theta(1)} = |G/N|.$$

If G/N is a p'-group, also prove that

$$\frac{\Phi_\mu(1)}{\Phi_\theta(1)} = \frac{\mu(1)}{\theta(1)}.$$

(8.7) If H is a subgroup of G and $\varphi \in \mathrm{IBr}(G)$, prove that $(\Phi_\varphi)_H$ is a nonnegative integer linear combination of $\{\Phi_\mu \mid \mu \in \mathrm{IBr}(H)\}$.

(8.8) Let $\mathcal{X}$ and $\mathcal{Y}$ be $\mathbb{C}$-similar irreducible representations of G with entries in S. Suppose that $\mathcal{X}^*$ is an irreducible F-representation of G. Prove that there exists a matrix $M \in \mathrm{GL}(n, S)$ such that $\mathcal{Y}M = M\mathcal{X}$.

(HINT: By using Corollary (2.9), first show that $\mathcal{Y}^*$ is also irreducible. Now, let P be a regular matrix in $\mathbf{K}$ such that $\mathcal{Y}P = P\mathcal{X}$. By using Lemma (2.5), show that we may choose P so that its entries are in $\mathbf{R}$ and $P^* \neq 0$. Use Schur's lemma to conclude that P^* is invertible. In this case, P^{-1} has its entries in S.)

(8.9) Suppose that $N \triangleleft G$ and let $\mathcal{Y}$ be a representation of N with entries in S such that $\mathcal{Y}^*$ is irreducible. If $\mathcal{Y}$ extends to G, prove that there exists an S-representation $\mathcal{X}$ of G extending $\mathcal{Y}$.

(HINT: Using Theorem (2.7), let $\mathcal{X}$ be a representation of G with entries in S such that $\mathcal{X}_N$ and $\mathcal{Y}$ are $\mathbb{C}$-similar. Then use Problem (8.8).)

(8.10) Suppose that $N \triangleleft G$ and let $\theta \in \mathrm{Irr}(N)$ be G-invariant. If $\theta^0 \in \mathrm{IBr}(N)$, we say that (G, N, θ) is an **ordinary-modular character triple**. Also, we say that two ordinary-modular character triples (G, N, θ) and (Γ, M, φ) are **isomorphic** if there exist an ordinary character triple isomorphism

$$(\sigma, \tau) : (G, N, \theta) \to (\Gamma, M, \varphi)$$

and a modular character triple isomorphism

$$(\mu, \tau) : (G, N, \theta^0) \to (\Gamma, M, \varphi^0)$$

such that for every $N \subseteq H \subseteq G$ and $\chi \in \mathrm{Ch}(H \mid \theta)$ we have that

$$\sigma_H(\chi)^0 = \mu_H(\chi^0).$$

We will say in this case that (σ, μ, τ) is an **isomorphism** $(G, N, \theta) \to (\Gamma, M, \varphi)$.

Now, suppose that (G, N, θ) is an ordinary-modular character triple and let $\alpha : G \to \Gamma$ be an onto homomorphism with $\ker(\alpha) \subseteq \ker(\theta)$. Let $M = \alpha(N)$ and let $\varphi \in \mathrm{Irr}(M)$ be the character corresponding to $\theta \in \mathrm{Irr}(N/\ker(\alpha))$. Prove that (G, N, θ) and (Γ, M, φ) are isomorphic as ordinary-modular character triples.

(8.11) Let (G, N, φ) be an ordinary-modular character triple. Let $\eta \in \mathrm{Irr}(G)$ be such that $(\eta_N \varphi)^0$ is irreducible. For $N \subseteq H \subseteq G$, define

$$\sigma_H : \mathrm{Ch}(H \mid \varphi) \to \mathrm{Ch}(H \mid \eta_N \varphi)$$

by $\sigma_H(\psi) = \psi\eta_H$. Also, define

$$\mu_H : \mathrm{Br}(H \,|\, \varphi^0) \to \mathrm{Br}(H \,|\, (\eta_N\varphi)^0)$$

by $\mu_H(\beta) = \beta(\eta^0)_H$. Let $i : G/N \to G/N$ be the identity map. Prove that

$$(\sigma, \mu, i) : (G, N, \varphi) \to (G, N, \eta_N\varphi)$$

is an isomorphism of ordinary-modular character triples.

(8.12) Suppose that (G, N, θ) is an ordinary-modular character triple. Prove that there exists an isomorphic ordinary-modular character triple (Γ, M, φ) where φ is linear and faithful.

(HINT: Mimic the proof of Theorem (8.28) using some of the previous problems.)

(8.13) Suppose that (G, N, θ) is an ordinary-modular character triple. Suppose that θ has p-defect zero. Prove that there exists an isomorphic ordinary-modular character triple (Γ, M, φ) where φ is linear, faithful and M is a p'-group.

(HINT: Use Problem (8.12) to find an ordinary-modular character triple (Γ, M, λ) isomorphic to (G, N, θ) with M central. Write $\lambda = \lambda_p\lambda_{p'}$, where λ_p is the p-part of λ (in the group $\mathrm{Irr}(M)$, for instance). Use Problem (3.10) to show that λ_p extends to Γ.)

(8.14) Let G be a group with a p-complement. Let q be a prime different from p. If q divides the degree of every nonlinear irreducible Brauer character of G, then G has a normal q-complement. Find an example where this is not true for $q = p$.

(HINT: Use Theorem (8.34) and Problem (2.8).)

(NOTE: If p does not divide $|G|$, this is a well-known result of J. Thompson. See Corollary (12.2) of [Isaacs, *Characters*].)

9. Blocks and Normal Subgroups

If G is a finite group, then $\text{Aut}(G)$ acts on the set of class functions defined on G (or G^0) by

$$\psi^\sigma(g) = \psi(g^{\sigma^{-1}}),$$

where $\sigma \in \text{Aut}(G)$, $\psi \in \text{cf}(G)$ (or $\text{cf}(G^0)$) and $g \in G$. This action permutes characters (both ordinary and modular) and, certainly, preserves irreducibility. In fact, since $(\chi^\sigma)^0 = (\chi^0)^\sigma$ for every $\chi \in \text{cf}(G)$, we clearly have that

$$d_{\chi\varphi} = d_{\chi^\sigma \varphi^\sigma}$$

for $\chi \in \text{Irr}(G)$ and $\varphi \in \text{IBr}(G)$ (Problem (2.1)). Consequently, if $\chi, \psi \in \text{Irr}(G)$, then χ and ψ are linked if and only if χ^σ and ψ^σ are linked. Furthermore, since $\varphi \in \text{IBr}(G)$ lies in a block B if and only if there is a $\chi \in \text{Irr}(B)$ such that $d_{\chi\varphi} \neq 0$, it easily follows that for every block B of G, $B^\sigma = \{\chi^\sigma \mid \chi \in B\}$ is another block of G with $\text{Irr}(B^\sigma) = \{\chi^\sigma \mid \chi \in \text{Irr}(B)\}$ and $\text{IBr}(B^\sigma) = \{\varphi^\sigma \mid \varphi \in \text{IBr}(B)\}$. We see, then, that $\text{Aut}(G)$ permutes the set $\text{Bl}(G)$. Furthermore, if D is a defect group of the block $B \in \text{Bl}(G)$, it is easy to check that D^σ is a defect group of B^σ (Problem (4.4)).

If $\sigma \in \text{Aut}(G)$, then σ also induces an automorphism of the R-algebra RG, where R is a ring, by setting

$$\left(\sum_{g \in G} a_g g\right)^\sigma = \sum_{g \in G} a_g g^\sigma.$$

Clearly,

$$\sigma(\mathbf{Z}(RG)) = \mathbf{Z}(RG).$$

It is straightforward to check that $(e_\chi)^\sigma = e_{\chi^\sigma}$ by using the formula for the idempotent e_χ (page 51). Therefore, $(f_B)^\sigma = f_{B^\sigma}$ and $(e_B)^\sigma = e_{B^\sigma}$. Furthermore, if we view σ as an automorphism of the F-algebra $\mathbf{Z}(FG)$, note that the composition of the algebra homomorphisms σ^{-1} and λ_B is λ_{B^σ} since

$$(\sigma^{-1} \cdot \lambda_B)(e_{B^\sigma}) = \lambda_B((e_{B^\sigma})^{\sigma^{-1}}) = \lambda_B(e_B) = 1.$$

If $N \triangleleft G$, the above discussion shows us that G acts on $\mathrm{Bl}(N)$ by conjugation. If $b \in \mathrm{Bl}(N)$ and $g \in G$, then b^g denotes the block b^{α_g}, where α_g is the automorphism of N that is conjugation by g. Hence, $\mathrm{Irr}(b^g) = \{\psi^g \,|\, \psi \in \mathrm{Irr}(b)\}$ and $\mathrm{IBr}(b^g) = \{\theta^g \,|\, \theta \in \mathrm{IBr}(b)\}$. We denote the **stabilizer** of b in G by $T(b) = \{g \in G \,|\, b^g = b\}$. If $\{b_1, \ldots, b_t\}$ is the G-orbit of b, we have that the idempotent $\sum_{i=1}^{t} f_{b_i}$ lies in $\mathbf{Z}(SG)$ (because the set $\{b_1, \ldots, b_t\}$ is G-invariant and each f_{b_i} lies in SN). Recall that $\{e_\chi \,|\, \chi \in \mathrm{Irr}(G)\}$ is the set of the primitive idempotents of $\mathbf{Z}(\mathbb{C}G)$ (see Theorem (2.12) of [Isaacs, *Characters*]) and Theorem (1.17.c)). Now, by Theorem (1.17.c) applied to $\mathbb{C}G$, the central idempotent $\sum_{i=1}^{t} f_{b_i} \in \mathbf{Z}(\mathbb{C}G)$ is a sum of e_χ for some $\chi \in \mathrm{Irr}(G)$. Hence, by Theorem (3.9), there exist uniquely determined blocks $B_1, \ldots, B_s \in \mathrm{Bl}(G)$ such that

$$\sum_{i=1}^{t} f_{b_i} = \sum_{i=1}^{s} f_{B_i} \, .$$

We say in this case that the block B_i **covers** b. We write $\{B_1, \ldots, B_s\} = \mathrm{Bl}(G \,|\, b)$ for the set of blocks of G which cover b.

Note that we also could have used the fact that $\sum_{i=1}^{t} e_{b_i}$ is an idempotent in $\mathbf{Z}(FG)$ to define the coverings of blocks (by applying Theorem (3.11)). However, if we write

$$\sum_{i=1}^{t} e_{b_i} = \sum_{i=1}^{s'} e_{B_i'}$$

for blocks $\{B_1', \ldots, B_{s'}'\}$ of G, then

$$\sum_{i=1}^{s'} e_{B_i'} = \sum_{i=1}^{t} e_{b_i} = \left(\sum_{i=1}^{t} f_{b_i}\right)^* = \left(\sum_{i=1}^{s} f_{B_i}\right)^* = \sum_{i=1}^{s} e_{B_i} \, ,$$

and necessarily

$$\{B_1, \ldots, B_s\} = \{B_1', \ldots, B_{s'}'\} \, .$$

So this yields the same notation of covering.

If R is a ring and $N \triangleleft G$, notice that the R-algebra $\mathbf{Z}(RG) \cap \mathbf{Z}(RN)$ is the R-span of $\{\hat{K} \,|\, K \in \mathrm{cl}(G) \text{ with } K \subseteq N\}$.

The first thing that we do is to characterize coverings of blocks in terms of characters. We need to prove the following.

(9.1) THEOREM. *Suppose that $N \triangleleft G$ and let $\chi \in \mathrm{Irr}(G)$ and $\theta \in \mathrm{Irr}(N)$. Then θ is an irreducible constituent of χ_N if and only if $\omega_\chi(\hat{K}) = \omega_\theta(\hat{K})$ for every $K \in \mathrm{cl}(G)$ contained in N.*

Proof. Let $\psi \in \mathrm{Irr}(N)$. First, we prove that $\omega_\theta(\hat{K}) = \omega_\psi(\hat{K})$ for every $K \in \mathrm{cl}(G)$ contained in N if and only if θ and ψ are G-conjugate. Suppose that $\psi = \theta^g$ for some $g \in G$. Then, for every $K \in \mathrm{cl}(G)$ contained in N, we have that

$$\omega_\psi(\hat{K}) = \omega_\psi((\hat{K})^g) = \omega_{\theta^g}((\hat{K})^g) = \omega_\theta(\hat{K}).$$

Conversely, since

$$\sum_{g \in G}(e_\theta)^g = \sum_{g \in G} e_{\theta^g} \in \mathbf{Z}(\mathbb{C}G) \cap \mathbf{Z}(\mathbb{C}N),$$

it follows that

$$0 \neq \omega_\theta\Big(\sum_{g \in G} e_{\theta^g}\Big) = \omega_\psi\Big(\sum_{g \in G} e_{\theta^g}\Big).$$

Thus, ψ and θ are G-conjugate (because $\omega_\psi(e_\mu) = \delta_{\psi\mu}$ for $\mu \in \mathrm{Irr}(N)$).

Now, by Clifford's theorem, write $\chi_N = e \sum_{i=1}^{r} \xi_i$, where $\{\xi_1, \ldots, \xi_r\}$ is the set of the distinct G-conjugates of $\xi \in \mathrm{Irr}(N)$. If K is a conjugacy class of G inside N, we have that

$$\omega_\xi(\hat{K}) = \frac{\xi(\hat{K})}{\xi(1)} = \frac{\sum_{i=1}^{r} \xi_i(\hat{K})}{r\xi(1)} = \frac{e\sum_{i=1}^{r} \xi_i(\hat{K})}{er\xi(1)} = \frac{\chi(\hat{K})}{\chi(1)} = \omega_\chi(\hat{K}).$$

By using the first part of the proof, the theorem follows. ∎

(9.2) THEOREM. *Suppose that $N \triangleleft G$. Let $b \in \mathrm{Bl}(N)$ and let $B \in \mathrm{Bl}(G)$. The following conditions are equivalent.*

(a) *B covers b;*

(b) *if $\chi \in B$, then every irreducible constituent of χ_N lies in a G-conjugate of b; and*

(c) *there is a $\chi \in B$ such that χ_N has an irreducible constituent in b.*

Proof. Let $\{b_1, \ldots, b_t\}$ be the G-orbit of b and write

$$\sum_{i=1}^{t} f_{b_i} = \sum_{i=1}^{s} f_{B_i},$$

where $\{B_1, \ldots, B_s\}$ is the set of blocks of G covering b. First, we prove the theorem for ordinary characters.

Assume that B covers b and let $\chi \in \mathrm{Irr}(B)$. Let $\psi \in \mathrm{Irr}(N)$ be an irreducible constituent of χ_N. By Theorem (9.1), we know that ω_χ and ω_ψ coincide in $\mathbf{Z}(\mathbb{C}G) \cap \mathbf{Z}(\mathbb{C}N)$. Hence,

$$1 = \omega_\chi\Big(\sum_{i=1}^{s} f_{B_i}\Big) = \omega_\psi\Big(\sum_{i=1}^{t} f_{b_i}\Big).$$

Therefore, ψ lies in some b_i (which is a G-conjugate of b). This proves that (a) implies (b). Since $\psi \in \mathrm{Irr}(b^x)$ if and only if $\psi^{x^{-1}} \in \mathrm{Irr}(b)$ for $x \in G$, it is clear that (b) implies (c).

Now, we assume that (c) holds for ordinary characters and we prove (a). Let $\chi \in \mathrm{Irr}(B)$ be such that χ_N has an irreducible constituent $\theta \in \mathrm{Irr}(b)$. By Theorem (9.1),

$$1 = \omega_\theta(\sum_{i=1}^{t} f_{b_i}) = \omega_\chi(\sum_{i=1}^{s} f_{B_i}).$$

Thus, there is an i with $B = B_i$, and we are done.

Finally, we prove the theorem for irreducible Brauer characters. Let $\varphi \in \mathrm{IBr}(B)$ and let $\theta \in \mathrm{IBr}(N)$ be an irreducible constituent of φ_N. If $\chi \in \mathrm{Irr}(B)$ is such that $d_{\chi\varphi} \neq 0$, then, since $(\chi_N)^0 = (\chi^0)_N$, it follows that there exists an irreducible constituent η of χ_N such that $d_{\eta\theta} \neq 0$. Then η and θ lie in the same block. By using this argument and the theorem for ordinary characters, it is clear that the theorem holds for Brauer characters. ∎

(9.3) COROLLARY. *Suppose that $N \triangleleft G$ and let $B \in \mathrm{Bl}(G)$. If B covers $b_1, b_2 \in \mathrm{Bl}(N)$, then b_1 and b_2 are G-conjugate.*

Proof. Let $\chi \in \mathrm{Irr}(B)$ and let $\theta \in \mathrm{Irr}(N)$ be a constituent of χ_N. By Theorem (9.2), we have that θ lies in a G-conjugate of b_1 and b_2, and the result follows. ∎

We may also characterize coverings of blocks going from N to G.

(9.4) THEOREM. *Suppose that $N \triangleleft G$ and let $b \in \mathrm{Bl}(N)$ be covered by $B \in \mathrm{Bl}(G)$. Given $\theta \in b$, there is a $\chi \in B$ over θ.*

Proof. First, we prove the theorem for ordinary characters. Suppose that $\theta \in \mathrm{Irr}(b)$. Let τ be any character in $\mathrm{Irr}(B)$ and let $\eta \in \mathrm{Irr}(b)$ be under τ (by Theorem (9.2)). Since η and θ lie in the same block, they are connected. To prove the theorem for ordinary characters, note that it suffices to show that if η is linked to some $\psi \in \mathrm{Irr}(N)$, then there exists a $\xi \in \mathrm{Irr}(B)$ over ψ. Suppose that $\varphi \in \mathrm{IBr}(N)$ is such that $d_{\eta\varphi} \neq 0 \neq d_{\psi\varphi}$. Then $(\tau_N)^0 = (\tau^0)_N$ is a Brauer character of N containing φ. It follows that there is an irreducible Brauer constituent μ of τ^0 lying over φ. Now, since ψ^0 contains φ, we may write $(\psi^G)^0 = (\psi^0)^G = \varphi^G + \Xi$, where Ξ is a Brauer character of G, or zero. Therefore, since μ is an irreducible constituent of φ^G by Corollary (8.7), we have that there is an irreducible constituent ξ of ψ^G such that ξ^0 contains μ. Then $\xi \in \mathrm{Irr}(B)$ because $\mu \in \mathrm{IBr}(B)$, ξ lies over ψ, and we are done.

To prove the theorem for Brauer characters, given $\theta \in \mathrm{IBr}(b)$, we choose $\gamma \in \mathrm{Irr}(b)$ such that $d_{\gamma\theta} \neq 0$. Then there exists a $\chi \in \mathrm{Irr}(B)$ over γ. Hence, some irreducible constituent of χ^0 lies over θ. This completes the proof of the theorem. ∎

Our next objective is to prove a useful characterization due to D. Passman of coverings of blocks by means of algebra homomorphisms.

(9.5) THEOREM (Passman). *Suppose that $N \triangleleft G$ and let $B \in \mathrm{Bl}(G)$ and $b \in \mathrm{Bl}(N)$. Then B covers b if and only if $\lambda_B(\hat{K}) = \lambda_b(\hat{K})$ for every $K \in \mathrm{cl}(G)$ contained in N.*

Proof.   Let $\{b_1, \ldots, b_t\}$ be the G-orbit of b and write

$$\sum_{i=1}^{t} e_{b_i} = \sum_{i=1}^{s} e_{B_i},$$

where $\{B_1, \ldots, B_s\} = \mathrm{Bl}(G \,|\, b)$.

Suppose first that $\lambda_B(\hat{K}) = \lambda_b(\hat{K})$ for every $K \in \mathrm{cl}(G)$ contained in N. Since $\sum_{i=1}^{t} e_{b_i} \in \mathbf{Z}(FG) \cap \mathbf{Z}(FN)$ (which, as we said, is the F-span of those conjugacy classes of G inside N), it follows that

$$\lambda_B\Big(\sum_{i=1}^{s} e_{B_i}\Big) = \lambda_B\Big(\sum_{i=1}^{t} e_{b_i}\Big) = \lambda_b\Big(\sum_{i=1}^{t} e_{b_i}\Big) = 1.$$

Hence, there is an i such that $B = B_i$. Thus, B covers b.

Conversely, suppose that B covers b. By Theorem (9.2), let $\chi \in \mathrm{Irr}(B)$ and let $\theta \in \mathrm{Irr}(b)$ be an irreducible constituent of χ_N. By Theorem (9.1), we have that

$$\omega_\chi(\hat{K}) = \omega_\theta(\hat{K})$$

for every conjugacy class K of G inside N. Now, $\lambda_B(\hat{K}) = \omega_\chi(\hat{K})^*$, $\lambda_b(\hat{K}) = \omega_\theta(\hat{K})^*$ and the proof of the theorem is complete. ∎

There is an easy corollary.

(9.6) COROLLARY. *Suppose that $N \triangleleft G$ with G/N a p-group. If $b \in \mathrm{Bl}(N)$, then there is a unique $B \in \mathrm{Bl}(G)$ covering b.*

Proof.   Let B and B' be blocks of G covering b and let K be a conjugacy class of G of p-regular elements. Then $K \subseteq N$ by hypothesis. By Passman's theorem, we have that

$$\lambda_B(\hat{K}) = \lambda_b(\hat{K}) = \lambda_{B'}(\hat{K}).$$

Since the idempotent e_B lies in the F-span of $\{\hat{K} \mid K \in \mathrm{cl}(G^0)\}$ (see Corollary (3.8)), it follows that

$$1 = \lambda_B(e_B) = \sum_{K \in \mathrm{cl}(G^0)} a_B(\hat{K})\lambda_B(\hat{K}) = \sum_{K \in \mathrm{cl}(G^0)} a_B(\hat{K})\lambda_{B'}(\hat{K}) = \lambda_{B'}(e_B).$$

Thus $B = B'$, as desired. ∎

Our next objective in this chapter is to analyse in more detail the first main theorem.

(9.7) THEOREM (Extended First Main Theorem). *If $B \in \mathrm{Bl}(G)$ has defect group P, then there exists a unique $\mathbf{N}_G(P)$-orbit of blocks $b \in \mathrm{Bl}(P\mathbf{C}_G(P))$ such that $b^G = B$. All these blocks b have defect group P and $b^{\mathbf{N}_G(P)}$ is the Brauer first main correspondent of B.*

First, we need a lemma.

(9.8) LEMMA. *Suppose that $N \triangleleft G$ and let $b \in \mathrm{Bl}(N)$ be such that b^G is defined. Then b^G covers b.*

Proof. By Passman's theorem, it suffices to check that λ_{b^G} and λ_b coincide on the conjugacy classes K of G contained in N. If K is such a class, then, since b^G is defined, we have that

$$\lambda_{b^G}(\hat{K}) = (\lambda_b)^G(\hat{K}) = \lambda_b(\widehat{K \cap N}) = \lambda_b(\hat{K})$$

and the lemma is proved. ∎

Suppose that $N \triangleleft G$ and let $b \in \mathrm{Bl}(N)$ be such that b^G is defined. If $g \in G$, note that $(b^g)^G$ is defined and equals b^G. This easily follows from the definition of induced blocks.

Proof of the Extended First Main Theorem. Let $B \in \mathrm{Bl}(G \mid P)$. By Theorem (4.14) we know that there exists a $b \in \mathrm{Bl}(P\mathbf{C}_G(P))$ such that $b^G = B$. Write $N = \mathbf{N}_G(P)$. We claim that b and b^N have defect group P. Since $P \triangleleft P\mathbf{C}_G(P)$, we have that P is contained in some defect group P_1 of b by Theorem (4.8). Now, P_1 is contained in some defect group P_2 of b^N (by Lemma (4.13)). Since $(b^N)^G = b^G = B$ by Problem (4.2), again by Lemma (4.13) we conclude that P_2 is contained in a G-conjugate of P. This proves the claim.

Now, if b_1 and b_2 are blocks of $P\mathbf{C}_G(P)$ with $(b_1)^G = (b_2)^G = B$, then $(b_1)^N$ and $(b_2)^N$ are blocks of N with defect group P inducing B. By the uniqueness part of the first main theorem, we have that $b_1^{\mathbf{N}_G(P)} = b_2^{\mathbf{N}_G(P)}$. Let us denote this block of $\mathbf{N}_G(P)$ by b_0. Now, by applying Lemma (9.8), we have that b_0 covers b_1 and b_2. By Corollary (9.3), we conclude that b_1

and b_2 are N-conjugate. Finally, if $b \in \mathrm{Bl}(P\mathbf{C}_G(P))$ induces B and $n \in N$, then $(b^n)^N = b^N$ by the remark preceding the proof of this theorem. Hence, $(b^n)^G = b^G$ by Problem (4.2) and the proof of the theorem is complete. ∎

If $B \in \mathrm{Bl}(G \,|\, P)$ and $b \in \mathrm{Bl}(P\mathbf{C}_G(P) \,|\, P)$ induces B, then b is called a **root** of B.

As we have seen, the extended first main theorem leads us to study blocks of $P\mathbf{C}_G(P)$ with defect group P, and this is what we do next. First, we study blocks of factor groups.

Recall that throughout these notes, we view $\mathrm{IBr}(G/N)$ and $\mathrm{Irr}(G/N)$ as subsets of $\mathrm{IBr}(G)$ and $\mathrm{Irr}(G)$, respectively. We already know (see the remark before Theorem (7.6)) that if $\bar{B} \in \mathrm{Bl}(G/N)$, then $\bar{B}$ is contained in a unique block $B \in \mathrm{Bl}(G)$. Note that, in general, a block B need not contain any block of G/N. It is clear that a block B of G contains a block of G/N if and only if some $\chi \in B$ (ordinary or modular) contains N in its kernel.

Now, write $\bar{G} = G/N$ and let

$$^{-}: RG \to R\bar{G}$$

be the natural R-algebra homomorphism (where R is an arbitrary ring). This is the homomorphism

$$\overline{\sum_{g \in G} r_g g} = \sum_{g \in G} r_g \bar{g}$$

for $r_g \in R$. It is clear that if $x \in \mathbf{Z}(RG)$, then $\bar{x} \in \mathbf{Z}(R\bar{G})$. Hence, if $B \in \mathrm{Bl}(G)$, then $\overline{e_B} = 0$ or $\overline{e_B}$ is an idempotent of $\mathbf{Z}(F\bar{G})$. In the latter case, we may write

$$\overline{e_B} = e_{\bar{B}_1} + \ldots + e_{\bar{B}_s}$$

for uniquely determined blocks $\bar{B}_1, \ldots, \bar{B}_s$ of $\bar{G}$ (by Theorem (3.11)). We say, then, that B **dominates** $\bar{B}_i$. Note that if $\bar{B} \in \mathrm{Bl}(G)$, then B dominates $\bar{B}$ if and only if the algebra homomorphism λ_B is the composition of $^{-}$: $\mathbf{Z}(FG) \to \mathbf{Z}(F\bar{G})$ and $\lambda_{\bar{B}}$.

Next, we wish to show that domination and inclusion of blocks are the same thing.

Consider the algebra homomorphism $^{-}: \mathbb{C}G \to \mathbb{C}\bar{G}$. If $\bar{\psi} \in \mathrm{Irr}(\bar{G})$ and $\psi \in \mathrm{Irr}(G)$ is the corresponding character of G (that is, if $\psi(g) = \bar{\psi}(\bar{g})$ for $g \in G$), it is clear that $\overline{e_\psi} = e_{\bar{\psi}}$ (just use the formula for the e_ψ). Now, we claim that if $\chi \in \mathrm{Irr}(G)$ does not contain N in its kernel, then $\overline{e_\chi} = 0$. To prove the claim it suffices to show that

$$\sum_{n \in N} \chi(gn) = 0$$

for $g \in G$. If $\mathcal{X}$ is a representation affording χ, it suffices to show that $\sum_{n \in N} \mathcal{X}(n) = 0$. This easily follows from Clifford's theorem for ordinary representations and Problem (2.1.a) of [Isaacs, *Characters*].

Now, let $\bar{\xi} \in \mathrm{Irr}(\bar{B})$, where $\bar{B}$ is a block of $\bar{G}$, and let B be an arbitrary block of G. Then B dominates $\bar{B}$ if and only if $\lambda_{\bar{B}}(\overline{e_B}) = 1$. This happens if and only if

$$\omega_{\bar{\xi}} \left(\sum_{\psi \in \mathrm{Irr}(B)} \overline{e_\psi} \right)^* = \omega_{\bar{\xi}} \left(\sum_{\substack{\psi \in \mathrm{Irr}(B) \\ N \subseteq \ker(\psi)}} e_{\bar{\psi}} \right)^* = 1 .$$

We deduce that B dominates $\bar{B}$ if and only if $\mathrm{Irr}(\bar{B}) \cap \mathrm{Irr}(B) \neq \emptyset$, which happens if and only if $\bar{B} \subseteq B$ (because every block of $\bar{G}$ is contained in a unique block of G). So we see that inclusion and domination are equivalent terms.

(9.9) THEOREM. *Let $N \lhd G$ and write $\bar{G} = G/N$.*

(a) Suppose that $\bar{B} \subseteq B$, where $\bar{B}$ is a block of $\bar{G}$ and B is a block of G. If $\bar{D}$ is a defect group of $\bar{B}$, then there is a defect group P of B such that $\bar{D} \subseteq PN/N$.

(b) If N is a p-group, then every block $B \in \mathrm{Bl}(G)$ contains a block $\bar{B} \in \mathrm{Bl}(\bar{G})$ such that $\delta(\bar{B}) = \{P N/N \mid P \in \delta(B)\}$.

(c) If N is a p'-group and $\bar{B} \subseteq B$, where $\bar{B}$ is a block of $\bar{G}$ and B is a block of G, then $\mathrm{Irr}(B) = \mathrm{Irr}(\bar{B})$, $\mathrm{IBr}(B) = \mathrm{IBr}(\bar{B})$ and $\delta(\bar{B}) = \{PN/N \mid P \in \delta(B)\}$.

Proof. First, we prove (a). Let $K = \mathrm{cl}(x)$ be a defect class of B and let $\bar{K} = \mathrm{cl}_{\bar{G}}(\bar{x}) = KN/N$ be the corresponding conjugacy class in $\bar{G}$. If for $y, z \in K$ we put $y \equiv z$ iff $yN = zN$, we have that $\bar{K} = \{x_1 N, \ldots, x_s N\}$, where $\{x_1, \ldots, x_s\}$ is a complete set of representatives of the equivalence classes in K under $\equiv$. Thus, $|\bar{K}| = s$. Furthermore, $K = (K \cap x_1 N) \cup \ldots \cup (K \cap x_s N)$. Since we have that $|K \cap x^g N| = |K \cap x N|$ for $g \in G$, it follows that

$$|\bar{K}| t = |K| ,$$

where $t = |K \cap xN|$. Thus,

$$\overline{K} = t \widehat{\overline{K}} .$$

Also, if $\mathbf{C}_{\bar{G}}(\bar{x}) = H/N$, then

$$t = \frac{|K|}{|\bar{K}|} = |H : \mathbf{C}_G(x)|$$

since $|K| = |G : \mathbf{C}_G(x)|$ and $|\bar{K}| = |\bar{G} : \mathbf{C}_{\bar{G}}(\bar{x})|$. Now,

$$0 \neq \lambda_B(\hat{K}) = \lambda_{\bar{B}}(\overline{\hat{K}}) = \lambda_{\bar{B}}(t^* \overline{\hat{K}}) = t^* \lambda_{\bar{B}}(\widehat{\bar{K}}),$$

and, from this equation, we conclude that p does not divide $|H : \mathbf{C}_G(x)|$. Also, by the min-max theorem, we see that $\bar{D} \subseteq_{\bar{G}} \bar{X}$, where $\bar{X}$ is a Sylow p-subgroup of $\mathbf{C}_{\bar{G}}(\bar{x})$. Now, if $P \in \mathrm{Syl}_p(\mathbf{C}_G(x))$, then $P \in \mathrm{Syl}_p(H)$ and thus $\bar{P} \in \mathrm{Syl}_p(H/N)$. Since P is a defect group of B, the proof of part (a) is complete.

Assume now that N is a p-group. By Lemma (2.32), we have that $N \subseteq \mathbf{O}_p(G) \subseteq \ker(\varphi)$ for every $\varphi \in \mathrm{IBr}(G)$. Since every block of $\bar{G}$ is contained in a unique block of G, note that if two irreducible Brauer characters of $\bar{G}$ lie in the same block of $\bar{G}$, then they lie in the same block when considered as Brauer characters of G. Hence, it is clear that we may write

$$\mathrm{IBr}(B) = \mathrm{IBr}(\overline{B_1}) \cup \ldots \cup \mathrm{IBr}(\overline{B_s}),$$

where $\{\overline{B_1}, \ldots, \overline{B_s}\}$ are the blocks of $\bar{G}$ contained in B. Now, by applying Corollary (3.17) we have that

$$\nu(|G|) - \mathrm{d}(B) = \min_{\varphi \in \mathrm{IBr}(B)} \{\nu(\varphi(1))\}$$

$$= \min_{j=1,\ldots,s} \{ \min_{\varphi \in \mathrm{IBr}(\overline{B_j})} \{\nu(\varphi(1))\}\} = \min_{j=1,\ldots,s} \{\nu(|G/N|) - \mathrm{d}(\overline{B_j})\}.$$

Therefore, it follows that there is a block $\overline{B_j}$ such that

$$\nu(|G|) - \mathrm{d}(B) = \nu(|G/N|) - \mathrm{d}(\overline{B_j}).$$

This shows that

$$\mathrm{d}(B) - \nu(|N|) = \mathrm{d}(\overline{B_j}).$$

If $\bar{D}$ is a defect group of $\overline{B_j}$, by part (a) we know that there is a defect group P of B such that $\bar{D} \subseteq P/N$ (recall that $N \subseteq P$ by Theorem (4.8)). Since $\mathrm{d}(B) - \nu(|N|) = \mathrm{d}(\overline{B_j})$, we conclude that $\bar{D} = P/N$, as desired.

Finally, assume that N is a p'-group and suppose that $\bar{B} \subseteq B$. If $\chi \in \mathrm{Irr}(B)$ with $N \subseteq \ker(\chi)$, we have that χ lies over 1_N. Hence, by Theorem (9.2), we have that B covers the block $\{1_N\}$. Therefore, every modular or ordinary character of B contains N in its kernel, again by Theorem (9.2). Thus, linking defines the same graph in B, whether it is considered in G or in $\bar{G}$. Hence, $\mathrm{Irr}(B) = \mathrm{Irr}(\bar{B})$ and also $\mathrm{IBr}(B) = \mathrm{IBr}(\bar{B})$. In particular, since p does not divide $|N|$, we have by definition that $\mathrm{d}(B) = \mathrm{d}(\bar{B})$. By part (a), the proof of the theorem is complete. ∎

Most of the results in the next theorem were needed in Chapter 7.

(9.10) THEOREM. *Suppose that G has a normal p-subgroup P such that $G/\mathbf{C}_G(P)$ is a p-group. Write $\bar{G} = G/P$. If $\bar{B} \in \mathrm{Bl}(\bar{G})$ and $B \in \mathrm{Bl}(G)$ is the unique block of G containing $\bar{B}$, then the map $\bar{B} \mapsto B$ is a bijection $\mathrm{Bl}(\bar{G}) \to \mathrm{Bl}(G)$. Also, $\mathrm{IBr}(\bar{B}) = \mathrm{IBr}(B)$, $\delta(\bar{B}) = \{D/P \mid D \in \delta(B)\}$ and the Cartan matrices of $\bar{B}$ and B are related by $C_B = |P|C_{\bar{B}}$.*

Proof. This is Theorem (7.6). For the equation involving the defect groups, Theorem (9.9.b) applies. ∎

As the reader probably has realized, the theorem above also gives us a relationship between projective indecomposable characters of G and $\bar{G}$:

$$\Phi_\varphi = |P|\Phi_{\bar\varphi}\,.$$

The general fact is the next theorem. Although it is not necessary for our purposes, perhaps this is a good place to prove it.

(9.11) THEOREM. *Let P be a normal p-subgroup of G, and write $\bar{G} = G/P$. Suppose that $g \in G^0$. If $\varphi \in \mathrm{IBr}(G)$ and $\bar\varphi \in \mathrm{IBr}(\bar{G})$ corresponds to φ, then*

$$\Phi_\varphi(g) = \frac{|\mathbf{C}_G(g)|}{|\mathbf{C}_{\bar{G}}(\bar{g})|}\Phi_{\bar\varphi}(\bar{g})\,.$$

Proof. If $\theta \in \mathrm{cf}(G^0)$, we define $\bar\theta \in \mathrm{cf}(\bar{G}^0)$ in the following way. First, it is clear that xP is a p'-element of G/P if and only if $x_p \in P$, which happens if and only if $xP = x_{p'}P$. So the natural map $G^0 \to \bar{G}^0$ given by $x \mapsto \bar{x}$ is surjective. Now, if $x, y \in G^0$ are such that $xP = yP$, by the Schur-Zassenhaus theorem it follows that $\langle x \rangle$ and $\langle y \rangle$ are P-conjugate. Hence, there is a $z \in P$ such that $x^z = y^i$ for some integer i. Since $yP = xP = x^zP = y^iP$, we deduce that $y^i = y$ and thus $x^z = y$. Hence, x and y are P-conjugate. We define $\bar\theta(\bar{x}) = \theta(x)$ for $x \in G^0$ and note that this is a well-defined map by the previous discussion. Then it is clear that $\bar\theta \in \mathrm{cf}(\bar{G}^0)$ and thus

$$\bar\theta = \sum_{\bar\varphi \in \mathrm{IBr}(\bar{G})} b_{\bar\varphi}\bar\varphi$$

for some complex numbers $b_{\bar\varphi}$. Also,

$$\theta = \sum_{\varphi \in \mathrm{IBr}(G)} a_\varphi\varphi$$

for some complex numbers a_φ. Since $\bar\varphi(\bar{x}) = \varphi(x)$ for $x \in G^0$, by the linear independence of $\mathrm{IBr}(G)$ we conclude that $a_\varphi = b_{\bar\varphi}$. In other words, by using Theorem (2.13), we have that

$$[\bar\theta, \Phi_{\bar\varphi}]^0 = b_{\bar\varphi} = a_\varphi = [\theta, \Phi_\varphi]^0\,.$$

Now, let g be a fixed element of G^0. We define a class function $\theta \in \mathrm{cf}(G^0)$ by setting

$$\theta(y) = \delta_{\mathrm{cl}(y),\mathrm{cl}(g^{-1})} .$$

By the first part of the proof,

$$\frac{\Phi_\varphi(g)}{|\mathbf{C}_G(g)|} = [\theta, \Phi_\varphi]^0 = [\bar\theta, \Phi_{\bar\varphi}]^0 = \frac{\Phi_{\bar\varphi}(\bar g)}{|\mathbf{C}_{\bar G}(\bar g)|} ,$$

as required. ∎

Now, we give a complete description of the blocks of $P\mathbf{C}_G(P)$ with defect group P.

(9.12) THEOREM. *Let P be a p-subgroup of G with $G = P\mathbf{C}_G(P)$ and let $B \in \mathrm{Bl}(G \,|\, P)$. Then there exists a unique $\theta \in \mathrm{Irr}(B)$ such that $P \subseteq \ker(\theta)$. Also, $\theta(1)_p = |G/P|_p$, so that θ has p-defect zero considered as a character of G/P. Furthermore, $\mathrm{IBr}(B) = \{\theta^0\}$. If $\xi \in \mathrm{Irr}(P)$ and we define $\theta_\xi(g) = \xi(g_p)\theta(g_{p'})$ for $g \in G$ with $g_p \in P$ and $\theta_\xi(g) = 0$ if $g_p \notin P$, then the map $\xi \mapsto \theta_\xi$ is a bijection $\mathrm{Irr}(P) \to \mathrm{Irr}(B)$. In particular, the set of heights of the irreducible characters in B is $\{\nu(\xi(1)) \,|\, \xi \in \mathrm{Irr}(P)\}$.*

Proof. Using Theorem (9.10), let $\bar B \in \mathrm{Bl}(G/P)$ be the unique block of G/P contained in B. We know by Theorem (9.9.b) that $\delta(\bar B) = \{X/P \,|\, X \in \delta(B)\}$ and thus $\bar B$ has defect zero. Therefore, by Theorem (3.18), we have that $\mathrm{Irr}(\bar B) = \{\theta\}$ and $\mathrm{IBr}(\bar B) = \{\theta^0\}$. Also, by Theorem (9.10), we have that $\mathrm{IBr}(B) = \{\theta^0\}$.

Now, if $\tau \in \mathrm{Irr}(B)$ has P in its kernel, then τ lies in some block of G/P which necessarily lies inside B. By Theorem (9.10), this block is $\bar B$ and therefore $\tau = \theta$.

Now, let $\chi \in \mathrm{Irr}(B)$ and write $\chi_P = e\xi$ for some $\xi \in \mathrm{Irr}(P)$. (Since $G = P\mathbf{C}_G(P)$, note that the characters of P are G-invariant.) By Corollary (5.9), we know that $\chi(g) = 0$ if $g_p \notin P$. Assume now that $g_p \in P$. Since $G/\mathbf{C}_G(P)$ is a p-group, note that $G^0 \subseteq \mathbf{C}_G(P)$. Hence, let $H = \langle g_{p'} \rangle \times P$. Write

$$\chi_H = \sum_{\psi \in \mathrm{Irr}(H)} a_\psi \psi$$

for some integers a_ψ. Since $\chi_P = e\xi$, it follows that every ψ with $a_\psi \neq 0$ is of the form $\lambda \times \xi$ for some $\lambda \in \mathrm{Irr}(\langle g_{p'} \rangle)$. Therefore, we may write $\chi_H = \alpha \times \xi$ for some character α of $\langle g_{p'} \rangle$. Hence, $\chi(g_{p'}) = \alpha(g_{p'})\xi(1)$ and we deduce that

$$\chi(g) = \alpha(g_{p'})\xi(g_p) = \frac{\chi(g_{p'})}{\xi(1)}\xi(g_p) .$$

Now, since $\mathrm{IBr}(B) = \{\theta^0\}$ and $\chi \in \mathrm{Irr}(B)$, we have that

$$\chi(g) = \frac{d_{\chi\theta^0}}{\xi(1)}\theta(g_{p'})\xi(g_p).$$

Since $G^0 \subseteq \mathbf{C}_G(P)$ observe that the map $P \times G^0 \to \{g \in G \mid g_p \in P\}$ given by $(x, y) \mapsto xy$ is a bijection. Now,

$$1 = [\chi, \chi] = \frac{1}{|G|}\Big(\frac{d_{\chi\theta^0}}{\xi(1)}\Big)^2 \sum_{x \in P, y \in G^0} \theta(y)\theta(y^{-1})\xi(x)\xi(x^{-1})$$

$$= \Big(\frac{d_{\chi\theta^0}}{\xi(1)}\Big)^2 \frac{1}{|G/P|} \sum_{y \in G^0} \theta(y)\theta(y^{-1}).$$

Since the map $G^0 \to (G/P)^0$ given by $y \mapsto yP$ is a bijection, and θ has defect zero considered as a character of G/P, by Theorem (3.18) we conclude that

$$\frac{1}{|G/P|} \sum_{y \in G^0} \theta(y)\theta(y^{-1}) = \frac{1}{|\bar{G}|} \sum_{\bar{y} \in \bar{G}^0} \bar{\theta}(\bar{y})\bar{\theta}(\bar{y}^{-1})$$

$$= \frac{1}{|\bar{G}|} \sum_{\bar{g} \in \bar{G}} \bar{\theta}(\bar{g})\bar{\theta}(\bar{g}^{-1}) = 1.$$

Therefore, $d_{\chi\theta^0} = \xi(1)$. Thus, given $\chi \in \mathrm{Irr}(B)$, there is a unique $\xi_\chi \in \mathrm{Irr}(P)$ such that $\chi(g) = \theta(g_{p'})\xi_\chi(g_p)$ whenever $g_p \in P$, while $\chi(g) = 0$ if $g_p \notin P$. It is clear, then, that the map $\chi \mapsto \xi_\chi$ from $\mathrm{Irr}(B) \to \mathrm{Irr}(P)$ is injective.

Now, since $\bar{\theta}$ has defect zero considered as character of $\bar{G}$, we have that $c_{\bar{\theta}^0\bar{\theta}^0} = 1$. By Theorem (9.10), we have that $c_{\theta^0\theta^0} = |P|$. Now,

$$|P| = c_{\theta^0\theta^0} = \sum_{\chi \in \mathrm{Irr}(B)} (d_{\chi\theta^0})^2 = \sum_{\chi \in \mathrm{Irr}(B)} \xi_\chi(1)^2 \leq \sum_{\mu \in \mathrm{Irr}(P)} \mu(1)^2 = |P|,$$

and we conclude that the map $\chi \mapsto \xi_\chi$ from $\mathrm{Irr}(B) \to \mathrm{Irr}(P)$ is bijective.

We have constructed the inverse of the map in the statement of this theorem. To prove the latter assertion note that

$$\nu(\theta_\xi(1)) = \nu(\xi(1)) + \nu(\theta(1)) = \nu(\xi(1)) + \nu(|G|) - \nu(|P|).$$

This proves that $\mathrm{height}(\theta_\xi) = \nu(\xi(1))$, as desired. ∎

If $B \in \mathrm{Bl}(G \mid P)$ and $b \in \mathrm{Bl}(P\mathbf{C}_G(P) \mid P)$ is a root of B, we have that there is a unique character $\theta \in \mathrm{Irr}(b)$ with $P \subseteq \ker(\theta)$ by Theorem (9.12).

This character, which satisfies the conclusions of Theorem (9.12) and which is uniquely defined by B and P up to $\mathbf{N}_G(P)$-conjugacy, is called the **canonical character** of B.

By the extended first main theorem and Theorem (9.12), we see that every block $B \in \mathrm{Bl}(G \mid P)$ is induced from a block b such that $\mathrm{l}(b) = 1$ and $\mathrm{k}(b) = |\mathrm{Irr}(P)|$. One of the main problems in block theory is to relate the modular invariants of a block B of G with the invariants of its Brauer first main correspondent.

There is a sort of converse of Theorem (9.12) due to A. Laradji.

(9.13) THEOREM (Laradji). *Suppose that D is a central p-subgroup of G and let $\chi \in \mathrm{Irr}(G)$ be such that $\chi(1)_p = |G : D|_p$. Then the block of χ has defect group D. Also, $\chi^0 \in \mathrm{IBr}(G)$ and $\chi(g) = 0$ if $g_p \notin D$.*

Proof. We define a class function ψ on G in the following way. We set $\psi(g) = 0$ if $g_p \notin D$ and $\psi(g) = \chi(g_{p'})$ if $g_p \in D$. We prove that $\psi \in \mathrm{Irr}(G)$ by using Brauer's characterization of characters. Let H be a subgroup of G of the form $H = P \times Q$, where P is a p-group and Q is a p'-group. Let $\xi \in \mathrm{Irr}(H)$ so that $\xi = \alpha \times \beta$, where $\alpha \in \mathrm{Irr}(P)$ and $\beta \in \mathrm{Irr}(Q)$. We wish to show that $[\psi_H, \xi] \in \mathbb{Z}$. Set $N = P \cap D \subseteq \mathbf{Z}(G)$. Notice that, by definition, ψ_H is zero off NQ and $\psi_{NQ} = 1_N \times \chi_Q$. Hence,

$$|H|[\psi_H, \xi] = |NQ|[\psi_{NQ}, \xi_{NQ}] = |NQ|[1_N \times \chi_Q, \alpha_N \times \beta]$$

and we deduce that

$$|H : NQ|[\psi_H, \xi] = |P : N|[\psi_H, \xi] \in \mathbb{Z}.$$

In particular, $[\psi_H, \xi] \in \mathbb{Q}$.

If $x \in G$, put

$$\omega(x) = \frac{|G : \mathbf{C}_G(x)|\chi(x)}{\chi(1)} \in \mathbf{R}.$$

Note that if $x \in Q$, then $PD \subseteq \mathbf{C}_G(x)$. Now,

$$\frac{|G : D|}{\chi(1)}|Q|[\psi_H, \xi] = \frac{|G : D|}{\chi(1)|P|}|H|[\psi_H, \xi] = \frac{|G : DP|}{|N|\chi(1)}|NQ|[1_N \times \chi_Q, \alpha_N \times \beta]$$

$$= \frac{|G : DP|}{|N|\chi(1)}|N|[1_N, \alpha_N]|Q|[\chi_Q, \beta] = [1_N, \alpha_N]\frac{|G : DP|}{\chi(1)}\sum_{x \in Q}\chi(x)\beta(x^{-1})$$

$$= [1_N, \alpha_N]\frac{|G : DP|}{\chi(1)}\sum_{x \in Q}\frac{|\mathbf{C}_G(x)|\chi(1)}{|G|}\omega(x)\beta(x^{-1})$$

$$= [1_N, \alpha_N] \sum_{x \in Q} |\mathbf{C}_G(x) : PD| \omega(x) \beta(x^{-1}) \in \mathbf{R}.$$

Since $\frac{|G:D|}{\chi(1)} |Q| [\psi_H, \xi] \in \mathbb{Q}$ (because $[\psi_H, \xi] \in \mathbb{Q}$), we deduce that

$$\frac{|G : D|}{\chi(1)} |Q| [\psi_H, \xi] \in \mathbb{Z}.$$

Now, by Ito's theorem (6.15) of [Isaacs, *Characters*] and by hypothesis, we have that $\frac{|G:D|}{\chi(1)}$ (and therefore $\frac{|G:D|}{\chi(1)} |Q|$) is an integer not divisible by p. Hence, using that $|P : N| [\psi_H, \xi] \in \mathbb{Z}$ and that $(\frac{|G:D|}{\chi(1)} |Q|, |P : N|) = 1$, we easily deduce that $[\psi_H, \xi] \in \mathbb{Z}$. Hence, by Brauer's characterization of characters we conclude that ψ is a generalized character of G.

Now, write $\chi_D = \chi(1)\lambda$, where $\lambda \in \mathrm{Irr}(D)$. By elementary character theory, we know that

$$\chi(x) = \lambda(x_p)\chi(x_{p'})$$

whenever $x_p \in D$. In particular, we have that

$$|\chi(x)| = |\chi(x_{p'})|$$

whenever $x_p \in D$. Now, since $\psi(1) > 0$ we have that

$$0 < [\psi, \psi] = \frac{1}{|G|} \sum_{\substack{x \in G \\ x_p \in D}} |\chi(x_{p'})|^2 = \frac{1}{|G|} \sum_{\substack{x \in G \\ x_p \in D}} |\chi(x)|^2 \le [\chi, \chi] = 1.$$

Thus, $[\psi, \psi] = 1$ and we conclude that $\psi \in \mathrm{Irr}(G)$, as claimed.

Now, since $\psi^0 = \chi^0$, it follows that ψ and χ lie in the same block B of G. Furthermore, note that $D \subseteq \ker(\psi)$ and $\psi(1)_p = \chi(1)_p = |G : D|_p$. Therefore, the block in G/D of ψ has defect zero. By Theorem (9.10), we deduce that $D \in \delta(B)$. Now, by Theorem (9.12) we have that $\chi^0 = \psi^0 \in \mathrm{IBr}(G)$. The last part follows from Corollary (5.9). ∎

Still there are many problems which we wish to study before concluding this chapter. One of them is to decide when $b \in \mathrm{Bl}(P\mathbf{C}_G(P) \,|\, P)$ is a root of b^G. For this problem (and others), we need Fong's theory on blocks and covering.

We start with the Clifford correspondence for blocks.

(9.14) THEOREM (Fong-Reynolds). *Let $N \triangleleft G$ and let $b \in \mathrm{Bl}(N)$. Let $T(b)$ be the stabilizer of b in G.*

(a) *The map $\mathrm{Bl}(T(b) \,|\, b) \to \mathrm{Bl}(G \,|\, b)$ given by $B \mapsto B^G$ is a bijection.*

(b) *If $B \in \mathrm{Bl}(T(b) \,|\, b)$, then $\mathrm{Irr}(B^G) = \{\psi^G \,|\, \psi \in \mathrm{Irr}(B)\}$ and $\mathrm{IBr}(B^G) = \{\varphi^G \,|\, \varphi \in \mathrm{IBr}(B)\}$. Also,*

$$d_{\chi\varphi} = d_{\chi^G \varphi^G}$$

for $\chi \in \mathrm{Irr}(B)$ and $\varphi \in \mathrm{IBr}(B)$.

(c) *If $B \in \mathrm{Bl}(T(b) \,|\, b)$, then every defect group of B is a defect group of B^G.*

(d) *If $B \in \mathrm{Bl}(T(b) \,|\, b)$ and $\chi \in \mathrm{Irr}(B)$, then*

$$\mathrm{height}(\chi) = \mathrm{height}(\chi^G).$$

Proof. First, notice that if $\theta \in b$, then $I_G(\theta) \subseteq T(b)$. This follows from the fact that, for $g \in G$, we have that $\theta^g \in b^g$ if and only if $\theta \in b$.

Let $B \in \mathrm{Bl}(T(b) \,|\, b)$ and let $\psi \in B$. We claim that ψ^G is irreducible. By Theorem (9.2), there is an irreducible constituent θ of ψ_N in b. Thus $I_G(\theta) \subseteq T(b)$ by the first paragraph. By the Clifford correspondence for ordinary and Brauer characters applied in $T(b)$, we have that $\psi = \tau^{T(b)}$ for some irreducible character τ of $I_G(\theta)$ over θ. Therefore, $\psi^G = \tau^G$ is an irreducible character of G which lies over θ. This proves the claim. In particular, if we choose $\psi \in \mathrm{Irr}(B)$, we deduce that the block B^G is defined (Corollary (6.2)), contains ψ^G and clearly covers b (Theorem (9.2)). Now, if $\psi \in \mathrm{IBr}(B)$, by applying Theorem (8.10) twice, we get

$$(\Phi_\psi)^G = ((\Phi_\tau)^{T(b)})^G = (\Phi_\tau)^G = \Phi_{\tau^G} = \Phi_{\psi^G}.$$

This shows that all irreducible characters $\eta \in \mathrm{Irr}(G)$ satisfying $d_{\eta\psi^G} \neq 0$ are of the form $\eta = \gamma^G$ for $\gamma \in \mathrm{Irr}(B)$. In particular, $\psi^G \in \mathrm{IBr}(B^G)$.

Next, we prove that induction defines surjective maps $\mathrm{Irr}(B) \to \mathrm{Irr}(B^G)$ and $\mathrm{IBr}(B) \to \mathrm{IBr}(B^G)$ for $B \in \mathrm{Bl}(T(b) \,|\, b)$. Suppose that $\eta \in \mathrm{Irr}(B^G)$. Let $\xi \in \mathrm{Irr}(B)$. Then η and ξ^G are linked. Since the irreducible constituents of $(\xi^G)^0 = (\xi^0)^G$ are irreducible Brauer characters induced from $\mathrm{IBr}(B)$, it follows that $d_{\eta\varphi^G} \neq 0$ for some $\varphi \in \mathrm{IBr}(B)$. Then η is an irreducible constituent of $(\Phi_\varphi)^G$ and we conclude that $\eta = \gamma^G$ for some $\gamma \in \mathrm{Irr}(B)$. Also, if $\delta \in \mathrm{IBr}(B^G)$, then $d_{\chi\delta} \neq 0$ for some $\chi \in \mathrm{Irr}(B^G)$. Then $\chi = \psi^G$ for some $\psi \in \mathrm{Irr}(B)$ by the first part of the claim. Now, $(\psi^G)^0 = (\psi^0)^G$, and we conclude that $\delta = \varphi^G$ for some $\varphi \in \mathrm{IBr}(B)$.

Now, let $B_1, B_2 \in \mathrm{Bl}(T(b) \,|\, b)$ and suppose that $\mu_1^G = \mu_2^G$, where $\mu_i \in \mathrm{IBr}(B_i)$ for $i = 1, 2$. We prove that $\mu_1 = \mu_2$. (The same argument proves the corresponding fact for ordinary characters μ_1 and μ_2.) Since $\mu_i^G \in B_i^G$

and these blocks cover b, by Theorem (9.2) we may find $\theta_i \in \mathrm{IBr}(b)$ such that θ_i lies under μ_i^G. Since θ_1 and θ_2 lie under μ_1^G for instance, by Clifford's theorem we deduce that $\theta_1 = \theta_2^g$ for some $g \in G$. Then $g \in T(b)$. Hence, μ_2 also lies over θ_1. Now, if $\tau_i \in \mathrm{IBr}(I_G(\theta_1))$ is the Clifford correspondent of μ_i over θ_1, it follows that $\tau_1^G = \mu_1^G = \mu_2^G = \tau_2^G$. Then $\tau_1 = \tau_2$ by the uniqueness in the Clifford correspondence and therefore $\mu_1 = \tau_1^{T(b)} = \tau_2^{T(b)} = \mu_2$, as desired. In particular, this, together with the claim in the previous paragraph, shows that induction defines bijections $\mathrm{Irr}(B) \to \mathrm{Irr}(B^G)$ and $\mathrm{IBr}(B) \to \mathrm{IBr}(B^G)$ for $B \in \mathrm{Bl}(T(b)\,|\,b)$. Also, this proves that the map $B \mapsto B^G$ from $\mathrm{Bl}(T(b)\,|\,b) \to \mathrm{Bl}(G\,|\,b)$ is injective. This map also is surjective since given a block B of G covering b and $\chi \in \mathrm{Irr}(B)$, by Theorem (9.2) it follows that χ lies over some $\theta \in \mathrm{Irr}(b)$. Then $I_G(\theta) \subseteq T(b)$ and by the Clifford correspondence $\chi = \mu^G$ for some $\mu \in \mathrm{Irr}(T(b))$ lying over θ. Hence, the block B_1 of μ covers b and necessarily induces B.

Now, if $B \in \mathrm{Bl}(T(b)\,|\,b)$, $\xi \in \mathrm{Irr}(B)$ and $\varphi \in \mathrm{IBr}(B)$, since $(\xi^G)^0 = (\xi^0)^G$ and the map induction is injective from $\mathrm{IBr}(B)$ to $\mathrm{IBr}(B^G)$, we deduce that

$$d_{\xi\varphi} = d_{\xi^G\varphi^G}.$$

This completes the proof of parts (a) and (b).

Now, we prove part (c). If $B \in \mathrm{Bl}(T(b)\,|\,b)$, then we have that $\mathrm{d}(B) = \mathrm{d}(B^G)$ because

$$\min\{(\psi^G(1))_p \,|\, \psi \in \mathrm{Irr}(B)\} = |G:T(b)|_p \min\{(\psi(1))_p \,|\, \psi \in \mathrm{Irr}(B)\}$$

and Definition (3.15). Since we always have that the every defect group of B is contained in some defect group of B^G (Lemma (4.13)), we have that every defect group of B is a defect group of B^G.

Finally, suppose that $B \in \mathrm{Bl}(T(b)\,|\,b)$ and let $\chi \in \mathrm{Irr}(B)$. Let P be a defect group of B, so that P is a defect group of B^G by part (c). Now,

$$\nu(\chi(1)) = \nu(|T(b)|) - \nu(|P|) + \mathrm{height}(\chi)$$

and therefore

$$\nu(\chi^G(1)) = \nu(|G:T(b)|) + \nu(\chi(1)) = \nu(|G|) - \nu(|T(b)|) + \nu(\chi(1))$$
$$= \nu(|G|) - \nu(|P|) + \mathrm{height}(\chi).$$

This proves part (d). $\blacksquare$

Suppose that $N \lhd G$ and $b \in \mathrm{Bl}(N)$. The key idea in Fong's theory of covering blocks is to consider the blocks $B \in \mathrm{Bl}(G\,|\,b)$ of maximal defect among the blocks in $\mathrm{Bl}(G\,|\,b)$, that is, with

$$\mathrm{d}(B) \geq \mathrm{d}(B')$$

for all $B' \in \mathrm{Bl}(G\,|\,b)$.

We need a rather technical lemma.

(9.15) LEMMA. Let $N \lhd G$, let $b \in \mathrm{Bl}(N)$ and let $B \in \mathrm{Bl}(G \,|\, b)$. Write (as we may)

$$\sum_{B' \in \mathrm{Bl}(G \,|\, b)} e_{B'} = \sum_{\substack{K \in \mathrm{cl}(G) \\ K \subseteq N}} u_b(\hat{K}) \hat{K}$$

for some $u_b(\hat{K}) \in F$. Then there exists a conjugacy class K of G contained in N such that $u_b(\hat{K}) \neq 0 \neq \lambda_B(\hat{K})$. If, in addition, $\mathrm{d}(B) \geq \mathrm{d}(B')$ for all $B' \in \mathrm{Bl}(G \,|\, b)$, then every such class K is such that $D_K \in \delta(B)$.

Proof. Recall that, by the definition of covering, it is possible to write

$$\sum_{B' \in \mathrm{Bl}(G \,|\, b)} e_{B'} = \sum_{\substack{K \in \mathrm{cl}(G) \\ K \subseteq N}} u_b(\hat{K}) \hat{K}$$

for some $u_b(\hat{K}) \in F$. Since B covers b, it follows that

$$1 = \lambda_B \Big(\sum_{B' \in \mathrm{Bl}(G \,|\, b)} e_{B'} \Big) = \sum_{\substack{K \in \mathrm{cl}(G) \\ K \subseteq N}} u_b(\hat{K}) \lambda_B(\hat{K}) .$$

We deduce that there is a $K \in \mathrm{cl}(G)$ contained in N such that

$$\lambda_B(\hat{K}) \neq 0 \neq u_b(\hat{K}) .$$

Consider now such a conjugacy class K and assume further that $\mathrm{d}(B) \geq \mathrm{d}(B')$ for all $B' \in \mathrm{Bl}(G \,|\, b)$. We prove that $D_K \in \delta(B)$. Since $\lambda_B(\hat{K}) \neq 0$, we have that $D_B \subseteq_G D_K$ by the min-max theorem. By using that

$$e_{B'} = \sum_{K \in \mathrm{cl}(G)} a_{B'}(\hat{K}) \hat{K} ,$$

we see that

$$u_b(\hat{K}) = \sum_{B' \in \mathrm{Bl}(G \,|\, b)} a_{B'}(\hat{K}) .$$

Hence, we deduce that there is a $B' \in \mathrm{Bl}(G \,|\, b)$ such that $a_{B'}(\hat{K}) \neq 0$. Again by the min-max theorem, we have that $D_K \subseteq_G D_{B'}$. Thus, $D_B \subseteq_G D_{B'}$. Since $\mathrm{d}(B) \geq \mathrm{d}(B')$, we have that D_B and $D_{B'}$ are G-conjugate. Thus, $D_K =_G D_B$, as desired. $\blacksquare$

(9.16) THEOREM (Fong). Let $N \lhd G$, let $b \in \mathrm{Bl}(N)$ and let $B \in \mathrm{Bl}(G \,|\, b)$. Then $\mathrm{d}(B) \geq \mathrm{d}(B')$ for all $B' \in \mathrm{Bl}(G \,|\, b)$ if and only if there is a conjugacy class K of G contained in N such that $\lambda_B(\hat{K}) \neq 0$ and $D_K \in \delta(B)$. Furthermore, in this case,

$$D_{B'} \subseteq_G D_B$$

for all $B' \in \mathrm{Bl}(G \,|\, b)$.

Proof. If $d(B) \geq d(B')$ for all $B' \in \mathrm{Bl}(G \mid b)$, then there is such a conjugacy class of G by Lemma (9.15).

Suppose now that K is a conjugacy class of G inside N such that $\lambda_B(\hat{K}) \neq 0$ with $D_K \in \delta(B)$. If B' is a block of G which covers b, by Passman's theorem we have that

$$\lambda_{B'}(\hat{K}) = \lambda_b(\hat{K}) = \lambda_B(\hat{K}) \neq 0.$$

Therefore, by the min-max theorem, $D_{B'} \subseteq_G D_K$. Since $D_K =_G D_B$ by hypothesis, we have that

$$D_{B'} \subseteq_G D_B$$

for all $B' \in \mathrm{Bl}(G \mid b)$. In particular, $d(B) \geq d(B')$ for all $B' \in \mathrm{Bl}(G \mid b)$. ∎

Next is one of the key results on covering and defect groups.

(9.17) THEOREM (Fong). *Let $N \triangleleft G$ and let $b \in \mathrm{Bl}(N)$ be G-invariant. Suppose that $B \in \mathrm{Bl}(G \mid b)$ is such that $d(B) \geq d(B')$ for all $B' \in \mathrm{Bl}(G \mid b)$. If P is a defect group of B, then p does not divide $|G : PN|$ and $P \cap N$ is a defect group of b. In particular,*

$$d(B) = d(b) + \nu(|G/N|).$$

Proof. Since b is G-invariant, using the notation of Lemma (9.15) we may write

$$e_b = \sum_{B' \in \mathrm{Bl}(G \mid b)} e_{B'} = \sum_{\substack{K \in \mathrm{cl}(G) \\ K \subseteq N}} u_b(\hat{K})\hat{K}$$

for some $u_b(\hat{K}) \in F$. If, as usual, we write

$$e_b = \sum_{L \in \mathrm{cl}(N)} a_b(\hat{L})\hat{L},$$

note that $u_b(\hat{K}) = a_b(\hat{L})$ whenever $L \subseteq K$.

By Lemma (9.15), there exists a conjugacy class K of G contained in N with $u_b(\hat{K}) \neq 0 \neq \lambda_B(\hat{K})$. Also, $D_K =_G P$. Let $x \in K$ be such that $P \in \mathrm{Syl}_p(\mathbf{C}_G(x))$ and let L be the conjugacy class of x in N so that $L \subseteq K$. Then K is the (disjoint) union of t different G-conjugates of L, where

$$t = \frac{|K|}{|L|} = \frac{|G : \mathbf{C}_G(x)|}{|N : \mathbf{C}_N(x)|} = |G : N\mathbf{C}_G(x)|.$$

Since b is G-invariant, notice that

$$\lambda_b(\hat{L}) = \lambda_b(\widehat{L^g})$$

for $g \in G$. Now, by Passman's theorem (and the previous remark), we have that

$$0 \neq \lambda_B(\hat{K}) = \lambda_b(\hat{K}) = t^* \lambda_b(\hat{L}).$$

It follows that $t = |G : N\mathbf{C}_G(x)|$ is a p'-number. Also, since P is a Sylow p-subgroup of $\mathbf{C}_G(x)$, we deduce that $|G : NP|$ is a p'-number. Finally, since $a_b(\hat{L}) = u_b(\hat{K})$ (by the first paragraph of the proof), we have that

$$a_b(\hat{L}) \neq 0 \neq \lambda_b(\hat{L}).$$

By definition, we conclude that L is a defect class for b. Now, since $P \cap N \in \text{Syl}_p(\mathbf{C}_N(x))$, we have that $P \cap N \in \delta(b)$, as desired. Finally, we have that PN/N is a Sylow p-subgroup of G/N and the second part is clear. $\blacksquare$

We assumed in the previous theorem that the block b is G-invariant. Note that the Fong-Reynolds correspondence $e \mapsto e^G$ carries the blocks e of the stabilizer $T(b)$ of b in G (covering b) of maximal defect onto the blocks e^G of G (covering b) of maximal defect. Also, since the defect groups of e are defect groups of e^G, we see that we may always apply Fong's theorem in $T(b)$.

There is a useful corollary.

(9.18) COROLLARY. *Suppose that $N \triangleleft G$. Let $b \in \text{Bl}(N)$ be G-invariant and let $B \in \text{Bl}(G \mid b)$ be such that $\text{d}(B) \geq \text{d}(B')$ for all $B' \in \text{Bl}(G \mid b)$. Let $\chi \in \text{Irr}(B)$ have height zero and let $\theta \in \text{Irr}(b)$ be a constituent of χ_N. Then θ has height zero in b and $|G : I_G(\theta)|$ and $[\chi_N, \theta]$ are not divisible by p.*

Proof. First, note that it is possible to choose such a θ by Theorem (9.2). By Clifford's theorem, write $\chi_N = e(\theta_1 + \ldots + \theta_t)$, where $\theta_1, \ldots, \theta_t$ are the different G-conjugates of θ. Hence,

$$\nu(|G|) - \text{d}(B) = \nu(\chi(1)) = \nu(et\theta(1))$$

$$= \nu(et) + \nu(\theta(1)) \geq \nu(et) + \nu(|N|) - \text{d}(b).$$

Since $\text{d}(B) = \text{d}(b) + \nu(|G/N|)$ by Theorem (9.17), it follows that $0 \geq \nu(et)$ and $\nu(\theta(1)) = \nu(|G|) - \text{d}(B) = \nu(|N|) - \text{d}(b)$. Thus, $e = [\chi_N, \theta]$ and $t = |G : I_G(\theta)|$ are p'-numbers and θ has height zero, as required. $\blacksquare$

If $N \triangleleft G$ and $B \in \text{Bl}(G)$, it is customary to say that B is **weakly regular** with respect to N if there is a conjugacy class K of G contained in N such that $\lambda_B(\hat{K}) \neq 0$ and $D_K \in \delta(B)$. By Fong's theory on covering, B is weakly regular if and only if B has maximal defect among the blocks which cover any block $b \in \text{Bl}(N)$ covered by B.

A block $B \in \text{Bl}(G)$ is **regular** with respect to $N \triangleleft G$ if $\lambda_B(\hat{K}) = 0$ for every conjugacy class K of G not contained in N. Clearly, regular blocks are weakly regular (consider any defect class of the block).

(9.19) THEOREM. *Suppose that $B \in \mathrm{Bl}(G)$ covers $b \in \mathrm{Bl}(N)$. Then B is regular with respect to N if and only if b^G is defined and $b^G = B$.*

Proof. Let $K \in \mathrm{cl}(G)$. If K is not contained in N, then

$$(\lambda_b)^G(\hat{K}) = \lambda_b(\widehat{K \cap N}) = 0 \,.$$

Hence, if $B = b^G$, then $\lambda_B = (\lambda_b)^G$ is zero on every conjugacy class K of G which is not contained in N. Therefore, B is regular.

Suppose now that B is regular. We prove that $\lambda_B(\hat{K}) = (\lambda_b)^G(\hat{K})$ for every conjugacy class K of G. Since B is regular, this is certainly true for conjugacy classes K which are not contained in N. On the other hand, if K is contained in N, then

$$(\lambda_b)^G(\hat{K}) = \lambda_b(\hat{K}) = \lambda_B(\hat{K})$$

by Passman's theorem. ∎

There is an easy (but useful) case when we know that blocks are regular with respect to some normal subgroup.

(9.20) LEMMA. *Let $B \in \mathrm{Bl}(G \,|\, P)$ and assume that $\mathbf{C}_G(P) \subseteq N \lhd G$. Then B is regular with respect to N.*

Proof. Let $K \in \mathrm{cl}(G)$ not be contained in N and suppose that $\lambda_B(\hat{K}) \neq 0$. Then, by the min-max theorem, there is an $x \in K$ such that $P \subseteq \mathbf{C}_G(x)$. Hence, $x \in \mathbf{C}_G(P) \subseteq N$, a contradiction. ∎

(9.21) COROLLARY. *Suppose that Q is a normal p-subgroup of G and let $b \in \mathrm{Bl}(Q\mathbf{C}_G(Q))$. Then b^G is the unique block of G which covers b.*

Proof. Let $N = Q\mathbf{C}_G(Q) \lhd G$. We know that b^G covers b by Lemma (9.8). (The block b^G is defined by Theorem (4.14).) Now, suppose that $B \in \mathrm{Bl}(G)$ covers b. Since $Q \lhd G$, it follows that Q is contained in every defect group P of B by Theorem (4.8). Hence, $\mathbf{C}_G(P) \subseteq \mathbf{C}_G(Q) \subseteq N$ and Lemma (9.20) applies. Thus, we deduce that B is regular with respect to N. By Theorem (9.19), $B = b^G$, as desired. ∎

We are finally ready to characterize when a block $b \in \mathrm{Bl}(P\mathbf{C}_G(P) \,|\, P)$ induces a block b^G of defect group P. This completes our analysis of the first main theorem.

(9.22) THEOREM. *Let $b \in \mathrm{Bl}(P\mathbf{C}_G(P) \,|\, P)$ and let $T(b)$ be the stabilizer of b in $\mathbf{N}_G(P)$. Then b^G has defect group P if and only if p does not divide $|T(b) : P\mathbf{C}_G(P)|$.*

Proof. First, we claim that b^G has defect group P if and only if $b^{\mathbf{N}_G(P)}$ has defect group P. If b^G has defect group P, we have that b has defect group P by the easy part of Theorem (9.7), for instance. Conversely, if $b^{\mathbf{N}_G(P)}$ has defect group P, then $(b^{\mathbf{N}_G(P)})^G$ has defect group P by the first main theorem. Since $b^G = (b^{\mathbf{N}_G(P)})^G$ (by Problem (4.2)), the claim is proven. So it is no loss to assume that $P \triangleleft G$.

Now, we claim that we may assume that b is G-invariant. First, observe that $b^{T(b)}$ and b^G cover b by Lemma (9.8). Also, note that $b^{T(b)}$, b^G and $(b^{T(b)})^G$ are defined by Theorem (4.14) and the Fong-Reynolds theorem. Since $(b^{T(b)})^G = b^G$ by Problem (4.2), we have that $b^{T(b)}$ is the Fong-Reynolds correspondent of b^G over b. Since $P \triangleleft G$, note that $b^{T(b)}$ has defect group P if and only if b^G has defect group P again by the Fong-Reynolds theorem. This proves the claim.

By Corollary (9.21), we know that b^G is the only block of G covering b. Now, by Fong's theorem (9.17), we know that if $Q \in \delta(b^G)$, then $Q \cap P\mathbf{C}_G(P) = P$ (because P is the unique defect group of b) and p does not divide $|G : Q\mathbf{C}_G(P)|$. Hence, we have that p does not divide $|G : P\mathbf{C}_G(P)|$ if and only if $Q = P$. ∎

With the hypotheses of Theorem (9.22) and using Theorem (9.12), note that $T(b) = I_{\mathbf{N}_G(P)}(\theta)$, where θ is the unique character in b with $P \subseteq \ker(\theta)$. Therefore, b^G has defect group P if and only if $|I_{\mathbf{N}_G(P)}(\theta) : P\mathbf{C}_G(P)|$ is a p'-number.

If $B \in \mathrm{Bl}(G \,|\, P)$ and $b \in \mathrm{Bl}(P\mathbf{C}_G(P) \,|\, P)$ is a root of B, then the p'-number $|T(b) : P\mathbf{C}_G(P)|$ is called the **inertial index** of B.

One of the famous deep conjectures in block theory is **Brauer's height zero conjecture.**

Height Zero Conjecture. Suppose that B is a block of G with defect group P. Then P is abelian if and only if $\mathrm{height}(\chi) = 0$ for all $\chi \in \mathrm{Irr}(B)$.

This conjecture is known to be true for p-solvable groups. This is an impressive theorem by D. Gluck and T. Wolf whose proof may be found in [Manz-Wolf].

With the amount of information that we already have on blocks, it is easy to prove that the conjecture is true for blocks with normal defect groups.

(9.23) THEOREM (Reynolds). *Suppose that B is a block of G with defect group $P \triangleleft G$. Then P is abelian if and only if $\mathrm{height}(\chi) = 0$ for all $\chi \in \mathrm{Irr}(B)$.*

Proof. We argue by induction on $|G|$. Let $b \in \text{Bl}(P\mathbf{C}_G(P) \,|\, P)$ be a root of B. Hence, $B = b^G$ and, by the Fong-Reynolds theorem (9.14.c,d), it is clear that we may assume that b is G-invariant. Hence, by Theorem (9.22) we deduce that $G/P\mathbf{C}_G(P)$ is a p'-group.

Now, we claim that $\{\text{height}(\chi) \,|\, \chi \in \text{Irr}(B)\} = \{\text{height}(\theta) \,|\, \theta \in \text{Irr}(b)\}$. Suppose that $\chi \in \text{Irr}(B)$. By Lemma (9.8), we have that B covers b. By Theorem (9.2), let $\theta \in \text{Irr}(b)$ lie under χ. By Corollary (11.29) of [Isaacs, *Characters*], we know that $\frac{\chi(1)}{\theta(1)}$ divides $|G : P\mathbf{C}_G(P)|$. Hence, $\chi(1)_p = \theta(1)_p$ and we deduce that $\text{height}(\chi) = \text{height}(\theta)$ because b and B have the same defect and $G/P\mathbf{C}_G(P)$ is a p'-group. The other inclusion follows similarly by using Theorem (9.4) instead of Theorem (9.2). This proves the claim.

By the claim and by the inductive hypothesis, we may assume that $G = P\mathbf{C}_G(P)$. In this case, we have that the set of heights of the irreducible characters in B is $\{\nu(\xi(1)) \,|\, \xi \in \text{Irr}(P)\}$ by Theorem (9.12). Since all the irreducible characters of P are linear if and only if P is abelian, the theorem follows. $\blacksquare$

Now, we prove a theorem which relates the defect groups of b and the defect groups of b^G in a special (but important) case.

If $H \subseteq G$, a block $b \in \text{Bl}(H \,|\, Q)$ is said to be **admissible** (with respect to G) if $\mathbf{C}_G(Q) \subseteq H$. Notice that this definition does not depend on the defect group of b that we choose since they are all H-conjugate. We claim that if b is admissible (with respect to G), then b^G is defined. This follows from the next argument. By Theorem (4.14) (or the extended first main theorem, if we wish), there is a block $b_0 \in \text{Bl}(Q\mathbf{C}_G(Q))$ such that $(b_0)^H = b$. Since $(b_0)^G$ is defined by Theorem (4.14), using Problem (4.2) we see that $b^G = ((b_0)^H)^G$ is also defined.

Finally, observe that if there is a p-subgroup P of G such that

$$P\mathbf{C}_G(P) \subseteq H \subseteq \mathbf{N}_G(P),$$

then every block b of H is admissible (with respect to G). This is true because $P \lhd H$ and hence, P is contained in every defect group Q of b. Thus, $\mathbf{C}_G(Q) \subseteq \mathbf{C}_G(P) \subseteq H$, as required.

(9.24) THEOREM. *Suppose that $b \in \text{Bl}(H \,|\, Q)$ is admissible (with respect to G). Then there is a defect group P of b^G such that $P \cap H = Q$. Moreover, $\mathbf{Z}(P) \subseteq \mathbf{Z}(Q)$.*

Proof. By Lemma (4.13), note that Q is contained in some defect group of b^G. Hence, if D_{b^G} is any defect group of b^G, then $\frac{|D_{b^G}|}{|Q|}$ is an integer. We will prove the theorem by double induction, first on $|G|$ and second on $\frac{|D_{b^G}|}{|Q|}$.

Let $b_Q \in \mathrm{Bl}(Q\mathbf{C}_G(Q)\,|\,Q)$ be a root of b. Then the block $(b_Q)^{\mathbf{N}_H(Q)}$ has defect group Q (because it is the Brauer first main correspondent of b by Theorem (9.7)). Also, $(b_Q)^{\mathbf{N}_H(Q)}$ is admissible (with respect to $\mathbf{N}_G(Q)$) since

$$\mathbf{C}_{\mathbf{N}_G(Q)}(Q) = \mathbf{C}_G(Q) \subseteq \mathbf{N}_H(Q)\,.$$

Suppose that $\mathbf{N}_G(Q) < G$. Arguing by induction on $|G|$, we conclude that there is a defect group X of $(b_Q)^{\mathbf{N}_G(Q)}$ such that $X \cap \mathbf{N}_H(Q) = Q$. Certainly, $Q \subseteq X$. If $X = Q$, then, by the first main theorem (and Problem (4.2)), we will have that $((b_Q)^{\mathbf{N}_G(Q)})^G = b^G$ has defect group Q, and we are done. So we may assume that $Q < X$. Now, the block $(b_Q)^{\mathbf{N}_G(Q)}$ is admissible with respect to G (by the last paragraph preceding the statement of this theorem) and induces b^G. By induction on $\frac{|D_b G|}{|Q|}$, there is a defect group P of b^G such that $P \cap \mathbf{N}_G(Q) = X$. Now, if $Q < P \cap H$, then

$$Q < \mathbf{N}_{P \cap H}(Q) = \mathbf{N}_G(Q) \cap P \cap H = X \cap H \cap \mathbf{N}_G(Q) = Q\,,$$

a contradiction. This contradiction implies that $Q = P \cap H$ and we are done in this case.

So we may assume that $Q \lhd G$. Now, by Corollary (9.21), we have that $(b_Q)^{T(b_Q)}$ is the unique block of $T(b_Q)$ which covers b_Q. Therefore, by Fong's theorem (9.17), if P is a defect group of $(b_Q)^{T(b_Q)}$, then $P \cap Q\mathbf{C}_G(Q) = Q$. Since $(T(b_Q) \cap H)/Q\mathbf{C}_G(Q)$ is a p'-group by Theorem (9.22) applied in H, we have that

$$P \cap H = P \cap (T(b_Q) \cap H) = P \cap Q\mathbf{C}_G(Q) = Q\,.$$

Now, $(b_Q)^{T(b_Q)}$ is the Fong-Reynolds correspondent of b^G over b_Q since $((b_Q)^{T(b_Q)})^G = b^G$ and $(b_Q)^{T(b_Q)}$ and $(b_Q)^G$ cover b_Q by Lemma (9.8). By the Fong-Reynolds theorem, we deduce that $P \in \delta(b^G)$.

To prove the second part of the theorem, note that

$$\mathbf{Z}(P) \subseteq \mathbf{C}_P(Q) = \mathbf{C}_P(Q) \cap H = \mathbf{C}_{P \cap H}(Q) = \mathbf{Z}(Q)\,.\qquad\blacksquare$$

Observe that the second part in Theorem (9.24) improves Theorem (5.18) for admissible blocks.

(9.25) COROLLARY. *If $b \in \mathrm{Bl}(H\,|\,Q)$ is admissible and b^G has abelian defect groups, then Q is a defect group of b^G.*

Proof. This is clear by Theorem (9.24). $\blacksquare$

Our next objective is to prove the following result.

(9.26) THEOREM (Knörr). *Let $b \in \text{Bl}(N \mid Q)$, where $N \lhd G$. If $B \in \text{Bl}(G)$ covers b, then there is a defect group P of B such that $P \cap N = Q$.*

Recall that we already obtained this theorem in the case where B was of maximal defect among $\text{Bl}(G \mid b)$ and b was G-invariant (Theorem (9.17)).

We need an easy lemma.

(9.27) LEMMA. *Suppose that $B \in \text{Bl}(G)$ covers $b \in \text{Bl}(N)$, where $N \lhd G$. Then $\text{d}(b) \leq \text{d}(B)$.*

Proof. Let $\theta \in \text{Irr}(b)$ be of height zero and let $\chi \in \text{Irr}(B)$ lie over θ (by Theorem (9.4)). By ordinary character theory (Corollary (11.29) of [Isaacs, *Characters*]), we know that $\frac{\chi(1)}{\theta(1)}$ divides $\frac{|G|}{|N|}$. Hence,

$$(\nu(|G|) - \text{d}(B) + \text{height}(\chi)) - (\nu(|N|) - \text{d}(b) + \text{height}(\theta)) \leq \nu(|G|) - \nu(|N|).$$

Since θ has height zero, it follows that $\text{d}(b) \leq \text{d}(b) + \text{height}(\chi) \leq \text{d}(B)$, as required. ∎

Proof of Theorem (9.26) (S. Gagola). We argue by double induction on $|G|$ and $|G : N|$. By the Fong-Reynolds theorem (and the inductive hypothesis), we may certainly assume that b is G-invariant.

We claim that G/N is simple. Suppose that M is a normal subgroup of G with $N < M < G$. Let $\mu \in \text{Irr}(B)$, let $\alpha \in \text{Irr}(M)$ be under μ and let $\beta \in \text{Irr}(N)$ be under α. Hence, β lies under μ and by Theorem (9.2) we have that $\beta \in \text{Irr}(b)$. Also, since α lies over β, again by Theorem (9.2) it follows that the unique block e of M which contains α covers b. Also, by the same argument, B covers e. By induction on $|G|$ and $|G : N|$, there is an $X \in \delta(e)$ such that $X \cap N = Q$, and there is a $Y \in \delta(B)$ such that $Y \cap M = X$. Now,

$$Y \cap N = (Y \cap M) \cap N = X \cap N = Q,$$

as desired.

Now, we claim that there exists $P \in \delta(B)$ such that $P \cap N \subseteq Q$. Using Lemma (9.15) (and its notation), if we write

$$e_b = \sum_{\substack{K \in \text{cl}(G) \\ K \subseteq N}} u_b(\hat{K}) \hat{K}$$

for some $u_b(\hat{K}) \in F$, then there exists a conjugacy class K of G inside N such that

$$u_b(\hat{K}) \neq 0 \neq \lambda_B(\hat{K}).$$

Note, in this case, that $u_b(\hat{K}) = a_b(\hat{L})$, whenever L is a conjugacy class of N inside K. Fix such a class L and let $x \in L \subseteq K$. Since $\lambda_B(\hat{K}) \neq 0$, by the min-max theorem there exists $P_1 \in \delta(B)$ such that P_1 is contained in a Sylow p-subgroup of $\mathbf{C}_G(x)$. Therefore, $P_1 \cap N$ is contained in a Sylow p-subgroup P_0 of $\mathbf{C}_N(x)$. Since $a_b(\hat{L}) \neq 0$, again by the min-max theorem we have that $(P_0)^n \subseteq Q$ for some $n \in N$. Thus, if $P = (P_1)^n \in \delta(B)$, then $P \cap N \subseteq Q$, as claimed.

Now, let $b_0 \in \mathrm{Bl}(\mathbf{N}_G(P) \,|\, P)$ be the Brauer first main correspondent of B. We claim that $(b_0)^{N\mathbf{N}_G(P)}$ covers b and has defect group P. First of all, observe that $(b_0)^{N\mathbf{N}_G(P)}$ is defined by Theorem (4.14). Now, by Lemma (4.13) we have that P is contained in some defect group X of $(b_0)^{N\mathbf{N}_G(P)}$. Also, because $((b_0)^{N\mathbf{N}_G(P)})^G = B$ (by Problem (4.2)), X is contained in some G-conjugate of P again by Lemma (4.13). This shows that $(b_0)^{N\mathbf{N}_G(P)}$ has defect group P. To finish the proof of the claim, let $\tau \in \mathrm{Irr}((b_0)^{N\mathbf{N}_G(P)})$. Since $((b_0)^{N\mathbf{N}_G(P)})^G = B$, by Corollary (6.4) there exists $\chi \in \mathrm{Irr}(B)$ over τ. Now, if θ is an irreducible constituent of τ_N, then θ lies under χ. By Theorem (9.2), we have that $\theta \in \mathrm{Irr}(b)$ (recall that b is G-invariant). Since τ lies over $\theta \in \mathrm{Irr}(b)$, by Theorem (9.2) it follows that $(b_0)^{N\mathbf{N}_G(P)}$ covers b, as claimed.

By induction on $|G|$ and the claim in the previous paragraph, we may assume that $N\mathbf{N}_G(P) = G$. Thus, $NP \lhd G$. Since G/N is simple, we have that $P \subseteq N$ or that $NP = G$. If $P \subseteq N$, then we have that $P = P \cap N \subseteq Q$. By Lemma (9.27), $\mathrm{d}(b) \leq \mathrm{d}(B)$ and we conclude that $P = Q$. In the other case, $NP = G$ and G/N is a p-group. In this case, by Corollary (9.6), B is the only block of G which covers b. By Fong's theorem (9.17), we have that $P \cap N \in \delta(b)$. Since $P \cap N \subseteq Q$, we have that $P \cap N = Q$, as required. $\blacksquare$

Next, we discuss the relationship between covering and the Brauer correspondence.

(9.28) THEOREM (Harris-Knörr). *Suppose that $N \lhd G$. Let $b \in \mathrm{Bl}(N \,|\, D)$ and let $b_0 \in \mathrm{Bl}(\mathbf{N}_N(D) \,|\, D)$ be the Brauer first main correspondent of b. Then the map $B \mapsto B^G$ from $\mathrm{Bl}(\mathbf{N}_G(D) \,|\, b_0) \to \mathrm{Bl}(G \,|\, b)$ is a defect group preserving bijection.*

If G and H are groups and $\sigma : G \to H$ is an isomorphism of groups, then σ induces an isomorphism of F-algebras $\sigma : \mathbf{Z}(FG) \to \mathbf{Z}(FH)$. If B is a block of G, then we define B^σ to be the unique block of H satisfying

$$\sigma^{-1} \cdot \lambda_B = \lambda_{B^\sigma}.$$

This is compatible with the definition that we made of B^σ at the beginning of this chapter. Of course, the characters and the defect groups of B are naturally mapped via σ to characters and defect groups of B^σ. We leave these details for the reader to check. Furthermore, if U is a subgroup of G and $b \in \mathrm{Bl}(U)$ is such that b^G is defined, then $(b^\sigma)^H$ is defined and equals $(b^G)^\sigma$. (See Problem (4.4).)

We will use the remark in the previous paragraph in the following form. If $U \subseteq V \subseteq G$ are subgroups of G, $g \in G$ and $b \in \mathrm{Bl}(U)$ is such that b^V is defined, then we have that $(b^g)^{V^g}$ is defined and equals $(b^V)^g$.

Proof of Theorem (9.28). We argue by induction on $|G|$. First of all notice that if B is a block of $\mathbf{N}_G(D)$, then B^G is defined by Theorem (4.14). Write $H = \mathbf{N}_G(D)$.

If $h \in H$, by the discussion before beginning this proof, we have that

$$(b_0^h)^N = (b_0^N)^h = b^h \,.$$

Also, b_0^h has defect group $D^h = D$. Thus, if $T(b)$ is the inertia group of b in G, by the uniqueness in the first main theorem we have that $T(b) \cap H$ is the inertia group in H of b_0. Now, by the Fong-Reynolds theorem, we have that induction of blocks defines defect group preserving bijections $i_1 : \mathrm{Bl}(H \cap T(b) \,|\, b_0) \to \mathrm{Bl}(H \,|\, b_0)$ and $i_2 : \mathrm{Bl}(T(b) \,|\, b) \to \mathrm{Bl}(G \,|\, b)$. Therefore, if $T(b) < G$, by the inductive hypothesis we have that induction of blocks defines a defect group preserving bijection $i_3 : \mathrm{Bl}(H \cap T(b) \,|\, b_0) \to \mathrm{Bl}(T(b) \,|\, b)$. By Problem (4.2), we easily see that the defect group preserving bijection

$$(i_1)^{-1} i_3 i_2 : \mathrm{Bl}(H \,|\, b_0) \to \mathrm{Bl}(G \,|\, b)$$

is induction of blocks. So it is no loss to assume that b is G-invariant. Since we have proved that $T(b) \cap H$ is the inertia subgroup of b_0 in H, it follows that b_0 is H-invariant.

Now, observe that H acts on the set $\Omega = \{Q \,|\, Q$ is a p-subgroup of G such that $Q \cap N = D\}$. Recall that Ω is nonempty since $D \in \Omega$. If $\mathcal{O}$ is a complete set of representatives of the orbits of the action of H on Ω, we claim that $Q_1, Q_2 \in \mathcal{O}$ are G-conjugate if and only if they are equal. If $Q_1^g = Q_2$ for some $g \in G$, then

$$D = Q_2 \cap N = Q_1^g \cap N = (Q_1 \cap N)^g = D^g \,,$$

and thus $g \in H$. The claim follows.

By Knörr's theorem (9.26) and the discussion in the previous paragraph, if B is a block of G (respectively, of H) covering b (respectively, b_0), it follows that B has a unique defect group in $\mathcal{O}$. Therefore, if for each p-subgroup Q of G we write

$$\mathrm{Bl}(G \,|\, Q, b) = \mathrm{Bl}(G \,|\, Q) \cap \mathrm{Bl}(G \,|\, b) \,,$$

then

$$\mathrm{Bl}(G \,|\, b) = \bigcup_{Q \in \mathcal{O}} \mathrm{Bl}(G \,|\, Q, b)$$

and
$$\mathrm{Bl}(H \mid b_0) = \bigcup_{Q \in \mathcal{O}} \mathrm{Bl}(H \mid Q, b_0)$$

are disjoint unions. Hence, to prove the theorem, it suffices to show that if $Q \in \mathcal{O}$, then induction defines a bijection $\mathrm{Bl}(H \mid Q, b_0) \to \mathrm{Bl}(G \mid Q, b)$.

Now, if $Q \in \mathcal{O}$, notice that $\mathbf{N}_G(Q) \subseteq H$ since $Q \cap N = D$. Hence, by Problem (4.3), we have that induction defines a bijection $\mathrm{Bl}(H \mid Q) \to \mathrm{Bl}(G \mid Q)$. Therefore, the theorem will be proved once we show that a block $B \in \mathrm{Bl}(H \mid Q)$ covers b_0 if and only if B^G covers b.

Since e_b is G-invariant and e_{b_0} is H-invariant, by the definition of covering we may write
$$e_b = \sum_{B \in \mathrm{Bl}(G \mid b)} e_B$$

and
$$e_{b_0} = \sum_{B \in \mathrm{Bl}(H \mid b_0)} e_B \,.$$

Therefore, a block $B \in \mathrm{Bl}(H)$ covers b_0 if and only if $\lambda_B(e_{b_0}) = 1$. Also, B^G covers b if and only if $\lambda_{B^G}(e_b) = 1$. Hence, it suffices to show that $\lambda_{B^G}(e_b) = \lambda_B(e_{b_0})$.

Let us write $\mathrm{Br}_D^G : \mathbf{Z}(FG) \to \mathbf{Z}(FH)$ for the Brauer homomorphism of G associated to D and let us write $\mathrm{Br}_D^N : \mathbf{Z}(FN) \to \mathbf{Z}(F(H \cap N))$ for the Brauer homomorphism of N associated to D. Recall that
$$\mathrm{Br}_D^G(\hat{K}) = \sum_{x \in K \cap \mathbf{C}_G(D)} x$$

for $K \in \mathrm{cl}(G)$ and
$$\mathrm{Br}_D^N(\hat{L}) = \sum_{x \in L \cap \mathbf{C}_N(D)} x$$

for $L \in \mathrm{cl}(N)$. Now, if $K \subseteq N$ is a conjugacy class of G, then $K \cap \mathbf{C}_G(D) = K \cap \mathbf{C}_N(D)$ and therefore
$$\mathrm{Br}_D^G(\hat{K}) = \mathrm{Br}_D^N(\hat{K}) \,.$$

Hence,
$$\mathrm{Br}_D^G(e_b) = \mathrm{Br}_D^N(e_b)$$

because $e_b \in \mathbf{Z}(FG) \cap \mathbf{Z}(FN)$. By the first main theorem, we have that $\mathrm{Br}_D^N(e_b) = e_{b_0}$ and thus
$$\mathrm{Br}_D^G(e_b) = e_{b_0} \,.$$

Since $\lambda_{B^G} = (\lambda_B)^G = \mathrm{Br}_D^G \lambda_B$ by Theorem (4.14), it follows that

$$\lambda_{B^G}(e_b) = \lambda_B(\mathrm{Br}_D^G(e_b)) = \lambda_B(e_{b_0}),$$

as desired. ∎

In 1979, J. Alperin and M. Broué introduced a fundamental idea which synthesized some of the basic results of block theory and group theory. Although we suggest the reader look at [Alperin, Book], we cannot resist writing down a proof of one of the main results there.

Let Q be a p-subgroup of G. A **subpair** of G is a pair (Q, b_Q), where b_Q is a block of $Q\mathbf{C}_G(Q)$. If $b \in \mathrm{Bl}(G)$, we say that (Q, b_Q) is a b-**subgroup** if $(b_Q)^G = b$. Also, a b-subgroup (Q, b_Q) is a **Sylow** b-**subgroup** if $Q \in \delta(b^G)$.

There is a natural containment between pairs. We write $(Q, b_Q) \lhd (P, b_P)$ if $Q \lhd P$, b_Q is invariant under conjugation by P and

$$(b_Q)^{P\mathbf{C}_G(Q)} = (b_P)^{P\mathbf{C}_G(Q)}.$$

In particular, note that if $(Q, b_Q) \lhd (P, b_P)$, then $(b_Q)^G = (b_P)^G$.
We write $(Q, b_Q) \subseteq (P, b_P)$ if there is a chain

$$(Q, b_Q) = (Q_0, b_0) \lhd (Q_1, b_1) \lhd \ldots \lhd (Q_n, b_n) = (P, b_P).$$

Note, again, that if $(Q, b_Q) \subseteq (P, b_P)$, then $(b_Q)^G = (b_P)^G$. Finally, if $b \in \mathrm{Bl}(G)$, (Q, b_Q) is a b-subgroup and $g \in G$, then $(Q, b_Q)^g = (Q^g, (b_Q)^g)$ is also a b-subgroup. (See the discussion before the proof of the Harris-Knörr theorem.)

(9.29) THEOREM (Alperin-Broué). *If b is a block of G, then every b-subgroup is contained in a Sylow b-subgroup and all the Sylow b-subgroups are G-conjugate.*

Proof. Suppose that (Q_1, b_1) and (Q_2, b_2) are Sylow b-subgroups. Then $Q_1, Q_2 \in \delta(b)$ and, thus, there exists a $g \in G$ such that $Q_1 = (Q_2)^g$. Hence, $(Q_2, b_2)^g = (Q_1, (b_2)^g)$. Now, $((b_2)^g)^G = (b_1)^G$. By the extended first main theorem (9.7), it follows that $(b_2)^g = (b_1)^n$ for some $n \in \mathbf{N}_G(Q_1)$. Then $(Q_2, b_2)^{gn^{-1}} = (Q_1, b_1)$, as desired.

To finish the proof, we assume that (Q, b_Q) is a maximal b-subgroup and we prove that $Q \in \delta(b)$. Write $b_0 = (b_Q)^{\mathbf{N}_G(Q)}$. Let T be the inertia group of b_Q in $\mathbf{N}_G(Q)$ and let $H/Q\mathbf{C}_G(Q) \in \mathrm{Syl}_p(T/Q\mathbf{C}_G(Q))$. Let P be a defect group of $(b_Q)^H$. Hence $Q \subseteq P$ by Theorem (4.8). By Corollary (9.21), it follows that $(b_Q)^H$ is the only block which covers b_Q. Therefore, Fong's theorem (9.17) applies and we deduce that $P \cap Q\mathbf{C}_G(Q)$ is a defect group of b_Q and $H = P\mathbf{C}_G(Q)$. Now, let b_P be a root of $(b_Q)^H$, so that b_P is a block of $PC_H(P)$ with defect group P which induces $(b_Q)^H$. Since $Q \subseteq P$, we see that $C_G(P) \subseteq C_G(Q) \subseteq H$. Therefore, (P, b_P) is a b-subgroup. By its construction, $(Q, b_Q) \lhd (P, b_P)$. Hence, by the maximality of (Q, b_Q), we deduce that $(Q, b_Q) = (P, b_P)$, and thus $H = QC_G(Q)$ and b_Q has defect group $Q = P \cap QC_G(Q)$. Then $T/QC_G(Q)$ is a p'-group, and we deduce that b has defect group Q by Theorem (9.22). ∎

The Alperin-Broué theory on subpairs led to the discovery of the Broué-Puig nilpotent blocks. Inspired by the Frobenius theorem on groups with a normal p-complement, a block $b \in \mathrm{Bl}(G)$ is said to be **nilpotent** if for every b-subgroup (Q, b_Q), $N_G((Q, b_Q))/C_G(Q)$ is a p-group, where $N_G((Q, b_Q))$ is the stabilizer of b_Q in $N_G(Q)$. If $b \in \mathrm{Bl}(G \mid P)$ is a nilpotent block, it is a remarkable theorem by M. Broué and L. Puig that $l(b) = 1$ and $k(b) = |\mathrm{Irr}(P)|$.

PROBLEMS

(9.1) If $N \lhd G$ and $B, B' \in \mathrm{Bl}(G)$, show that B and B' cover the same block of N if and only if there exists a sequence $\chi_1, \ldots, \chi_s$ of irreducible characters, where $\chi_1 \in \mathrm{Irr}(B)$ and $\chi_s \in \mathrm{Irr}(B')$, such that either χ_i and χ_{i+1} belong to the same block or $(\chi_i)_N$ and $(\chi_{i+1})_N$ have an irreducible constituent in common. Prove the same fact for irreducible Brauer characters.

(9.2) If A is a p-group acting on G and fixing $B \in \mathrm{Bl}(G)$, prove that there is a $\chi \in \mathrm{Irr}(B)$ fixed by A.

(9.3) Suppose that A is a group acting on G. If B is a p-block of positive defect, prove that B does not consist of A-conjugates of some $\theta \in \mathrm{Irr}(B)$.

(HINT: Consider the character $\sum_{\chi \in \mathrm{Irr}(B)} \chi(1)\chi$, weak block orthogonality and Corollary (9.18) in the semidirect product GA.)

(9.4) Let N be a normal subgroup of G, let b be a G-invariant block of N and let $B \in \mathrm{Bl}(G \mid P)$ be covering b. If $\tilde{b}$ is the unique block of PN covering b, prove that $\tilde{b}$ has defect group P.

(9.5) Suppose that $N \lhd G$. Let $b \in \mathrm{Bl}(N)$ and $B \in \mathrm{Bl}(G)$. Show that B covers b if and only if $e_b e_B \neq 0$.

(9.6) Let B be a block of G. A subsection (x, b) associated to B is **major** if the defect of b is the defect of B. Given a p-element x of G and a block $B \in \mathrm{Bl}(G \mid D)$, prove that there exists a major subsection (x, b) associated to B if and only if x is G-conjugate to some element of $Z(D)$. If D is abelian, prove that every subsection associated to B is major.

(NOTE: One of the key results concerning major subsections is that if (x, b) is a major subsection associated to B and $\chi \in \mathrm{Irr}(B)$, then $\chi^{(x,b)} \neq 0$. This result, due to Brauer, should be compared with Theorem (5.21).)

(9.7) Let $B \in \mathrm{Bl}(G \,|\, D)$ and let $b \in \mathrm{Bl}(D\mathbf{C}_G(D))$ be a root of B. Let Ω be a complete set of representatives of $T(b)$-conjugacy classes in $\mathbf{Z}(D)$, where $T(b)$ is the stabilizer of b in $\mathbf{N}_G(D)$. Prove that

$$\{(x, b^{\mathbf{C}_G(x)})\}_{x \in \Omega}$$

is a complete set of representatives of the G-conjugacy classes of the major subsections belonging to B.

(9.8) Let $N \triangleleft G$ such that G/N is a p-group. Suppose that $\theta \in \mathrm{Irr}(N)$ is G-invariant and has p-defect zero. Prove that there exists a subgroup P of G such that $G = PN$ and $P \cap N = 1$.

10. Characters and Blocks in p-Solvable Groups

We start this chapter with a remarkable result.

(10.1) THEOREM (Fong-Swan). *Suppose that G is p-solvable. If $\varphi \in \mathrm{IBr}(G)$, then there exists a $\chi \in \mathrm{Irr}(G)$ such that $\chi^0 = \varphi$.*

The proof we give here is a variation of a proof by M. Isaacs due to T. Wolf. First, we need a lemma.

(10.2) LEMMA. *Suppose that $N \lhd G$ and G/N is a p'-group. Let $\psi \in \mathrm{Irr}(N)$ be such that $\psi^0 \in \mathrm{IBr}(N)$ and $I_G(\psi) = I_G(\psi^0)$. Then the map $\chi \mapsto \chi^0$ from $\mathrm{Irr}(G \,|\, \psi) \to \mathrm{IBr}(G \,|\, \psi^0)$ is a bijection.*

Proof. We argue by induction on $|G|$. Let $T = I_G(\psi) = I_G(\psi^0)$ and assume that $T < G$. By induction, we have that the map $\eta \mapsto \eta^0$ is a bijection from $\mathrm{Irr}(T \,|\, \psi) \to \mathrm{IBr}(T \,|\, \psi^0)$. Now, by using the Clifford correspondence for ordinary and Brauer characters and the fact that $(\eta^G)^0 = (\eta^0)^G$, it is clear that we may assume that ψ is G-invariant.

Write $\psi^0 = \theta$ and put

$$\psi^G = \sum_{\xi \in \mathrm{Irr}(G \,|\, \psi)} a_\xi \xi$$

and

$$\theta^G = \sum_{\varphi \in \mathrm{IBr}(G \,|\, \theta)} b_\varphi \varphi \,.$$

(Of course, we are using Corollary (8.7), which states that the set of irreducible constituents of θ^G is $\mathrm{IBr}(G \,|\, \theta)$.)

Let $\chi \in \mathrm{Irr}(G \,|\, \psi)$. Then $\chi_N = a_\chi \psi$ by Frobenius reciprocity. Since $\mathbf{I}(\theta^G, \varphi) = \mathbf{I}(\theta, \varphi_N)$ (Corollary (8.7)) and using the fact that the set $\mathrm{IBr}(G)$ is linearly independent, we get

$$a_\chi \theta = (\chi_N)^0 = (\chi^0)_N = \sum_{\varphi \in \mathrm{IBr}(G)} d_{\chi\varphi} \varphi_N = \sum_{\varphi \in \mathrm{IBr}(G \,|\, \theta)} d_{\chi\varphi} b_\varphi \theta \,.$$

We conclude that

$$a_\chi = \sum_{\varphi \in \mathrm{IBr}(G \mid \theta)} d_{\chi\varphi} b_\varphi$$

and that $d_{\chi\varphi} = 0$ for $\varphi \notin \mathrm{IBr}(G \mid \theta)$. Now, since $(\psi^G)^0 = (\psi^0)^G = \theta^G$ and

$$(\psi^G)^0 = \sum_{\xi \in \mathrm{Irr}(G \mid \psi)} a_\xi \xi^0 = \sum_{\xi \in \mathrm{Irr}(G \mid \psi)} a_\xi \left(\sum_{\varphi \in \mathrm{IBr}(G)} d_{\xi\varphi} \varphi \right)$$

$$= \sum_{\varphi \in \mathrm{IBr}(G)} \left(\sum_{\xi \in \mathrm{Irr}(G \mid \psi)} d_{\xi\varphi} a_\xi \right) \varphi ,$$

it follows that

$$\sum_{\xi \in \mathrm{Irr}(G \mid \psi)} d_{\xi\varphi} a_\xi = b_\varphi$$

for all $\varphi \in \mathrm{IBr}(G \mid \theta)$. Therefore,

$$a_\chi = \sum_{\varphi \in \mathrm{IBr}(G \mid \theta)} d_{\chi\varphi} \left(\sum_{\xi \in \mathrm{Irr}(G \mid \psi)} d_{\xi\varphi} a_\xi \right)$$

$$= \sum_{\xi \in \mathrm{Irr}(G \mid \psi)} \left(\sum_{\varphi \in \mathrm{IBr}(G \mid \theta)} d_{\chi\varphi} d_{\xi\varphi} \right) a_\xi = \sum_{\xi \in \mathrm{Irr}(G \mid \psi)} \mathbf{I}(\chi^0, \xi^0) a_\xi .$$

It follows that $\mathbf{I}(\chi^0, \chi^0) = 1$ and $\mathbf{I}(\chi^0, \xi^0) = 0$ for $\chi \neq \xi \in \mathrm{Irr}(G \mid \psi)$. This shows that $\chi^0 \in \mathrm{IBr}(G \mid \theta)$ for $\chi \in \mathrm{Irr}(G \mid \psi)$ and that the map $\chi \mapsto \chi^0$ from $\mathrm{Irr}(G \mid \psi) \to \mathrm{IBr}(G \mid \theta)$ is injective.

Finally, suppose that $\varphi \in \mathrm{IBr}(G \mid \theta)$. Then

$$0 \neq b_\varphi = \sum_{\xi \in \mathrm{Irr}(G \mid \psi)} d_{\xi\varphi} a_\xi$$

and we conclude that there is a $\xi \in \mathrm{Irr}(G \mid \psi)$ such that $d_{\xi\varphi} \neq 0$. Since $\xi^0 \in \mathrm{IBr}(G)$ by the first part, it follows that $\xi^0 = \varphi$, as desired. $\blacksquare$

We will obtain the Fong-Swan theorem as a special case ($N = 1$) of the following result. For this, we will need a well-known fact about ordinary characters: if $N \triangleleft G$ and $\theta \in \mathrm{Irr}(N)$ is a G-invariant character such that $(|G : N|, \theta(1)\mathrm{o}(\theta)) = 1$, then θ extends to G and there is a unique extension χ of θ to G such that $\mathrm{o}(\chi) = \mathrm{o}(\theta)$ (Corollary (8.16) of [Isaacs, *Characters*]). The fact that θ extends to G is also a consequence of Theorem (8.23) (since $\mathrm{Irr}(G) = \mathrm{IBr}(G)$ whenever G is a p'-group). Once the existence has been proven, the uniqueness of such an extension is not difficult to establish.

(10.3) THEOREM (Wolf). *Suppose that G/N is p-solvable and let $\psi \in \operatorname{Irr}(N)$ be such that $\psi^0 \in \operatorname{IBr}(N)$ and $I_G(\psi) = I_G(\psi^0)$. Assume further that p does not divide $o(\psi)\psi(1)$. Then, given $\varphi \in \operatorname{IBr}(G \mid \psi^0)$, there exists a $\chi \in \operatorname{Irr}(G \mid \psi)$ such that $\chi^0 = \varphi$.*

Proof. We argue by induction on $|G/N|$. Let $T = I_G(\psi) = I_G(\psi^0)$. Suppose that $\eta \in \operatorname{IBr}(T \mid \psi)$ is the Clifford correspondent of φ over ψ^0. If $T < G$, then, using induction, we obtain a $\xi \in \operatorname{Irr}(T \mid \psi)$ such that $\xi^0 = \eta$. Now, $(\xi^G)^0 = (\xi^0)^G = \eta^G = \varphi$ and the theorem is proven in this case. Hence, we may assume that ψ is G-invariant.

Now, let M/N be a chief factor of G. If M/N is a p'-group, then by Lemma (10.2) we know that $\gamma \mapsto \gamma^0$ defines a bijection $\operatorname{Irr}(M \mid \psi) \to \operatorname{IBr}(M \mid \psi^0)$. In particular, for $\gamma \in \operatorname{Irr}(M \mid \psi)$, we note that $I_G(\gamma) = I_G(\gamma^0)$ since the above map is injective and ψ is G-invariant. Since φ lies over ψ^0, we see that φ lies over γ^0 for some $\gamma \in \operatorname{Irr}(M \mid \psi)$. Now, write $\gamma_N = d\psi$ for some integer d. Since $\frac{\gamma(1)}{\psi(1)}$ divides $|M : N|$, we see that $\gamma(1)$ is a p'-number. Also, observe that $\det(\gamma)_N = \det(\psi)^d$. Hence,

$$\det(\gamma)^{|M:N|o(\psi)} = 1$$

and we deduce that $o(\gamma)$ divides $|M/N|o(\psi)$, a p'-number. In this case, the theorem follows by induction on $|G : N|$.

Finally, we assume that M/N is a p-group. By Green's theorem (8.11), we have that $\operatorname{IBr}(M \mid \psi^0) = \{\delta\}$, where $\delta_N = \psi^0$. In particular, δ is G-invariant. Now, let $\tau \in \operatorname{Irr}(M)$ be the unique extension of ψ with $o(\tau) = o(\psi)$ (using Corollary (6.28) or Corollary (8.16) of [Isaacs, *Characters*]). By uniqueness, note that τ is G-invariant. Also, we have that $(\tau^0)_N = (\tau_N)^0 = \psi^0$. Hence, we see that τ^0 is irreducible (otherwise, τ^0 is the sum of two Brauer characters and so is $(\tau^0)_N = \psi^0 \in \operatorname{IBr}(N)$, a contradiction). Thus, $\tau^0 \in \operatorname{IBr}(M \mid \psi^0)$ and therefore $\tau^0 = \delta$. Since φ necessarily lies over δ, the theorem follows by induction on $|G : N|$. ∎

Proof of the Fong-Swan Theorem. Just set $N = 1$ in Theorem (10.3). ∎

(There is an alternative proof of the Fong-Swan theorem which may be obtained by using character triples. See Problem (10.11).)

(10.4) COROLLARY. *Suppose that G is p-solvable. Then $\operatorname{IBr}(G) = \{\chi^0 \mid \chi \in \operatorname{Irr}(G)$ is such that χ^0 is not of the form $\alpha^0 + \beta^0$ for characters α, β of $G\}$.*

Proof. This is clear by the Fong-Swan theorem. ∎

In particular, we see that for p-solvable groups the set $\operatorname{IBr}(G)$ is uniquely defined. But more than that is true. Observe that the set $\operatorname{IBr}(G)$ can be easily determined from the character table of G by using Corollary (10.4).

(10.5) COROLLARY. *Suppose that G is p-solvable and let $\varphi \in \mathrm{IBr}(G)$. If $\sigma \in \mathrm{Gal}(\mathbb{Q}_{|G|}/\mathbb{Q})$, then $\varphi^\sigma \in \mathrm{IBr}(G)$.*

Proof. By the Fong-Swan theorem, let $\chi \in \mathrm{Irr}(G)$ be such that $\chi^0 = \varphi$. If $(\chi^\sigma)^0 = \alpha^0 + \beta^0$ for some characters α and β of G, then

$$\varphi = \chi^0 = (\alpha^{\sigma^{-1}})^0 + (\beta^{\sigma^{-1}})^0.$$

Now, $(\alpha^{\sigma^{-1}})^0$ and $(\beta^{\sigma^{-1}})^0$ are Brauer characters of G because $\alpha^{\sigma^{-1}}$ and $\beta^{\sigma^{-1}}$ are ordinary characters of G. This is not possible since φ is irreducible. By Corollary (10.4), $\varphi^\sigma = (\chi^\sigma)^0 \in \mathrm{IBr}(G)$. ∎

In general, given a p-solvable group G and $\varphi \in \mathrm{IBr}(G)$, there might be different $\chi, \xi \in \mathrm{Irr}(G)$ such that $\chi^0 = \varphi = \xi^0$. (As an easy example, if $\lambda \in \mathrm{Irr}(G)$ is a linear character of p-power order, then $\lambda^0 = 1_{G^0}$.) It is a nontrivial theorem of M. Isaacs that, in p-solvable groups, given $\varphi \in \mathrm{IBr}(G)$, it is possible to choose a canonical $\chi \in \mathrm{Irr}(G)$ such that $\chi^0 = \varphi$. In fact, such a canonical character χ behaves well with respect to normal subgroups; that is, if $N \triangleleft G$ and ψ is an irreducible constituent of χ_N then $\psi^0 \in \mathrm{IBr}(N)$ and ψ is the canonical lift of ψ^0. Next, we prove this result for odd primes p. (The prime $p = 2$ poses some extra difficulties.)

Recall that $\chi \in \mathrm{Irr}(G)$ is **p-rational** if χ has its values in $\mathbb{Q}_{|G|_{p'}}$. By elementary Galois theory, note that $\chi \in \mathrm{Irr}(G)$ is p-rational if and only if $\chi^\sigma = \chi$ for every $\sigma \in \mathrm{Gal}(\mathbb{Q}_{|G|}/\mathbb{Q}_{|G|_{p'}})$.

(10.6) THEOREM (Isaacs). *Suppose that G is p-solvable, where p is odd. Given $\varphi \in \mathrm{IBr}(G)$, then there exists a unique p-rational $\chi \in \mathrm{Irr}(G)$ such that $\chi^0 = \varphi$. Furthermore, if $N \triangleleft G$ and ψ is an irreducible constituent of χ_N, then ψ is p-rational and $\psi^0 \in \mathrm{IBr}(N)$.*

(This theorem is not true for $p = 2$. Just take G to be any elementary abelian 2-group.)

We need an extendibility theorem which plays the role Green's theorem has in the proof of Theorem (10.3).

(10.7) THEOREM (Isaacs). *Let $N \triangleleft G$ be such that G/N is a p-group for $p \neq 2$. Suppose that $\theta \in \mathrm{Irr}(N)$ is invariant in G and p-rational. Then θ^G has a unique p-rational irreducible constituent χ. Furthermore, $\chi_N = \theta$.*

Proof. This is Theorem (6.30) of [Isaacs, *Characters*]. ∎

A useful remark when dealing with p-rational characters of a group and its subgroups is the following. If H is a subgroup of G, since $\mathbb{Q}_{|H|} \subseteq \mathbb{Q}_{|G|}$ it easily follows that χ^σ is defined for characters χ of H and $\sigma \in \mathrm{Gal}(\mathbb{Q}_{|G|}/\mathbb{Q})$. In fact, since restriction defines an isomorphism

$$\mathrm{Gal}(\mathbb{Q}_{|G|}/\mathbb{Q}_{|G|_{p'}}) \cong \mathrm{Gal}(\mathbb{Q}_{|G|_p}/\mathbb{Q})$$

and $\mathbb{Q}_{|H|_p} \subseteq \mathbb{Q}_{|G|_p}$, by elementary Galois theory we see that every $\mu \in \mathrm{Gal}(\mathbb{Q}_{|H|}/\mathbb{Q}_{|H|_{p'}})$ can be extended to some $\sigma \in \mathrm{Gal}(\mathbb{Q}_{|G|}/\mathbb{Q}_{|G|_{p'}})$. We deduce that $\chi \in \mathrm{Irr}(H)$ is p-rational if and only if $\chi^\sigma = \chi$ for every $\sigma \in \mathrm{Gal}(\mathbb{Q}_{|G|}/\mathbb{Q}_{|G|_{p'}})$.

We also need a lemma.

(10.8) LEMMA. *Let $N \lhd G$ be such that G/N is a p-group for $p \neq 2$. Suppose that $\theta \in \mathrm{Irr}(N)$ is p-rational. Then θ^G has a unique p-rational irreducible constituent χ. Furthermore, suppose that $\theta^0 \in \mathrm{IBr}(N)$ and that $I_G(\theta) = I_G(\theta^0)$. Then $\chi^0 \in \mathrm{IBr}(G)$.*

Proof. Let $T = I_G(\theta)$. By Theorem (10.7), we know that there exists a unique p-rational character $\eta \in \mathrm{Irr}(T|\theta)$. Also, $\eta_N = \theta$. Let $\sigma \in \mathrm{Gal}(\mathbb{Q}_{|G|}/\mathbb{Q}_{|G|_{p'}})$. Since $\eta^\sigma = \eta$, it follows that $(\eta^G)^\sigma = \eta^G$. This proves that $\eta^G = \chi$ is p-rational. Suppose now that $\mu \in \mathrm{Irr}(G|\theta)$ is p-rational. Since both μ and θ are fixed by σ, we deduce that the Clifford correspondent ξ of μ over θ is fixed by σ. Hence, we deduce that ξ is a p-rational character over θ. Therefore, $\xi = \eta$ and thus $\mu = \xi^G = \eta^G = \chi$. The first part of the lemma follows.

Assume now that $\theta^0 \in \mathrm{IBr}(N)$ and that $I_G(\theta) = I_G(\theta^0)$. Since $\eta_N = \theta$, we have that $(\eta^0)_N = (\eta_N)^0 = \theta^0 \in \mathrm{IBr}(N)$. Thus, η^0 is irreducible because it extends an irreducible Brauer character. Also, $\eta^0 \in \mathrm{IBr}(I_G(\theta^0)|\theta^0)$. By the Clifford correspondence for Brauer characters, we have that $(\eta^0)^G \in \mathrm{IBr}(G)$. Hence, $\chi^0 = (\eta^G)^0 = (\eta^0)^G \in \mathrm{IBr}(G)$, as desired. ∎

Proof of Theorem (10.6). We argue by induction on $|G|$.

Let $\chi \in \mathrm{Irr}(G)$ be p-rational and such that $\chi^0 \in \mathrm{IBr}(G)$. If $N \lhd G$ and θ is an irreducible constituent of χ_N, we prove by induction on $|N|$ that θ is p-rational and $\theta^0 \in \mathrm{IBr}(N)$.

Let $\mathcal{G} = \mathrm{Gal}(\mathbb{Q}_{|G|}/\mathbb{Q}_{|G|_{p'}})$ and let $V = \{g \in G \,|\, \theta^g = \theta^\sigma \text{ for some } \sigma \in \mathcal{G}\}$, which is a subgroup of G. Let T be the inertia group of θ in G, so that $T \subseteq V$. Also, let $\psi \in \mathrm{Irr}(T)$ be the Clifford correspondent of χ over θ. Write $\eta = \psi^V \in \mathrm{Irr}(V)$.

We claim that η is p-rational with $\eta^0 \in \mathrm{IBr}(V)$. The second part is clear since $(\eta^0)^G = (\eta^G)^0 = \chi^0 \in \mathrm{IBr}(G)$ and thus, η^0 must be irreducible. To prove that η is p-rational, we observe that, by two applications of the Clifford correspondence (one in V and the other one in G), induction defines a bijection $\mathrm{Irr}(V|\theta) \to \mathrm{Irr}(G|\theta)$. Now, suppose that $\sigma \in \mathcal{G}$. Since $\chi = \chi^\sigma$

it follows by Clifford's theorem that θ and θ^σ are G-conjugate. Therefore, θ and θ^σ are V-conjugate. Hence, η and η^σ lie in $\mathrm{Irr}(V \mid \theta)$ and both induce χ. We conclude that $\eta^\sigma = \eta$ and this proves the claim.

If $V < G$, arguing by induction on $|G|$, we have that the theorem is true for V. Since $\eta \in \mathrm{Irr}(V)$ is p-rational with $\eta^0 \in \mathrm{IBr}(V)$, $N \lhd V$, and θ is an irreducible constituent of η_N, we deduce that θ is p-rational and $\theta^0 \in \mathrm{IBr}(N)$, as desired. Therefore, we may assume that $V = G$.

Now, we claim that $T = G$. Given $g \in G$, by the definition of $V = G$, there exists a $\sigma \in \mathcal{G}$ such that $\theta^g = \theta^\sigma$. Since $I_G(\theta^\sigma) = I_G(\theta)$, we conclude that $T^g = T$. Hence, $T \lhd G$. Also, note that ψ^g lies over $\theta^g = \theta^\sigma$ and induces χ. Since $\chi = \chi^\sigma$, we also have that ψ^σ induces χ and certainly lies over θ^σ. By the uniqueness in the Clifford correspondence, we conclude that $\psi^g = \psi^\sigma$. Hence, $(\psi^0)^g = (\psi^0)^\sigma = \psi^0$ (since σ fixes $\mathbb{Q}_{|G|_{p'}}$). We deduce that the Brauer character ψ^0 is G-invariant. However, $\chi^0 = (\psi^G)^0 = (\psi^0)^G$ is irreducible. By Problem (8.3), $T = G$ and the claim is proved.

Now, we may write $\chi_N = e\theta$ and in this case it is clear that θ is p-rational. Since we may certainly assume that $N > 1$, let N/M be a chief factor of G. Let $\xi \in \mathrm{Irr}(M)$ be under θ (and therefore under χ). By induction on $|N|$, we know that ξ is p-rational and $\xi^0 \in \mathrm{IBr}(M)$. Also, by induction on $|G|$, we know that ξ is the unique p-rational lifting of ξ^0. In particular, $I_G(\xi) = I_G(\xi^0)$. If N/M is a p'-group, we conclude that $\theta^0 \in \mathrm{IBr}(N)$ by Lemma (10.2). If N/M is a p-group, we conclude that $\theta^0 \in \mathrm{IBr}(N)$ by Lemma (10.8).

Next, given $\varphi \in \mathrm{IBr}(G)$, we prove that there exists a unique p-rational $\chi \in \mathrm{Irr}(G)$ such that $\chi^0 = \varphi$. Let $U \lhd G$ be a maximal normal subgroup of G and let $\mu \in \mathrm{IBr}(U)$ be an irreducible constituent of φ_U. By induction, there exists a unique p-rational $\tau \in \mathrm{Irr}(U)$ such that $\tau^0 = \mu$. By uniqueness, note that $I_G(\tau) = I_G(\tau^0)$.

First suppose that G/U is a p-group. By Lemma (10.8), there exists a unique p-rational character $\chi \in \mathrm{Irr}(G \mid \tau)$. Furthermore, χ^0 is irreducible (and lies over $\tau^0 = \mu$). By Green's theorem (8.11), we have that $\mathrm{IBr}(G \mid \tau^0) = \{\varphi\}$ and thus we conclude that $\chi^0 = \varphi$. Also, if δ is another p-rational lifting of φ, by the first part of this proof we know that the irreducible constituents of δ_U are p-rational and lift irreducible Brauer characters of U. Since φ lies over μ, by uniqueness we see that δ lies over τ. Hence, $\delta = \chi$ because χ is the unique p-rational character over τ.

Suppose now that G/U is a p'-group. By Lemma (10.2), we know that the map $\chi \mapsto \chi^0$ from $\mathrm{Irr}(G \mid \tau) \to \mathrm{IBr}(G \mid \tau^0)$ is a bijection. Now, if $\sigma \in \mathcal{G}$ and $\chi \in \mathrm{Irr}(G \mid \tau)$, we have that $(\chi^\sigma)^0 = (\chi^0)^\sigma = \chi^0$. By the uniqueness in Lemma (10.2), we conclude that $\chi^\sigma = \chi$. This proves that every irreducible character over τ is p-rational. Since $\varphi \in \mathrm{IBr}(G \mid \tau^0)$, we deduce that there exists a p-rational $\chi \in \mathrm{Irr}(G \mid \tau)$ such that $\chi^0 = \varphi$. Also, if $\delta^0 = \varphi$ and

$\delta \in \mathrm{Irr}(G)$ is p-rational, by the first part of the theorem we know that every irreducible constituent of δ_U is p-rational and lifts an irreducible Brauer character of U. Since φ_U contains τ^0, by the uniqueness of the p-rational lifting applied in U we see that δ_U contains τ. Hence, $\delta = \chi$ by Lemma (10.2). This completes the proof of the theorem. ∎

Next, we devote ourselves to studying the relationship between Brauer characters and the characters of a p-complement of G.

(10.9) THEOREM. *Suppose that G is p-solvable and let H be a p-complement of G. Then the map $\varphi \mapsto \varphi_H$ from $\{\varphi \in \mathrm{IBr}(G) \,|\, \varphi(1) \text{ has } p'\text{-degree}\} \to \mathrm{Irr}(H)$ is a well-defined injective map.*

We need a lemma.

(10.10) LEMMA. *Suppose that G is p-solvable and let $N \triangleleft G$ be such that G/N is a p'-group. Let H be a p-complement of G and assume that $\theta \in \mathrm{IBr}(N)$ is such that $\theta_{H \cap N} \in \mathrm{Irr}(H \cap N)$. Then restriction defines a bijection $\mathrm{IBr}(G \,|\, \theta) \to \mathrm{Irr}(H \,|\, \theta_{N \cap H})$.*

Proof. First of all, recall that if φ is a Brauer character, then φ_H is an ordinary character of H. Also, if φ and γ are Brauer characters with $\varphi_H = \gamma_H$, then $\varphi = \gamma$ because every p-regular element lies in some G-conjugate of H (and φ and γ are class functions).

Since G/N is a p'-group, we have that $G = NH$.

Let $\varphi \in \mathrm{IBr}(G \,|\, \theta)$. We prove that $\varphi_H \in \mathrm{Irr}(H)$ using induction on $|G|$. Let $T = I_G(\theta)$. We claim that

$$T \cap H = I_H(\theta_{N \cap H}).$$

If $t \in H$ fixes θ, it is clear that t fixes $\theta_{N \cap H}$. Conversely, if $t \in H$ fixes $\theta_{N \cap H}$, then θ and θ^t coincide on $H \cap N$, a p-complement of N. By the first paragraph of this proof, we deduce that $\theta^t = \theta$. This proves the claim.

Now, let $\eta \in \mathrm{IBr}(T \,|\, \theta)$ be the Clifford correspondent of φ over θ. If $T < G$, by induction we have that $\eta_{T \cap H} \in \mathrm{Irr}(T \cap H)$. (Note that $T \cap H$ is a p-complement of T.) Now, since $\eta_{T \cap H}$ lies over $\theta_{N \cap H}$, we deduce that $(\eta_{T \cap H})^H \in \mathrm{Irr}(H)$ using the Clifford correspondence. By Mackey's theorem (Problem (8.5)), we have that

$$\varphi_H = (\eta^G)_H = (\eta_{N \cap H})^H$$

is irreducible. Hence, we may assume that θ is G-invariant. Thus, $\theta_{N \cap H}$ is H-invariant.

Now, let $\delta \in \mathrm{Irr}(H)$ be under φ. Since $\varphi_N = e\theta$ for some integer e, $\varphi_{N \cap H} = e\theta_{N \cap H}$ and we may also write $\delta_{N \cap H} = f\theta_{N \cap H}$ for some integer f. Since p does not divide $|G/N|$, by Corollary (8.7) (and Frobenius reciprocity) we may write

$$\theta^G = e\varphi + \Delta \quad \text{and} \quad (\theta_{N\cap H})^H = f\delta + \Psi,$$

where Δ is a Brauer character of G (or zero), Ψ is a character of H (or zero) and $\mathbf{I}(\Delta, \varphi) = 0 = [\Psi, \delta]$. Also, write

$$\varphi_H = u\delta + \Xi,$$

where $u = [\varphi_H, \delta]$ and Ξ is a character of H (or zero). Now, by Mackey's theorem

$$e\varphi_H + \Delta_H = (\theta^G)_H = (\theta_{N\cap H})^H = f\delta + \Psi$$

and we conclude that $eu \leq f$. Also,

$$e\theta_{N\cap H} = \varphi_{H\cap N} = u\delta_{H\cap N} + \Xi_{N\cap H} = uf\theta_{H\cap N} + \Xi_{N\cap H}$$

and therefore $uf \leq e$. Hence, $eu^2 = (eu)u \leq fu \leq e$ and we deduce that $u = 1$ and $e = f$. Now,

$$\varphi(1) = e\theta(1) = f\theta(1) = \delta(1),$$

and we see that $\varphi_H = \delta$ is irreducible.

By the first paragraph of the proof, our map is clearly injective. It remains to show that it is onto. Let $\xi \in \mathrm{Irr}(H \mid \theta_{N\cap H})$. By Frobenius reciprocity, we have that ξ is an irreducible constituent of $(\theta_{N\cap H})^H$. By Corollary (8.7), write $\theta^G = \sum_{\varphi \in \mathrm{IBr}(G \mid \theta)} a_\varphi \varphi$ for some integers a_φ. Now,

$$(\theta_{N\cap H})^H = (\theta^G)_H = \sum_{\varphi \in \mathrm{IBr}(G \mid \theta)} a_\varphi \varphi_H.$$

Since $\varphi_H \in \mathrm{Irr}(H)$ for every $\varphi \in \mathrm{IBr}(G \mid \theta)$, we deduce that $\xi = \varphi_H$ for some $\varphi \in \mathrm{IBr}(G \mid \theta)$. ∎

Proof of Theorem (10.9). We argue by induction on $|G|$. Let $N \triangleleft G$ be a maximal normal subgroup and let $\theta \in \mathrm{IBr}(N)$ be an irreducible constituent of φ_N. By Clifford's theorem, we have that $\theta(1)$ divides $\varphi(1)$ and thus θ has p'-degree. By induction, we have that $\theta_{N\cap H} \in \mathrm{Irr}(N \cap H)$.

Suppose first that G/N is a p-group. Hence, $H \subseteq N$. Then, by Swan's theorem (8.22), we have that $\frac{\varphi(1)}{\theta(1)}$ divides $|G : N|$. Since $\varphi(1)$ has p'-degree, we conclude that $\varphi_N = \theta$. Since $\varphi_H = \varphi_{H\cap N} = \theta_{H\cap N} = \theta_H$ is irreducible, we are done in this case.

Suppose now that G/N is a p'-group. Then $\varphi_H \in \mathrm{Irr}(H)$ by applying Lemma (10.10). The injectivity of the map is obvious. (See the first paragraph of the proof of Lemma (10.10).) ∎

Brauer characters of p'-degree are important because of Huppert's theorem, among other reasons.

(10.11) THEOREM (Huppert). *Suppose that G is p-solvable. If $\varphi \in$
$\mathrm{IBr}(G)$, then there exists a subgroup J of G and $\gamma \in \mathrm{IBr}(J)$ of p'-degree
such that $\gamma^G = \varphi$.*

We derive Huppert's theorem from an argument of Isaacs.

(10.12) LEMMA (Isaacs). *Suppose that G is p-solvable. If $\varphi \in \mathrm{IBr}(G)$
does not have p'-degree, then there exists a normal subgroup N of G
such that the irreducible constituents of φ_N have p'-degree and are not
G-invariant.*

Proof. Let N be a normal subgroup of G maximal such that the irreducible
constituents of φ_N have p'-degree. We prove that if $\theta \in \mathrm{IBr}(N)$ is an
irreducible constituent of φ_N, then θ is not G-invariant.

Suppose that θ is G-invariant. By hypothesis, we know that $N < G$.
Let $M/N > 1$ be a chief factor of G, so that M/N is either a p-group or a
p'-group. Let $\eta \in \mathrm{IBr}(M)$ lie under φ. As θ is G-invariant, we have that η
lies over θ.

If M/N is a p-group, by Green's theorem, we have that $\eta_N = \theta$. Thus,
η has p'-degree and this contradicts the maximality of N. On the other
hand, if M/N is a p'-group, by Dade's theorem (8.30), we have that $\frac{\eta(1)}{\theta(1)}$
divides $|M : N|$. Hence, $\eta(1)$ has p'-degree and this contradicts again the
maximality of N. ∎

Proof of Theorem (10.11) We may certainly assume that φ does not
have p'-degree. If $\varphi = \tau^G$ for some $\tau \in \mathrm{IBr}(U)$, where $U < G$, then the result
follows from induction on $|G|$. Otherwise, by the Clifford correspondence
we have that φ_N is a multiple of a single $\theta \in \mathrm{IBr}(N)$ for every $N \lhd G$. But
this contradicts Lemma (10.12). ∎

(NOTE: It is a nontrivial fact that in a p-solvable group G the number
of irreducible Brauer characters of p'-degree coincides with $|\mathrm{Irr}(\mathbf{N}_G(P)/P)|$,
where P is a Sylow p-subgroup of G.)

Our next objective is an important theorem of P. Fong.

(10.13) THEOREM (Fong). *Suppose that G is p-solvable and let H
be a p-complement of G. If $\varphi \in \mathrm{IBr}(G)$, then there exists a character α of
H such that $\alpha^G = \Phi_\varphi$. Furthermore, every such α is irreducible and has
degree $\varphi(1)_{p'}$.*

(10.14) COROLLARY (Fong's Dimensional Formula). *If G is p-
solvable and $\varphi \in \mathrm{IBr}(G)$, then*

$$\Phi_\varphi(1) = |G|_p \varphi(1)_{p'}.$$

Proof. This is clear by Theorem (10.13). ∎

To prove Fong's theorem (10.13), we need some lemmas.

(10.15) LEMMA. *Suppose that G is p-solvable. Let H be a p-complement of G and assume that $\varphi \in \mathrm{IBr}(G)$ has p'-degree. If $\mu \in \mathrm{IBr}(G)$ lies over φ_H, then $\mu = \varphi$.*

Proof. We argue by induction on $|G|$. By Theorem (10.9), let $\alpha = \varphi_H \in \mathrm{Irr}(H)$. Let N be a maximal normal subgroup of G. Let $\xi \in \mathrm{Irr}(N \cap H)$ be an irreducible constituent of $\alpha_{N \cap H}$. Since φ lies over ξ, there exists an irreducible constituent θ of φ_N which lies over ξ. Now, by Clifford's theorem we have that $\theta(1)$ divides $\varphi(1)$ and thus θ has p'-degree. By Theorem (10.9), we have that $\theta_{N \cap H}$ is irreducible. Hence, $\theta_{N \cap H} = \xi$. Since μ lies over α, we have that μ lies over ξ, so there exists an irreducible constituent τ of μ_N which lies over ξ. By the inductive hypothesis, we have that $\tau = \theta$.

If G/N is a p-group, then φ and μ are irreducible Brauer characters of G lying over θ. By Green's theorem, we deduce that $\varphi = \mu$.

Suppose now that G/N is a p'-group. By Lemma (10.10), we know that restriction defines a bijection from $\mathrm{IBr}(G \mid \theta)$ onto $\mathrm{Irr}(H \mid \theta_{N \cap H})$. Since μ lies over θ, we deduce that μ_H is irreducible. Hence, $\mu_H = \alpha = \varphi_H$. Therefore, $\mu = \varphi$, as required. ∎

The next lemma should be compared with Corollary (2.17).

(10.16) LEMMA. *Suppose that G is p-solvable. If χ is a character of G vanishing off p-regular elements, then χ is a nonnegative integer linear combination of $\{\Phi_\varphi \mid \varphi \in \mathrm{IBr}(G)\}$.*

Proof. By Theorem (2.13), write

$$\chi = \sum_{\varphi \in \mathrm{IBr}(G)} a_\varphi \Phi_\varphi$$

for some $a_\varphi \in \mathbb{C}$. Then, again by Theorem (2.13), we have that $a_\varphi = [\chi, \varphi]^0$. Now let $\psi \in \mathrm{Irr}(G)$ be such that $\psi^0 = \varphi$ by the Fong-Swan theorem. Then

$$[\chi, \varphi]^0 = [\chi, \psi]$$

is a nonnegative integer, as required. ∎

Proof of Theorem (10.13). We argue by induction on $|G|$ that there exists some character α of H with degree $\varphi(1)_{p'}$ inducing Φ_φ. (If this is the case, notice that every character β of H with $\beta^G = \Phi_\varphi$ necessarily has degree $\varphi(1)_{p'}$ because $\beta^G(1)_p = |G|_p$ and $\Phi_\varphi(1) = \alpha^G(1) = |G|_p\varphi(1)_{p'}$.)

Suppose first that φ does not have p'-degree. Then, by Isaacs' lemma (10.12), there exists a normal subgroup $N \triangleleft G$ such that the irreducible constituents of φ_N are not G-invariant. Let $\theta \in \mathrm{IBr}(N)$ be one of them. Set $T = I_G(\theta)$. By elementary group theory, there exists a $g \in G$ such that $T \cap H^g$ is a p-complement of T. Therefore, by replacing θ by $\theta^{g^{-1}}$ (and T by $T^{g^{-1}}$), we may assume that $T \cap H$ is a p-complement of T.

Now, if $\mu \in \mathrm{IBr}(T)$ is the Clifford correspondent of φ over θ, we have that $(\Phi_\mu)^G = \Phi_\varphi$ by Theorem (8.10). By induction, there exists a character γ in $H \cap T$ of degree $\mu(1)_{p'}$ such that $\gamma^T = \Phi_\mu$. Now, if $\alpha = \gamma^H$, then

$$\alpha^G = \gamma^G = (\gamma^T)^G = (\Phi_\mu)^G = \Phi_\varphi.$$

Also,

$$\alpha(1) = |H : T \cap H|\gamma(1) = |G : T|_{p'}\mu(1)_{p'} = \varphi(1)_{p'}.$$

So we may assume that φ has p'-degree. We claim that the ordinary character φ_H of H induces Φ_φ. If $\chi \in \mathrm{Irr}(G)$, we have that

$$\chi^0 = \sum_{\psi \in \mathrm{IBr}(G)} d_{\chi\psi}\psi.$$

Hence,

$$\chi_H = \sum_{\psi \in \mathrm{IBr}(G)} d_{\chi\psi}\psi_H.$$

By Lemma (10.15),

$$[\chi_H, \varphi_H] = d_{\chi\varphi}.$$

Now, by Frobenius reciprocity, we have that

$$[\chi, (\varphi_H)^G] = d_{\chi\varphi}.$$

Hence,

$$(\varphi_H)^G = \sum_{\chi \in \mathrm{Irr}(G)} d_{\chi\varphi}\chi = \Phi_\varphi$$

and the claim is proved. Since $\varphi(1) = \varphi_H(1) = \varphi(1)_{p'}$, the proof by induction is completed.

Finally, we prove that if α is a character of H with $\alpha^G = \Phi_\varphi$, then $\alpha \in \mathrm{Irr}(H)$. If α is not irreducible, then $\alpha = \gamma + \beta$ for some characters

γ, β of H. Therefore, $\Phi_\varphi = \gamma^G + \beta^G$. Now Lemma (10.16) applies and contradicts the fact that the characters Φ_φ are linearly independent. $\blacksquare$

If $\varphi \in \mathrm{IBr}(G)$ and $\alpha \in \mathrm{Irr}(H)$ induces Φ_φ, we say that α is a **Fong character** associated with φ.

Next, we give two characterizations of Fong characters.

(10.17) THEOREM. *Suppose that G is p-solvable and let H be a p-complement of G. Let $\varphi \in \mathrm{IBr}(G)$. Then $\alpha \in \mathrm{Irr}(H)$ is a Fong character associated with φ if and only if*

$$[\mu_H, \alpha] = \delta_{\mu\varphi}$$

for every $\mu \in \mathrm{IBr}(G)$.

Proof. If α is a Fong character associated with φ, then

$$[\mu_H, \alpha] = [\hat{\mu}_H, \alpha] = [\hat{\mu}, \alpha^G] = [\mu, \alpha^G]^0 = [\mu, \Phi_\varphi]^0 = \delta_{\mu\varphi},$$

using Theorem (2.13). Conversely, suppose that $[\mu_H, \alpha] = \delta_{\mu\varphi}$ for every $\mu \in \mathrm{IBr}(G)$ and let $\chi \in \mathrm{Irr}(G)$. Then

$$[\alpha^G, \chi] = [\alpha, \chi_H] = \sum_{\mu \in \mathrm{IBr}(G)} d_{\chi\mu}[\alpha, \mu_H] = d_{\chi\varphi}.$$

Thus, $\alpha^G = \Phi_\varphi$, as required. $\blacksquare$

In particular, notice that if α is a character of a p-complement H of G with $\alpha^G = \Phi_\varphi$, then α is an irreducible constituent of φ_H by Theorems (10.13) and (10.17).

(10.18) THEOREM. *Suppose that G is p-solvable and let H be a p-complement of G. If $\varphi \in \mathrm{IBr}(G)$, then the Fong characters associated with φ are the irreducible constituents of φ_H of smallest possible degree. This degree is $\varphi(1)_{p'}$.*

Proof. If $\varphi \in \mathrm{IBr}(G)$ and $\alpha \in \mathrm{Irr}(H)$, notice that α is an irreducible constituent of φ_H if and only if

$$0 \neq [\varphi_H, \alpha] = [\hat{\varphi}_H, \alpha] = [\hat{\varphi}, \alpha^G] = [\varphi, \alpha^G]^0 \neq 0.$$

Now, by Theorem (2.13) we may write

$$\alpha^G = \sum_{\mu \in \mathrm{IBr}(G)} [\alpha^G, \mu]^0 \Phi_\mu \tag{1}$$

where each $[\alpha^G, \mu]^0$ is a nonnegative integer by Lemma (10.16). Therefore,

$$\alpha(1) = \sum_{\mu \in \mathrm{IBr}(G)} [\alpha^G, \mu]^0 \mu(1)_{p'} \tag{2}$$

by Fong's dimensional formula.

By Theorems (10.13) and (10.17), we already know that φ_H has an irreducible constituent of degree $\varphi(1)_{p'}$. Also, if α is any irreducible constituent of φ_H, by (2) and the first paragraph, we see that

$$\alpha(1) \geq [\alpha^G, \varphi]^0 \varphi(1)_{p'} \geq \varphi(1)_{p'}.$$

It remains to show that if α is an irreducible constituent of φ_H of degree $\varphi(1)_{p'}$, then α is a Fong character associated with φ. By (2) and the first paragraph, we deduce that $[\alpha^G, \varphi]^0 = 1$ and $[\alpha^G, \mu]^0 = 0$ if $\varphi \neq \mu \in \mathrm{IBr}(G)$. Hence, by (1) we have that $\alpha^G = \Phi_\varphi$, and the theorem is proved. ∎

We close this analysis of Brauer characters and p-complements in p-solvable groups with a theorem of M. Isaacs. Recall that $\chi \in \mathrm{Irr}(G)$ is **quasiprimitive** if for every $N \triangleleft G$ we have that χ_N is a multiple of a single irreducible character of N.

(10.19) THEOREM (Isaacs). *Suppose that G is p-solvable. Let H be a p-complement of G. If $\alpha \in \mathrm{Irr}(H)$ is quasiprimitive, then $\alpha^G = \Phi_\varphi$ for some $\varphi \in \mathrm{IBr}(G)$.*

Proof. Let $\varphi \in \mathrm{IBr}(G)$ be over α. If φ has p'-degree, then $\alpha = \varphi_H$ and $\alpha^G = \Phi_\varphi$ by Theorem (10.18), for instance. So we may assume that φ does not have p'-degree. By Lemma (10.12), there exists $N \triangleleft G$ such that the irreducible constituents of φ_N have p'-degree and are not G-invariant.

Now, let $\nu \in \mathrm{Irr}(N \cap H)$ be under α and let $\theta \in \mathrm{IBr}(N)$ be over ν and under φ. Since θ has p'-degree, we have that $\nu = \theta_{N \cap H}$ by Theorem (10.9). Now, since α is quasiprimitive, we have that $\alpha_{N \cap H} = e\nu$ for some integer e. Hence, ν is H-invariant and by the easy part of Theorem (10.9) we have that θ is H-invariant. Therefore $H \subseteq T \subseteq G$, where $T = I_G(\theta) < G$. Arguing by induction on $|G|$, we have that $\alpha^T = \Phi_\mu$, for some $\mu \in \mathrm{IBr}(T)$. Now μ lies over α by Theorem (10.17) and therefore over ν. Hence we may find $\eta \in \mathrm{IBr}(N)$ lying over ν and under μ. By Lemma (10.15), we have that $\eta = \theta$ and thus, by the Clifford correspondence, we have that μ^G is irreducible. Now, by Theorem (8.10), we have that $(\Phi_\mu)^G = \Phi_{\mu^G}$. Hence

$$\alpha^G = (\alpha^T)^G = (\Phi_\mu)^G = \Phi_\varphi$$

and we are done. ∎

Now we come back to blocks.

The main tool used when studying blocks of p-solvable groups is Fong's theorem (10.20). This, together with the Fong-Reynolds theorem and covering theory, completely describes the blocks of p-solvable groups. Recall that if G is a p'-group, then the p-blocks of G are the singletons $\{\chi\}$ for $\chi \in \operatorname{Irr}(G)$.

(10.20) THEOREM (Fong). *Suppose that G is p-solvable and let $\theta \in \operatorname{Irr}(\mathbf{O}_{p'}(G))$ be G-invariant. Then there is a unique p-block B of G covering $\{\theta\}$. Also, $\operatorname{Irr}(B) = \operatorname{Irr}(G \mid \theta)$ and $\operatorname{IBr}(B) = \operatorname{IBr}(G \mid \theta)$. Furthermore, the defect groups of B are the Sylow p-subgroups of G.*

Proof. Let B be a block of G which covers $\{\theta\}$. By Theorem (9.2) we have that $\operatorname{Irr}(B) \subseteq \operatorname{Irr}(G \mid \theta)$ and $\operatorname{IBr}(B) \subseteq \operatorname{IBr}(G \mid \theta)$. If we prove that $\operatorname{Irr}(B) = \operatorname{Irr}(G \mid \theta)$, then we will have that B is the unique block of G covering θ. Thus, $\operatorname{IBr}(B) = \operatorname{IBr}(G \mid \theta)$ since every irreducible Brauer character lying over θ belongs to some block which covers $\{\theta\}$.

We prove that $\operatorname{Irr}(B) = \operatorname{Irr}(G \mid \theta)$ and $\delta(B) = \operatorname{Syl}_p(G)$ using induction on $|G|$.

Write $N = \mathbf{O}_{p'}(G)$. We may certainly assume that p divides $|G|$ and we choose $M \triangleleft G$ such that $N \subseteq M$ and M is a maximal normal subgroup of G containing $\mathbf{O}_{p'p}(G)$, if possible.

If B_1 is a block of G covering $\{\theta\}$ and b_1 is a block of M covered by B_1, we claim that b_1 covers $\{\theta\}$. To see this, let $\chi \in B_1$ be such that χ_M has an irreducible constituent $\eta \in b_1$ (which exists by Theorem (9.2.c)). Since χ_N is a multiple of θ by Theorem (9.2), we deduce that η_N is also a multiple of θ. Hence, b_1 covers $\{\theta\}$, again by Theorem (9.2). This proves the claim.

Now, let $b \in \operatorname{Bl}(M)$ be covered by B, so that b covers $\{\theta\}$ by the previous claim. Using induction, we have that $\operatorname{Irr}(b) = \operatorname{Irr}(M \mid \theta)$ and that the defect groups of b are the Sylow p-subgroups of M. In particular, b is the unique block of M which covers $\{\theta\}$. Now, if B_1 is a block of G covering $\{\theta\}$, by the claim in the previous paragraph, it follows that B_1 covers b. We see that it suffices to show that B is the only block of G covering b and that $\delta(B) = \operatorname{Syl}_p(G)$.

If G/M is a p-group, we have that B is the unique block of G covering b by Corollary (9.6). Also, since b is the unique block covering $\{\theta\}$, by uniqueness we have that b is G-invariant. Now, by Fong's theorem (9.17), we have that if $P \in \delta(B)$, then $P \cap M \in \delta(b)$ and $PM = G$. Hence, $P \cap M$ is a Sylow p-subgroup of M. Thus

$$|G|_p = |G : M||M|_p = |P : P \cap M||P \cap M| = |P|$$

and the theorem is proved in this case.

So we may assume that G/M is a p'-group (and hence, we may assume that M contains $\mathbf{O}_{p'p}(G)$). Since the defect groups of b are the Sylow p-subgroups of M, there exists a character $\xi \in \mathrm{Irr}(b)$ of p'-degree (that is, of height zero in b). Since B covers b, by Theorem (9.4) there exists an $\alpha \in \mathrm{Irr}(B)$ over ξ. In particular, since $\frac{\alpha(1)}{\xi(1)}$ divides $|G : M|$ we deduce that α has p'-degree. Thus, the defect of B is $\nu(|G|)$, and we have that the defect groups of B are the Sylow p-subgroups of G. Now, if B' is another block covering b, again by Theorem (9.4) we may choose $\beta \in \mathrm{Irr}(B')$ over ξ. (As before, β has p'-degree.) It is clear that to prove the theorem, it suffices to show that α and β lie in the same block. For this, we will see that if K is a conjugacy class of G and $x \in K$, then p divides in S the algebraic integer

$$\omega = |K|(\frac{\alpha(x)}{\alpha(1)} - \frac{\beta(x)}{\beta(1)}).$$

If this is the case, then $\omega^* = 0$ and we will have that

$$\omega_\alpha(\hat{K})^* = \omega_\beta(\hat{K})^*$$

for every conjugacy class K of G, proving that α and β lie in the same p-block of G. If $K \subseteq M$, by Theorem (9.1) we have that

$$\omega_\alpha(\hat{K}) = \omega_\xi(\hat{K}) = \omega_\beta(\hat{K})$$

and thus $\omega = 0$ in this case.

So we may assume that $x \notin M$. Since α and β have p'-degree, note that $\frac{\alpha(x)}{\alpha(1)}$ and $\frac{\beta(x)}{\beta(1)}$ are actually in our ring S. Therefore, it suffices to prove that $|K|$ is divisible by p. If not, $\mathbf{C}_G(x)$ contains a Sylow p-subgroup of G and in particular, we have that $xN \in \mathbf{C}_{G/N}(\mathbf{O}_p(G/N))$. Hence, by the Hall-Higman lemma 1.2.3, we have that $\mathbf{C}_{G/N}(\mathbf{O}_p(G/N)) \subseteq \mathbf{O}_p(G/N) \subseteq M/N$. We deduce that $x \in M$, a contradiction which proves the theorem. ∎

The importance of Fong's theorem on blocks of p-solvable groups is that it gives a path between block theory and ordinary character theory. It is a fact that a large fraction of modular representation theory of p-solvable groups can be translated to ordinary representation theory. Fong's theorem on blocks and the Fong-Swan theorem are examples of this.

Let us see how to use Fong's theorem.

(10.21) THEOREM. *Suppose that G is p-solvable and let B be a p-block of G with defect group P. If $\chi \in \mathrm{Irr}(B)$, then*

$$\mathrm{height}(\chi) \le \nu(|P : \mathbf{Z}(P)|).$$

Proof. We argue by induction on $|G|$. Write $N = \mathbf{O}_{p'}(G)$ and let $\theta \in \mathrm{Irr}(N)$ be an irreducible constituent of χ_N. Hence, the block $\{\theta\}$ of N is covered by B by Theorem (9.2). Let T be the inertia group of θ and let $\psi \in \mathrm{Irr}(T)$ be the Clifford correspondent of χ over θ. Of course, T is the inertia group of the block $\{\theta\}$. By the Fong-Reynolds theorem, there is a G-conjugate D of P, such that D is the defect group of the block of ψ. Also, $\mathrm{height}(\chi) = \mathrm{height}(\psi)$. Since $|P : \mathbf{Z}(P)| = |D : \mathbf{Z}(D)|$, by induction we may assume that θ is G-invariant.

By Fong's theorem (10.20), we have that P is a Sylow p-subgroup of G and thus $\mathrm{height}(\chi) = \nu(\chi(1))$.

If $\mathbf{O}_{p'p}(G) = G$, then θ extends to G (by Corollary (6.28) of [Isaacs, *Characters*], for instance) and therefore, by Gallagher's theorem (Corollary (6.17) of [Isaacs, *Characters*]), we have that $\chi = \hat{\theta}\beta$, where $\hat{\theta} \in \mathrm{Irr}(G)$ extends θ and $\beta \in \mathrm{Irr}(G)$ has N in its kernel. Therefore, $\beta_P \in \mathrm{Irr}(P)$. Also, we have that $\beta(1)$ divides $|P : \mathbf{Z}(P)|$ (by Ito's theorem (6.15) of [Isaacs, *Characters*]). In this case,

$$\mathrm{height}(\chi) = \nu(\beta(1)) \leq \nu(|P : \mathbf{Z}(P)|)$$

and we are done.

Now, let M be a maximal normal subgroup of G containing $\mathbf{O}_{p'p}(G)$, let $\xi \in \mathrm{Irr}(M)$ be under χ and let b be the block of ξ. Since $\mathbf{O}_{p'}(M) = N$, note that $P \cap M \in \mathrm{Syl}_p(M)$ is a defect group of b by Fong's theorem (10.20). By induction, we have that

$$\nu(\xi(1)) \leq \nu(|P \cap M : \mathbf{Z}(P \cap M)|)\,.$$

Also, by applying the Hall-Higman's lemma 1.2.3, notice that

$$\mathbf{Z}(P)\mathbf{O}_{p'}(G)/\mathbf{O}_{p'}(G) \subseteq \mathbf{C}_G(\mathbf{O}_{p'p}(G)/\mathbf{O}_{p'}(G))$$

$$\subseteq \mathbf{O}_{p'p}(G)/\mathbf{O}_{p'}(G) \subseteq M/\mathbf{O}_{p'}(G)\,,$$

and thus

$$\mathbf{Z}(P) \subseteq \mathbf{Z}(P \cap M) \subseteq P \cap M \subseteq P\,.$$

If G/M is a p'-group, then $P \subseteq M$. Note that since $\frac{\chi(1)}{\xi(1)}$ divides $|G : M|$, we have that $\chi(1)_p = \xi(1)_p$. Hence, $\mathrm{height}(\chi) = \mathrm{height}(\xi)$ and the theorem follows by induction. So we may assume that G/M is a p-group. Hence, $|G/M| = p$ and by Corollary (6.19) of [Isaacs, *Characters*] we have two more possibilities, namely, $\chi_M = \xi$ and $\xi^G = \chi$. If $\chi_M = \xi$, then we have that

$$\nu(\chi(1)) = \nu(\xi(1)) \leq \nu(|P \cap M : \mathbf{Z}(P \cap M)|) \leq \nu(|P : \mathbf{Z}(P)|)\,.$$

If $\xi^G = \chi$, we have that

$$\nu(\chi(1)) = \nu(\xi(1)) + 1 \leq \nu(|P \cap M : \mathbf{Z}(P \cap M)|) + 1 \leq \nu(|P : \mathbf{Z}(P)|)$$

since $P \cap M < P$ and $\mathbf{Z}(P) \subseteq \mathbf{Z}(P \cap M)$. This finishes the proof of the theorem. ∎

Notice that Theorem (10.21) implies that, for p-solvable groups, if a block B has abelian defect group, then all irreducible characters in B have height zero.

The next result is another application of Fong's theorem on blocks.

(10.22) THEOREM. *Suppose that G is p-solvable and let B be a block of G of defect d. If $\varphi, \mu \in \mathrm{IBr}(B)$, then*

$$c_{\varphi\mu} \leq p^d .$$

Proof. By the Fong-Reynolds theorem and Fong's theorem on blocks (and arguing as in Theorem (10.21)), we may assume that $d = \nu(|G|)$. Also, by Theorem (2.22) we know that

$$c_{\varphi\mu} \leq \Phi_1(1) .$$

Now, by Fong's dimensional formula, $\Phi_1(1) = |G|_p$ and we are done. ∎

(NOTE: Theorem (10.22) is known to be false for arbitrary groups.)

As we mentioned in Chapter 3, K. Harada has conjectured that the p-blocks are the minimal subsets $U \subseteq \mathrm{Irr}(G)$ such that $\sum_{\chi \in U} \chi(1)\chi$ vanishes off p-regular elements.

(10.23) THEOREM (Kiyota-Okuyama). *Suppose that G is p-solvable. If $U \subseteq \mathrm{Irr}(G)$ is such that $\sum_{\chi \in U} \chi(1)\chi \in \mathrm{vcf}(G)$, then U is a union of p-blocks.*

Proof. As we are going to see, the fact that G is p-solvable is only used to apply the Fong-Swan theorem.

Write $\Delta_U = \sum_{\chi \in U} \chi(1)\chi$. Since $\Delta_U \in \mathrm{vcf}(G)$, by Theorem (2.13) we may write

$$\Delta_U = \sum_{\varphi \in \mathrm{IBr}(G)} a_\varphi \Phi_\varphi ,$$

where a_φ are complex numbers. Let $V = \{\varphi \in \mathrm{IBr}(G) \mid a_\varphi \neq 0\}$.

We claim that if $\theta \in V$, then $a_\theta = \theta(1)$. By the Fong-Swan theorem, let $\psi \in \mathrm{Irr}(G)$ be such that $\psi^0 = \theta \in V$. Now, by using Theorem (2.13), notice that

$$[\Delta_U, \psi] = [\Delta_U, \theta]^0 = \sum_{\varphi \in \mathrm{IBr}(G)} a_\varphi [\Phi_\varphi, \theta]^0 = a_\theta \neq 0 \,.$$

Therefore, $a_\theta = [\Delta_U, \psi] = \psi(1) = \theta(1)$, as claimed.

Now, if $\chi \in \mathrm{Irr}(G)$, then we have that

$$[\Delta_U, \chi] = \frac{1}{|G|} \sum_{g \in G^0} \Delta_U(g) \chi(g^{-1})$$

$$= \frac{1}{|G|} \sum_{g \in G^0} \left(\sum_{\theta \in V} \theta(1) \Phi_\theta(g) \right) \left(\sum_{\varphi \in \mathrm{IBr}(G)} d_{\chi\varphi} \varphi(g^{-1}) \right)$$

$$= \sum_{\varphi \in \mathrm{IBr}(G)} \left(d_{\chi\varphi} \sum_{\theta \in V} \theta(1) \left(\frac{1}{|G|} \sum_{g \in G^0} \Phi_\theta(g) \varphi(g^{-1}) \right) \right)$$

$$= \sum_{\varphi \in \mathrm{IBr}(G)} \left(d_{\chi\varphi} \sum_{\theta \in V} \theta(1) [\Phi_\theta, \varphi]^0 \right) = \sum_{\varphi \in V} d_{\chi\varphi} \varphi(1) \leq \chi(1) \,.$$

If $\chi \in U$, then we have that $[\Delta_U, \chi] = \chi(1)$. Therefore we conclude that $d_{\chi\varphi} = 0$ if $\varphi \notin V$. On the other hand, if $\chi \notin U$, then $0 = [\Delta_U, \chi] = \sum_{\varphi \in V} d_{\chi\varphi} \varphi(1)$ and we obtain that $d_{\chi\varphi} = 0$ if $\varphi \in V$.

Suppose finally that $\chi \in U$ and that χ and $\psi \in \mathrm{Irr}(G)$ have an irreducible Brauer constituent μ in common. Since $d_{\chi\mu} \neq 0$, we have that $\mu \in V$. Since $d_{\psi\mu} \neq 0$, this forces $\psi \in U$. This proves that U is a union of p-blocks. ∎

PROBLEMS

(10.1) Suppose that G is p-solvable and let $N \triangleleft G$ be such that $G = JN$, where $J \subseteq G$ has p-power index. Assume that $\theta \in \mathrm{IBr}(N)$ is such that $\theta_{J \cap N} \in \mathrm{IBr}(J \cap N)$. Prove that restriction defines a bijection

$$\mathrm{IBr}(G \,|\, \theta) \to \mathrm{Irr}(J \,|\, \theta_{J \cap N}) \,.$$

(10.2) Suppose that G is p-solvable and let $\mu \in \mathrm{IBr}(K)$ be such that $\mu^G \in \mathrm{IBr}(G)$, where K is a subgroup of G. Prove that μ^G is the only irreducible Brauer character of G over μ.

(10.3) Suppose that G is p-solvable and let H be a p-complement of G. Prove that χ has p-defect zero if and only if χ is induced from a character of H.

(10.4) Let $\chi \in \mathrm{Irr}(G)$ have p-defect zero. Prove that χ is a $\mathbb{Z}$-linear combination of characters of G induced from p'-subgroups of G.

(HINT: Use Solomon's theorem (8.10) of [Isaacs, *Characters*] which asserts that 1_G is a $\mathbb{Z}$-linear combination of characters of the form $(1_H)^G$ for some solvable subgroups H of G. Use Fong's theorem (10.13).)

(NOTE: W. Willems has proved that if B is a block of defect group D and $\chi \in \mathrm{Irr}(B)$, then χ is a $\mathbb{Z}$-linear combination of characters of G induced from subgroups H such that the Sylow p-subgroups of H are contained in G-conjugates of D. See [Willems, 1979].)

(10.5) Suppose that G is p-solvable and let $\chi \in \mathrm{Irr}(G)$ have p-defect zero. If $\chi_{\mathbf{O}_{p'}(G)}$ is a multiple of some irreducible character, prove that G is a p'-group.

(10.6) Suppose that G is p-solvable, with p odd. Let $\chi \in \mathrm{Irr}(G)$ be p-rational such that $\chi^0 \in \mathrm{IBr}(G)$. Show that $\mathbf{O}_p(G) \subseteq \ker(\chi)$.

(10.7) Suppose that G has odd order. Prove that the principal block is the unique selfdual block of G.

(10.8) Suppose that G is p-solvable. If p does not divide $\varphi(1)$ for every $\varphi \in \mathrm{IBr}(G)$, then G has a normal Sylow p-subgroup. Find an example where this is not true for primes $q \neq p$.

(NOTE: As we said in Chapter 2, this result is true for every finite group G.)

(10.9) Suppose that G is p-solvable and let $\mu \in \mathrm{IBr}(K)$ be such that $\mu^G = \varphi \in \mathrm{IBr}(G)$, where K is a subgroup of G. Prove that $(\Phi_\mu)^G = \Phi_\varphi$.

(10.10) Suppose that G is p-solvable and let H be a p-complement of G. If $\lambda \in \mathrm{Irr}(H)$ is linear, show that $\lambda^G = \Phi_\varphi$ for some $\varphi \in \mathrm{IBr}(G)$ with p-power degree.

(10.11) Prove the Fong-Swan theorem by using character triple isomorphisms.

(HINT: Argue by induction on $|G : \mathbf{O}_{p'}(G)|$. Use Problem (8.13) and the fact that $\mathbf{O}_p(G) \subseteq \ker(\varphi)$ for every $\varphi \in \mathrm{IBr}(G)$.)

(10.12) (Nagao) The following is known as the k(GV)-conjecture:

k(GV)-**Conjecture.** Let V be an elementary abelian p-group on which a p'-group G acts faithfully and irreducibly. Then the number of conjugacy classes k(GV) of the semidirect product GV is less than or equal to $|V|$.

If G is p-solvable, prove that the k(GV)-conjecture is equivalent to the Brauer's k(B)-conjecture.

(HINT: Use the Fong-Reynolds theorem, Theorem (10.20) and Problem (11.10) of [Isaacs, *Characters*] to deduce that Brauer's $k(B)$-conjecture for p-solvable groups is equivalent to $k(G/\mathbf{O}_{p'}(G)) \leq |G|_p$. Finally, use an inductive argument and that $k(G) \leq k(G/N)k(N)$ for $N \triangleleft G$.)

(10.13) The character table of $G = \mathrm{GL}(2,3)$ is

$|\langle x \rangle|$	1	2	2	4	3	6	8	8
$|\mathrm{cl}(x)|$	1	1	12	6	8	8	6	6
χ_1	1	1	1	1	1	1	1	1
χ_2	1	1	-1	1	1	1	-1	-1
χ_3	2	2	0	2	-1	-1	0	0
χ_4	3	3	1	-1	0	0	-1	-1
χ_5	3	3	-1	-1	0	0	1	1
χ_6	2	-2	0	0	-1	1	$i\sqrt{2}$	$-i\sqrt{2}$
χ_7	2	-2	0	0	-1	1	$-i\sqrt{2}$	$i\sqrt{2}$
χ_8	4	-4	0	0	1	-1	0	0

Using the Fong-Swan theorem, show that the Brauer character table of G for $p = 3$ is

$|\langle x \rangle|$	1	2	2	4	8	8
$|\mathrm{cl}(x)|$	1	1	12	6	6	6
φ_1	1	1	1	1	1	1
φ_2	1	1	-1	1	-1	-1
φ_3	3	3	1	-1	-1	-1
φ_4	3	3	-1	-1	1	1
φ_5	2	-2	0	0	$i\sqrt{2}$	$-i\sqrt{2}$
φ_6	2	-2	0	0	$-i\sqrt{2}$	$i\sqrt{2}$

(10.14) Suppose that G is p-solvable. Let $\varphi \in \mathrm{IBr}(G)$ and suppose that $N \triangleleft G$ is such that the irreducible constituents of φ_N have p'-degree. If $\chi = \alpha^G$ for some $\alpha \in \mathrm{IBr}(U)$, prove that $|NU : U|$ is not divisible by p.

(HINT: Use induction on $|N|$. Problem (10.1) is relevant.)

(10.15) Suppose that G is p-solvable. Let $\alpha \in \mathrm{IBr}(U)$ and $\beta \in \mathrm{IBr}(V)$ be such that $\alpha^G = \varphi = \beta^G$, where U and V are subgroups of G and α and β have p'-degree. If P_1 is a Sylow p-subgroup of U and P_2 is a Sylow p-subgroup of V, prove that P_1 and P_2 are G-conjugate. Each element in the G-conjugacy class of P_1, for instance, is called a **vertex** of φ. By Huppert's theorem, note that every $\varphi \in \mathrm{IBr}(G)$ has associated a unique conjugacy class of vertices. If P is a vertex of φ, $|G|_p = p^a$, $|P| = p^b$ and $\varphi(1)_p = p^c$, deduce that $a - b = c$.

(HINT: Use induction on $|G|$. If $\varphi(1)$ is a p'-number, then the result is clear. If not, then use Lemma (10.12) and Problem (10.14).)

(10.16) Let G be a p-solvable group and let H be a p-complement of G. Suppose that $\alpha \in \mathrm{Irr}(H)$ satisfies $\alpha(x) = \alpha(y)$ whenever $x, y \in H$ are G-conjugate. Show that there exists $\chi \in \mathrm{Irr}(G)$ such that $\chi_H = \alpha$.

11. Groups with Sylow p-Subgroups of Order p

One of the deepest results in modular representation theory is the description of the blocks with a cyclic defect group. This theory was begun by R. Brauer in the early forties with the study of groups with a Sylow p-subgroup of order p and (a year before that) with the general study of the blocks of defect one. Brauer's achievements are of great importance for character theory and have many consequences.

Many years later (in 1967), J. Thompson used Green's new results on indecomposable modules to produce another proof of some of Brauer's earlier theorems. Afterwards, E. C. Dade was able to give a full description of the blocks with cyclic defect groups.

Here, we will restrict ourselves to the analysis of the p-blocks of the groups having a Sylow p-subgroup of order p; probably, one of the most important cases. To do that, we will need to study the fundamental paper of Brauer on blocks of defect one. (After this is done, it is not very difficult to give a complete description of the blocks of defect one.) Brauer's results, however, cannot be obtained by purely character theoretic methods. We will need to use representations, Schur indices and some algebraic number theory.

We start with the statement for the principal block case (so that the reader can immediately see how powerful these results are).

(11.1) THEOREM. *Suppose that $|G|_p = p$, where p is odd. Let $P \in \mathrm{Syl}_p(G)$ and write $N = \mathbf{N}_G(P)$, $C = \mathbf{C}_G(P)$, $e = |N : C|$ and $t = \frac{p-1}{e}$. Let B_0 be the principal block of G.*

(a) The number of ordinary irreducible characters in B_0 is $e + t$. They can be labelled as $\chi_1, \ldots, \chi_e, \theta_1, \ldots, \theta_t$. The characters $\chi_1, \ldots, \chi_e$ (called **nonexceptional**) *are p-rational and have degrees*

$$\chi_i(1) \equiv \epsilon_i \bmod p$$

for uniquely defined $\epsilon_i = \pm 1$ and $1 \leq i \leq e$. The characters $\theta_1, \ldots, \theta_t$ (called **exceptional** *if $t \geq 2$) have the same value on p-regular elements.*

Also, there is a unique $\epsilon = \pm 1$ such that

$$\theta_j(1) \equiv \epsilon e \bmod p$$

for $1 \le j \le t$.
   (b)

$$\epsilon(\theta_1)^0 = \sum_{i=1}^{e} \epsilon_i(\chi_i)^0 .$$

   (c) *The number of irreducible Brauer characters in B_0 is e. Also, $d_{\chi\varphi} \le 1$ for $\chi \in \mathrm{Irr}(B_0)$ and $\varphi \in \mathrm{IBr}(B_0)$.*
   (d) *If $1 \ne x \in P$ and $y \in C^0 = \mathbf{O}_{p'}(N)$, then $\chi_i(xy) = \epsilon_i$ for $1 \le i \le e$.*
   (e) *There exist representatives $\lambda_1, \ldots, \lambda_t$ of the action of N on $\mathrm{Irr}(P) - \{1_P\}$ such that*
$$\theta_j(xy) = \epsilon(\lambda_j \times 1_{\mathbf{O}_{p'}(N)})^N(x)$$

for $1 \le j \le t$, $1 \ne x \in P$ and $y \in C^0$.
   (f) *Assume in addition that $\mathbf{C}_G(P) = P$. Then $\chi(1)_p = |G|_p$ for every $\chi \in \mathrm{Irr}(G) - B_0$.*

As we see, Theorem (11.1) gives very important information on the character table of groups which are divisible by some prime to the first power only. (As an important example of this class of groups, the reader should be aware of the fact that if G is a sporadic simple group, then there exists a prime p such that $|G|_p = p$.)

Assume the notation of Theorem (11.1) and suppose that G does not have a normal p-complement. (By using Burnside's transfer theorem, it is very easy to check that this happens if and only if $e > 1$.) Hence, $e \not\equiv 1 \bmod p$ because $e \le p - 1$. We say that $\chi \in \mathrm{Irr}(B_0)$ is of + **type** if its degree is congruent with 1 or $-e$ modulo p. Also, we say that $\chi \in \mathrm{Irr}(B_0)$ is of − **type** if its degree is congruent with -1 or e modulo p. (With our hypothesis and Theorem (11.1), observe that every irreducible character in B_0 is exactly of + type or of − type.) Theorem (11.1.b) implies that the sum of all the degrees of plus type equals the sum of all the degrees of minus type, except that the t exceptional characters count as one character only for this calculation.

Let us start with an application.

(11.2) COROLLARY. *Suppose that G is a simple group of order 7920 (that is, $G = M_{11}$). Then $|\mathrm{Irr}(G)| = 10$ and the degrees of the irreducible characters of G are $1, 10, 10, 10, 11, 16, 16, 44, 45$ and 55.*

Proof. Write $7920 = 2^4 \cdot 3^2 \cdot 5 \cdot 11$.

By elementary but somewhat tedious group theoretic calculations, it is possible to prove the following facts. If $p = 11$, the Sylow p-subgroups of G are selfcentralizing and the size of their normalizer is 55. If $p = 5$, the Sylow p-subgroups of G are also selfcentralizing and the size of their normalizer is 20.

Now, we calculate the degrees of the irreducible characters in the principal 11-block of G. Using the notation of Theorem (11.1), we have that $e = |N : P| = 5$ and $t = 2$ for $p = 11$. Therefore, there are seven irreducible characters in the principal 11-block of G. Two of them are exceptional and five nonexceptional. (One nonexceptional character is the trivial character.) The exceptional characters have degree congruent with 5 or 6 mod 11. (The ones with degree congruent with 6 are of $+$ type, and the others are of $-$ type.) The nonexceptional characters have degrees congruent with 1 or 10 modulo 11. Of course, all of them have degree not divisible by 11. By using the fact that the squares of the degrees are less than $|G| = 7920$, we conclude that the only possible $+$ type degrees are $\{1, {}^*6^*, 12, 45\}$. (We write ${}^*a^*$ to indicate that a is a degree corresponding to an exceptional character.) Similarly, the only possible $-$ type degrees are $\{{}^*5^*, 10, {}^*16^*, {}^*60^*\}$. (We stress that we have discarded the possible $+$ type exceptional degree ${}^*72^*$ because $t = 2$ and $72^2 + 72^2 > |G|$.)

We start by assuming that 5 is the exceptional degree. We have to find an integer solution to the equation

$$5 + 10a = 1 + 12b + 45c$$

for $0 \le a, b, c$ and $a + b + c = 4$. Working modulo 5, we see that $b = 2$. We find that $a = 2, b = 2, c = 0$ is a solution. Hence

$$1, 5, 5, 10, 10, 12, 12 \tag{1}$$

is a possibility for the degrees of the irreducible characters in the principal 11-block. (We will see afterwards that this solution is not possible.)

Suppose now that 6 is the exceptional degree. Then we have to find a solution to the equation

$$10a = 1 + 6 + 12b + 45c$$

for $0 \le a, b, c$ and $a + b + c = 4$. Working modulo 5, we see that $b = 4$, which is not possible.

Suppose now that 60 is the exceptional degree. The only solution to the equation

$$60 + 10a = 1 + 12b + 45c$$

for $0 \leq a, b, c$ and $a + b + c = 4$ is $a = 1, c = 1$ and $b = 2$. However, the sum of the squares of the degrees $(2 \cdot 60^2 + 45^2 + 2 \cdot 12^2 + 10^2 + 1 > |G|)$ is too large.

Finally, we assume that 16 is the exceptional degree. There is a unique solution for

$$16 + 10a = 1 + 12b + 45c$$

for $0 \leq a, b, c$ and $a + b + c = 4$. The unique solution is $a = 3, b = 0$ and $c = 1$. This gives the possibility

$$1, 10, 10, 10, 16, 16, 45 \qquad (2)$$

for the degrees of the irreducible ordinary characters in the principal 11-block of G.

Next, we work with $p = 5$ and $e = 4$. Hence, the principal 5-block of G consists of $e + 1 = 5$ irreducible ordinary characters (one of them, the trivial character). In this case, $t = 1$ and there are no exceptional characters. By discarding the degrees which are too large, we obtain that the possible + type degrees are $\{1, 6, 11, 16, 36, 66\}$. Similarly, we have that the possible − type degrees are $\{4, 9, 24, 44\}$. By Theorem (11.1.e), we have that every character outside the principal 11-block has degree divisible by 11. The possible degrees $4, 6, 9, 24$ and 36 should correspond to characters outside the 11-principal block since they do not appear as possibilities in (1) or (2). Since they are not divisible by 11, we conclude that they are not irreducible character degrees of G. Hence, the + type possible degrees are $\{1, 11, 16, 66\}$ and the possible − type degree is just $\{44\}$. We have to find a solution to the equation

$$1 + 11a + 16b + 66c = 44d$$

for $a + b + c + d = 4$ and $0 \leq a, b, c, d$. Working modulo 11, we have that $b = 2$. Also, $a = 1 = d$. Hence, the irreducible character degrees of the principal 5-block are uniquely determined:

$$1, 11, 16, 16, 44.$$

Now, by applying Theorem (11.e) for $p = 5$, we can exclude the possibility (1) for the principal 11-block since 12 is not divisible by 5.

We conclude that G has nine irreducible characters of degrees 1, 10, 10, 10, 11, 16, 16, 44 and 45. By Theorem (11.1.e), any other irreducible character of G has degree divisible by 55. Now,

$$|G| - 1 - 3 \cdot 10^2 - 11^2 - 2 \cdot 16^2 - 44^2 - 45^2 = 55^2$$

and we deduce that there are ten irreducible characters of G of degrees

$$1, 10, 10, 10, 11, 16, 16, 44, 45, 55,$$

as desired. ∎

The blocks of defect one for $p = 2$ are not difficult to describe.

(11.3) THEOREM. *Suppose that $p = 2$ and let B be a block of defect one. Let $P = \{1, x\} \in \delta(B)$. Then $\mathrm{Irr}(B) = \{\chi_1, \chi_2\}$, $\mathrm{IBr}(B) = \{\varphi\}$ and the decomposition matrix of B is*

$$D_B = \begin{pmatrix} 1 \\ 1 \end{pmatrix}.$$

If $\theta \in \mathrm{Irr}(\mathbf{C}_G(x))$ is the canonical character of B, then there are $\epsilon_i = \pm 1$ such that

$$\chi_i(xy) = \epsilon_i \theta(y)$$

for all $y \in \mathbf{C}_G(x)^0$ and $i = 1, 2$. Also, $\epsilon_1 \epsilon_2 = -1$.

Proof. By the Brauer-Feit theorem (3.27), we know that

$$\mathrm{k}(B) \leq \frac{1}{4} 2^2 + 1 = 2.$$

Since B does not have defect zero, by Theorem (3.18) we have that $\mathrm{k}(B) = 2$, $\mathrm{l}(B) < 2$ and the first part of the theorem is complete.

By Theorem (3.26), we know that the greatest elementary divisor of the 1×1 Cartan matrix C_B is 2. Therefore, $C_B = (2)$ and this easily determines the decomposition matrix. In particular, note that $(\chi_1)^0 = (\chi_2)^0$.

Now, let $b \in \mathrm{Bl}(\mathbf{C}_G(x))$ be a root of B. Since $\mathbf{N}_G(P) = \mathbf{C}_G(x)$ and all the roots of B are $\mathbf{N}_G(P)$-conjugate, by the extended first main theorem (9.7) it follows that b is the unique block of $\mathbf{C}_G(x)$ inducing B. Furthermore, by Theorem (9.12), we know that $\mathrm{IBr}(b) = \{\theta^0\}$, where $\theta \in \mathrm{Irr}(b)$ is the canonical character of B. Since b is the Brauer first main correspondent of B, we have that $\{1, x\} \in \delta(b)$. Hence, $c_{\theta^0 \theta^0} = 2$ again by Theorem (3.26).

By Corollary (5.8), we see that for p-regular y in $\mathbf{C}_G(x)$, we have

$$\chi_i(xy) = \epsilon_i \theta(y),$$

where ϵ_i is the generalized decomposition number $d^x_{\chi_i \theta^0}$. Since the order of x is 2, we have that $\epsilon_i \in \mathbb{Z}$. (See the remark after the proof of Lemma (5.1).) By Lemma (5.13.b), we have that

$$\epsilon_1^2 + \epsilon_2^2 = c_{\theta^0 \theta^0} = 2.$$

However, $\epsilon_1 \neq \epsilon_2$ since $\chi_1 \neq \chi_2$, both characters are zero on elements whose p-part is not trivial and not G-conjugate to x (Corollary (5.9)) and $(\chi_1)^0 = (\chi_2)^0$. Hence, we deduce that $\epsilon_1 \epsilon_2 = -1$, as desired. ∎

With the same notation as in the previous theorem, note that the character

$$\epsilon_1 \chi_1 + \epsilon_2 \chi_2$$

is zero on p-regular elements. Also, the values of the characters χ_i on p-singular elements can be obtained locally. The same type of results hold for general p-blocks of defect one for every prime p, although these are much harder to obtain.

When studying blocks of defect one, the deepest part is to prove Theorem (11.4) below.

Two irreducible characters $\chi, \psi \in \mathrm{Irr}(G)$ are p-**conjugate** if there exists a $\sigma \in \mathrm{Gal}(\mathbb{Q}_n/\mathbb{Q}_m)$ such that $\chi^\sigma = \psi$, where $n = |G|$ and $m = |G|_{p'}$. Since $\sigma \in \mathrm{Gal}(\mathbb{Q}_n/\mathbb{Q}_m)$ fixes χ if and only if it fixes (elementwise) the field $\mathbb{Q}_m(\chi)$ (that is, the smallest subfield of $\mathbb{C}$ containing $\mathbb{Q}_m$ and the values $\{\chi(g) \mid g \in G\}$), by elementary Galois theory we deduce that the number r_χ of p-conjugate characters to a given character χ is

$$r_\chi = |\mathrm{Gal}(\mathbb{Q}_n/\mathbb{Q}_m) : \mathrm{Gal}(\mathbb{Q}_n/\mathbb{Q}_m(\chi))| = |\mathbb{Q}_m(\chi) : \mathbb{Q}_m| \,.$$

Note that if χ and ψ are p-conjugate, then $\chi^0 = \psi^0$. In this case, χ and ψ lie in the same p-block and their corresponding rows in the decomposition matrix are exactly the same. We see that if B is a p-block, then $\mathrm{Gal}(\mathbb{Q}_n/\mathbb{Q}_m)$ acts on $\mathrm{Irr}(B)$. We may then write $\mathrm{Irr}(B) = \mathcal{O}(\xi_1) \cup \ldots \cup \mathcal{O}(\xi_\omega)$, where $\mathcal{O}(\xi)$ is the set of p-conjugates of $\xi \in \mathrm{Irr}(B)$.

(11.4) THEOREM (Brauer). *Let B be a p-block of G of defect one, where p is odd.*

(a) If $\chi \in \mathrm{Irr}(B)$, then $d_{\chi\varphi} \leq 1$ for all $\varphi \in \mathrm{IBr}(G)$.

(b) Let $\{\xi_1, \ldots, \xi_\omega\}$ be a complete set of representatives for the classes of p-conjugate characters in $\mathrm{Irr}(B)$. If $\varphi \in \mathrm{IBr}(B)$, then there exist exactly two different indices i and j such that φ is an irreducible constituent of $(\xi_i)^0$ and $(\xi_j)^0$.

The proof of Theorem (11.4) requires the use of representations and some algebraic number theory.

Next, we devote ourselves to developing the results that we are going to need. Most of what we will use are standard facts on Dedekind domains which can be found in [Isaacs, *Algebra*]. In addition, we will use a deeper theorem about the algebraic integers in a cyclotomic extension. Recall that **R** is the ring of algebraic integers in $\mathbb{C}$ and that we have fixed a maximal ideal M of **R** containing $p\mathbf{R}$.

(11.5) THEOREM. *Suppose that $\mathbb{Q}_n = \mathbb{Q}(\xi)$, where ξ is a primitive nth root of unity. Then $\mathbf{R} \cap \mathbb{Q}_n = \mathbb{Z}[\xi]$.*

Proof. See, for instance, Theorem (21.13) of [Curtis-Reiner]. ∎

(11.6) NOTATION.   Let $n > 1$ be an integer. Suppose that $n = p^c m$, where p does not divide m and $c > 0$. We write $\mathbf{R}_n = \mathbf{R} \cap \mathbb{Q}_n$ for the algebraic integers in $\mathbb{Q}_n$. Let ρ be a primitive p^cth root of unity, so that $\mathbb{Q}_n = \mathbb{Q}_m(\rho)$. Let $\mathcal{G} = \mathrm{Gal}(\mathbb{Q}_n/\mathbb{Q}_m)$. Note that $M_m = \mathbf{R}_m \cap M$ is a prime ideal of $\mathbf{R}_m$ containing p. Also, $M_n = \mathbf{R}_n \cap M$ is a prime ideal of $\mathbf{R}_n$ containing p with $M_n \cap \mathbf{R}_m = M_m$.

Recall that $\mathbf{R}_m$ and $\mathbf{R}_n$ are Dedekind domains (Corollary (28.22) of [Isaacs, *Algebra*]) and that in a Dedekind domain every nonzero prime ideal is maximal. Therefore, M_m and M_n are maximal ideals of $\mathbf{R}_m$ and $\mathbf{R}_n$, respectively.

We are interested in the associated localized rings. Write

$$S_n = \{\frac{\alpha}{\beta} \mid \alpha \in \mathbf{R}_n, \beta \in \mathbf{R}_n - M\},$$

and

$$S_m = \{\frac{\alpha}{\beta} \mid \alpha \in \mathbf{R}_m, \beta \in \mathbf{R}_m - M\}.$$

These are local rings with maximal ideals

$$\mathcal{P}_n = \{\frac{\alpha}{\beta} \mid \alpha \in M_n, \beta \in \mathbf{R}_n - M\}$$

and

$$\mathcal{P}_m = \{\frac{\alpha}{\beta} \mid \alpha \in M_m, \beta \in \mathbf{R}_m - M\}.$$

Note that $S_m \subseteq S_n \subseteq S$. Therefore, we may restrict the ring homomorphism $* : S \to F$ to S_n (or S_m). It is clear that the kernel of $* : S_n \to F$ is $\mathcal{P}_n$.

Using Theorem (11.5), and the fact that a primitive nth root of unity ν may be written as $\nu = \xi\rho$, where ξ is a primitive mth root of unity, it easily follows that

$$\mathbf{R}_n = \mathbf{R}_m[\rho].$$

Finally, note that if $\sigma \in \mathrm{Gal}(\mathbb{Q}_n/\mathbb{Q})$, then σ is a ring automorphism of $\mathbf{R}_n$.

(11.7) THEOREM.   *Assume Notation (11.6). Then M_n is the unique prime ideal of $\mathbf{R}_n$ such that $M_n \cap \mathbf{R}_m = M_m$. In particular, M_n and $\mathcal{P}_n$ are $\mathcal{G}$-invariant.*

Proof. Let $I = \mathbf{R}_n M_m + \mathbf{R}_n(1 - \rho)$. We prove that I is the unique prime ideal of $\mathbf{R}_n$ with $I \cap \mathbf{R}_m = M_m$. In this case, $I = M_m$ and the theorem will be proved.

Let $\Phi_{p^c}(x)$ be the cyclotomic polynomial of p^cth roots of unity. Thus,

$$\Phi_{p^c}(x) = \prod_{\substack{1 \le k \le p^c \\ (k,p)=1}} \left(x - \rho^k\right).$$

By elementary field theory, recall that $\Phi_{p^c}(x) = \Phi_p(x^{p^{c-1}})$, where $\Phi_p(x) = 1 + x + \ldots + x^{p-1}$. Hence,

$$p = \prod_{\substack{1 \le k \le p^c \\ (k,p)=1}} \left(1 - \rho^k\right).$$

Now, let J be any prime ideal of $\mathbf{R}_n$ with $\mathbf{R}_m \cap J = M_m$. Then

$$\prod_{\substack{1 \le k \le p^c \\ (k,p)=1}} \left(1 - \rho^k\right) = p \in J.$$

Therefore, there exists a k with $1 \le k \le p^c$ and $(k,p) = 1$ such that $1 - \rho^k \in J$. Note that $(\rho^k)^l = \rho$ for some nonnegative integer l since $\langle \rho \rangle = \langle \rho^k \rangle$. Write $\rho^k = \delta$ so that $\delta \in \mathbf{R}_n$ and $1 - \delta \in J$. Thus,

$$1 - \rho = 1 - \delta^l = (1 - \delta)(\delta^{l-1} + \ldots + \delta + 1) \in J.$$

Hence, $I \subseteq J$. We see that to prove the theorem it suffices to check that I is a maximal ideal of $\mathbf{R}_n$ with $I \cap \mathbf{R}_m = M_m$. Note that $I \subseteq M$ since $I^* = 0$ (because $(1 - \rho)^* = 0$ and $M_m \subseteq M$). Hence, $M_m \subseteq I \subseteq M$ and we conclude that $M_m = I \cap \mathbf{R}_m$. Now, consider the map $x \mapsto x + I$ from $\mathbf{R}_m \to \mathbf{R}_n/I$. Since $\mathbf{R}_n = \mathbf{R}_m[\rho]$ and $\rho + I = 1 + I$, it follows that the map is onto. The kernel of the map is $\mathbf{R}_m \cap I = M_m$. Hence, we see that $\mathbf{R}_m/M_m$ and $\mathbf{R}_n/I$ are isomorphic as rings. Since $\mathbf{R}_m/M_m$ is a field, $\mathbf{R}_n/I$ is also a field, proving that I is maximal.

Since every $\sigma \in \mathcal{G}$ induces a ring automorphism of $\mathbf{R}_n$ fixing $\mathbf{R}_m$, by uniqueness, we have that M_n is $\mathcal{G}$-invariant. By definition, it follows that $\mathcal{P}_n$ is also $\mathcal{G}$-invariant. $\blacksquare$

(11.8) THEOREM. *Assume Notation (11.6). We have that*

$$S_n = S_m[\rho].$$

Therefore, $S_n \cap \mathbb{Q}_m = S_m$ and $\mathcal{P}_n \cap \mathbb{Q}_m = \mathcal{P}_m$.

Proof. Let $\Phi_{p^c}(x)$ be the p^cth cyclotomic polynomial and suppose that t is its degree. We know that $\Phi_{p^c}(x)$ is the minimal polynomial of ρ over $\mathbb{Q}$. In addition, by elementary Galois theory, $\Phi_{p^c}(x)$ is also the minimal polynomial of ρ over $\mathbb{Q}_m$. Hence, $\{1, \rho, \rho^2, \ldots, \rho^{t-1}\}$ is a $\mathbb{Q}_m$-basis of $\mathbb{Q}_n = \mathbb{Q}_m[\rho]$. Therefore, each element of $S_m[\rho]$ is expressible in a unique way as an S_m-linear combination of $1, \rho, \rho^2, \ldots, \rho^{t-1}$ because $S_m \subseteq \mathbb{Q}_m$ and $\Phi_{p^c}(x) \in \mathbb{Z}[x]$.

It is clear that $S_m[\rho] \subseteq S_n$ since $S_m \subseteq S_n$ and $\rho \in \mathbf{R}_n \subseteq S_n$. Since $\mathbf{R}_n = \mathbf{R}_m[\rho] \subseteq S_m[\rho]$ and $S_n = \{\frac{\alpha}{\beta} \mid \alpha \in \mathbf{R}_n, \beta \in \mathbf{R}_n - M\}$, it suffices to show that if $\beta \in \mathbf{R}_n - M$, then $\beta^{-1} \in S_m[\rho]$. Since M_n is $\mathcal{G}$-invariant by Theorem (11.7), it follows that $\beta^\sigma \in \mathbf{R}_n - M$ for all $\sigma \in \mathcal{G}$. Now, $\gamma = \prod_{\sigma \in \mathcal{G}} \beta^\sigma \in \mathbb{Q}_m$ because it is fixed by every $\sigma \in \mathcal{G}$. Therefore,

$$\gamma = \prod_{\sigma \in \mathcal{G}} \beta^\sigma \in \mathbf{R}_n \cap \mathbb{Q}_m - M = \mathbf{R}_m - M.$$

Also,

$$\beta^{-1} = \frac{1}{\gamma} \left(\prod_{\sigma \in \mathcal{G} - \{1\}} \beta^\sigma \right).$$

Since $\beta \in \mathbf{R}_n = \mathbf{R}_m[\rho]$, it follows that $\beta^\sigma \in \mathbf{R}_m[\rho]$ for all $\sigma \in \mathcal{G}$. Hence, $\prod_{\sigma \in \mathcal{G} - \{1\}} \beta^\sigma \in \mathbf{R}_m[\rho]$ and thus

$$\beta^{-1} = \frac{1}{\gamma} \left(\prod_{\sigma \in \mathcal{G} - \{1\}} \beta^\sigma \right) \in S_m[\rho],$$

as desired.

Now, since an element $a_0 + a_1\rho + \ldots + a_{t-1}\rho^{t-1}$ with $a_i \in S_m \subseteq \mathbb{Q}_m$ lies in $\mathbb{Q}_m$ if and only if $a_i = 0$ for $i > 1$, we have that $S_m[\rho] \cap \mathbb{Q}_m = S_m$, as required.

Finally, note that $\mathcal{P}_n \cap \mathbb{Q}_m = \mathcal{P}_n \cap S_n \cap \mathbb{Q}_m = \mathcal{P}_n \cap S_m$. So it suffices to show that $\mathcal{P}_n \cap S_m = \mathcal{P}_m$. It is clear that $\mathcal{P}_m \subseteq \mathcal{P}_n \cap S_m$. The other inclusion follows because $\mathcal{P}_m$ is the unique maximal ideal of S_m. ∎

(11.9) THEOREM. *Assume Notation (11.6). Then there exist different prime ideals* $P_1, \ldots, P_s$ *of* $\mathbf{R}_m$ *such that*

$$p\mathbf{R}_m = \prod_{i=1}^{s} P_i.$$

Proof. Let ξ be a primitive mth root of unity and consider the mth cyclotomic polynomial

$$\Phi_m(x) = \prod_{\substack{1 \le k \le m \\ (k,m)=1}} (x - \xi^k) \in \mathbb{Z}[x].$$

Then

$$\Phi_m(x)^* = \prod_{\substack{1 \le k \le m \\ (k,m)=1}} (x - (\xi^*)^k) \in \mathbb{Z}_p[x].$$

By Lemma (2.1), it follows that all the powers $(\xi^*)^k$ are distinct for distinct k with $1 \le k \le m$. Therefore, $\Phi_m(x)^*$ has no multiple roots and we may write

$$\Phi_m(x)^* = f_1(x) \dots f_s(x),$$

where $f_i(x) \in \mathbb{Z}_p[x]$ is an irreducible polynomial in $\mathbb{Z}_p[x]$ and $f_i \ne f_j$ for $i \ne j$. By the Chinese remainder theorem, we conclude that

$$\mathbb{Z}_p[x]/(\Phi_m(x)^*) \cong \mathbb{Z}_p[x]/(f_1(x)) \oplus \dots \oplus \mathbb{Z}_p[x]/(f_s(x)),$$

as rings. Note that each $F_i = \mathbb{Z}_p[x]/(f_i(x))$ is a field since $f_i(x)$ is irreducible. Since the ring $F_1 \oplus \dots \oplus F_s$ has s different maximal ideals

$$I_i = F_1 \oplus \dots \oplus F_{i-1} \oplus F_{i+1} \oplus \dots \oplus F_s$$

satisfying $0 \subseteq I_1 I_2 \dots I_s \subseteq I_1 \cap I_2 \cap \dots \cap I_s = 0$, it is clear that in order to prove the theorem, it suffices to check that the ring $\mathbf{R}_m/p\mathbf{R}_m$ is isomorphic to $\mathbb{Z}_p[x]/(\Phi_m(x)^*)$. To do that we use Theorem (11.5).

Let us write $\mathbf{R}_m/p\mathbf{R}_m = \mathbb{Z}[\xi]/p\mathbb{Z}[\xi]$. We define $\theta : \mathbb{Z}_p[x] \to \mathbb{Z}[\xi]/p\mathbb{Z}[\xi]$ in the following way. If $f(x)^* \in \mathbb{Z}_p[x]$ (where $f(x) \in \mathbb{Z}[x]$), we set

$$\theta(f(x)^*) = f(\xi) + p\mathbb{Z}[\xi].$$

Note that if $f(x)^* = g(x)^*$ then $f(\xi) - g(\xi) \in p\mathbb{Z}[\xi]$ and therefore the map is well defined. Also, it is clear that θ is an onto ring homomorphism. So it suffices to show that the kernel of θ is $(\Phi_m(x)^*)$. Suppose that $f(x) \in \mathbb{Z}[x]$ is such that $f(\xi) \in p\mathbb{Z}[\xi]$. We want to show that $f(x)^* \in (\Phi_m(x)^*)$. In this case, there exists a polynomial $g(x) \in \mathbb{Z}[x]$ such that $f(\xi) - pg(\xi) = 0$. We may certainly replace $f(x)$ by $f(x) - pg(x)$ and assume that $f(\xi) = 0$. Since the leading coefficient of $\Phi_m(x)$ is 1, we can apply the division algorithm to $f(x)$ and $\Phi_m(x)$ in $\mathbb{Z}[x]$. Hence, there are integer polynomials $d(x)$ and $r(x)$ with $f(x) = d(x)\Phi_m(x) + r(x)$, where $r(x) = 0$ or $r(x)$ has smaller degree than $\Phi_m(x)$. Since $r(\xi) = f(\xi) - d(\xi)\Phi_m(\xi) = 0$ and $\Phi_m(x)$ is the minimal polynomial of ξ over $\mathbb{Q}$, it follows that $r(x) = 0$. Thus, $\Phi_m(x)$ divides $f(x)$ in $\mathbb{Z}[x]$ and $f(x)^* \in (\Phi_m(x)^*)$, as required. ∎

In the next theorem we are going to use two standard facts on a Dedekind domain D. First, we will use that every nonzero ideal I of D can be written in a unique way as a product of prime ideals of D (Theorem (29.3) of [Isaacs, *Algebra*]). Also, we will use the fact that if $A \subseteq B$ are ideals of a Dedekind domain D, then there exists another ideal C of D such that $A = BC$. (With the standard notation on Dedekind domains, it suffices to take $C = B^{-1}A \subseteq D$ and apply that $BB^{-1} = D$ (Theorem (29.6) of [Isaacs, *Algebra*]).)

(11.10) THEOREM. *Assume Notation (11.6). If r is any positive integer, then*
$$(\mathcal{P}_m)^r \cap \mathbb{Q} = \{q \in \mathbb{Q} \,|\, \nu(q) \geq r\}.$$

Proof. It is easy to check that
$$(\mathcal{P}_m)^r = \{\frac{\alpha}{\beta} \,|\, \alpha \in (M_m)^r, \beta \in \mathbf{R}_m - M\}.$$

Also, it is clear that
$$\{q \in \mathbb{Q} \,|\, \nu(q) \geq r\} \subseteq (\mathcal{P}_m)^r \cap \mathbb{Q}.$$

Suppose that
$$\frac{a}{b} = \frac{\alpha}{\beta}$$
where $\alpha \in (M_m)^r, \beta \in \mathbf{R}_m - M$ and $a, b \in \mathbb{Z}$ are such that $(a, b) = 1$. We want to prove that p^r divides a. If u is an integer not divisible by p, note that $u\beta \in \mathbf{R}_m - M$. Hence, by replacing β by $a_{p'}\beta$, we may assume that $a = p^s$ for some nonnegative integer s. Our aim is to show that $s \geq r$. Now, $p^s\beta = b\alpha \in (M_m)^r$. If $x \in \mathbf{R}_m$, denote by $(x) = \mathbf{R}_m x$ the principal ideal generated by x. Then
$$(p)^s(\beta) \subseteq (M_m)^r.$$

Write $(p) = P_1 \ldots P_t$, where, by Theorem (11.9), $P_1, \ldots, P_t$ are distinct prime ideals of $\mathbf{R}_m$. Also, write $(\beta) = Q_1 \ldots Q_v$, where $Q_1, \ldots, Q_v$ are prime ideals of $\mathbf{R}_m$. Note that $Q_i \neq M_m$ for all i because $\beta \in \mathbf{R}_m - M$. Then
$$(P_1)^s \ldots (P_t)^s Q_1 \ldots Q_v \subseteq (M_m)^r.$$

By the comments preceding the statement of this theorem, we conclude that there exists an ideal I of $\mathbf{R}_m$ such that
$$(P_1)^s \ldots (P_t)^s Q_1 \ldots Q_v = I(M_m)^r.$$

By the uniqueness of the decomposition of the products of prime ideals in Dedekind domains, there exists a unique $1 \leq j \leq t$ such that $P_j = M_m$ and we may conclude that $r \leq s$, as desired. ∎

If $K/\mathbb{Q}$ is a finite degree extension of $\mathbb{Q}$, then $\mathbf{R} \cap K$ is a Dedekind domain (Corollary (28.22) of [Isaacs, *Algebra*]) and thus $M \cap K$ is a maximal ideal of $\mathbf{R} \cap K$ (because it is a prime ideal). In particular, $(\mathbf{R} \cap K)^* \cong \mathbf{R} \cap K / M \cap K$ is a subfield of F.

Suppose that G is a finite group and write $|G| = p^a m$, where $|G|_{p'} = m$. Then $F_m = (\mathbf{R}_m)^*$ is a field which contains a primitive mth root of unity (by Lemma (2.1)). Since the polynomial $x^{p^a m} - 1 = (x^m - 1)^{p^a} \in F_m[x]$ splits into linear factors in the field F_m, it follows that F_m is a splitting field for G by Corollary (9.15) of [Isaacs, *Characters*]. Let us fix $\{\mathcal{Y}_1, \ldots, \mathcal{Y}_l\}$, a complete set of irreducible F_m-representations of G, so that $\{\mathcal{Y}_1, \ldots, \mathcal{Y}_l\}$ is a complete set of irreducible F-representations of G (Corollary (9.8) of [Isaacs, *Characters*]).

(11.11) THEOREM. *Suppose that p is an odd prime number. Let G be a group and write $|G| = p^a m$, where p does not divide m. Let $\chi \in \mathrm{Irr}(G)$. Then there exists a representation $\mathcal{X}$ of G affording χ satisfying the following conditions:*

(a) $\mathcal{X}$ has its entries in S_χ, where $S_\chi = \{\frac{\alpha}{\beta} \mid \alpha \in \mathbf{R}_\chi, \beta \in \mathbf{R}_\chi - M\}$ and $\mathbf{R}_\chi = \mathbf{R} \cap \mathbb{Q}_m(\chi)$;

(b) the F-representation $\mathcal{X}^*$ has the form

$$\mathcal{X}^* = \begin{pmatrix} \mathcal{Y}_{i_1} & * & \cdots & * \\ 0 & \mathcal{Y}_{i_2} & \cdots & * \\ \vdots & \vdots & \ddots & \vdots \\ 0 & 0 & \cdots & \mathcal{Y}_{i_s} \end{pmatrix}$$

for some not necessarily distinct indices $1 \leq i_j \leq l$.

Proof. We have that $\mathbb{Q}_m \subseteq \mathbb{Q}_m(\chi) \subseteq \mathbb{Q}_{p^a m}$. By a result of P. Fong (Corollary (10.13) of [Isaacs, *Characters*]), we have that the Schur index $m_{\mathbb{Q}_m}(\chi) = 1$. Now, by Corollary (10.2.d) of [Isaacs, *Characters*] we conclude that χ is afforded by a $\mathbb{Q}_m(\chi)$-representation. By Problem (2.12), we have that χ can be afforded by an S_χ-representation $\mathcal{X}$.

We show that it is possible to choose a representation $\mathcal{X}$ which also satisfies condition (b) of the theorem. Let $(\mathbf{R}_\chi)^* = F_0 = (S_\chi)^*$, a subfield of F. Note that $F_m \subseteq F_0$. (In fact, $F_0 = F_m$ but we do not need that.) Hence, we have that $\{\mathcal{Y}_1, \ldots, \mathcal{Y}_l\}$ is a complete set of irreducible F_0-representations of G (by Corollary (9.8) of [Isaacs, *Characters*]). Now, by the discussion preceding Theorem (1.20), there exists a $Q \in \mathrm{GL}(r, F_0)$, where r is the degree of $\mathcal{X}$, such that

$$Q\mathcal{X}^*Q^{-1} = \begin{pmatrix} \mathcal{Y}_{i_1} & * & \cdots & * \\ 0 & \mathcal{Y}_{i_2} & \cdots & * \\ \vdots & \vdots & \ddots & \vdots \\ 0 & 0 & \cdots & \mathcal{Y}_{i_s} \end{pmatrix}$$

for some indices $1 \le i_j \le l$. Let P be a matrix with entries in $\mathbf{R}_\chi$ such that $P^* = Q$. Since $\det(P)^* = \det(P^*) = \det(Q) \ne 0$, it follows that $\det(P) \in \mathbf{R}_\chi - M$. Thus, $P^{-1} \in \mathrm{GL}(r, S_\chi)$. Hence, if we replace $\mathcal{X}$ by $P\mathcal{X}P^{-1}$, we have that $P\mathcal{X}P^{-1}$ has its entries in S_χ and $(P\mathcal{X}P^{-1})^* = P^*\mathcal{X}^*(P^{-1})^* = Q\mathcal{X}^*Q^{-1}$ has the desired form. ∎

In the proof of Brauer's theorem (11.4), we will use the well-known **Schur relations**. If $\mathcal{X}$ and $\mathcal{Y}$ are irreducible $\mathbb{C}$-representations of G and we write $\mathcal{X}(g) = (x_{ij}(g))$ and $\mathcal{Y}(g) = (y_{ij}(g))$ for $g \in G$, we have that

$$\sum_{g \in G} x_{ij}(g) y_{kl}(g^{-1}) = 0$$

if $\mathcal{X}$ and $\mathcal{Y}$ are not similar. If $\mathcal{X} = \mathcal{Y}$ and d is the degree of $\mathcal{X}$, then

$$\sum_{g \in G} x_{ij}(g) x_{kl}(g^{-1}) = \delta_{il}\delta_{jk}\frac{|G|}{d}.$$

(See Problem (2.20) of [Isaacs, *Characters*].)

Proof of Theorem (11.4) We fix ρ, a primitive pth root of unity. Let $m = |G|_{p'}$ and set $n = pm$. Let $\mathcal{G} = \mathrm{Gal}(\mathbb{Q}_m(\rho)/\mathbb{Q}_m)$, so that $|\mathcal{G}| = p - 1$. Also, $\mathbb{Q}_m(\rho) = \mathbb{Q}_n$. We use Notation (11.6). Write $|G|_p = p^a$.

Suppose that $\chi \in \mathrm{Irr}(B)$. First of all, note that $\chi(1)_p = p^{a-1}$ because $\chi(1)_p = p^{a-1+h}$ where $h = \mathrm{height}(\chi) \ge 0$ and, on the other hand, $\chi(1)_p < |G|_p = p^a$ (since the block B does not have defect zero).

We claim that if $\chi \in \mathrm{Irr}(B)$, then $\mathbb{Q}_m(\chi) \subseteq \mathbb{Q}_m(\rho) = \mathbb{Q}_n$. Let $g \in G$. We want to show that $\chi(g) \in \mathbb{Q}_m(\rho)$. This is clear if g is p-regular because in this case $\chi(g) \in \mathbb{Q}_m$. So we assume that p divides the order of g. By Corollary (5.9), we have that $\chi(g) = 0$ if the order of g_p is bigger than p. Hence, we may assume that the order of g_p is p. In this case, the generalized decomposition numbers $d^{g_p}_{\chi\mu}$ lie in $\mathbb{Z}[\rho]$ (see the definition of the generalized decomposition numbers after Lemma (5.1)). By Lemma (5.1), we see that

$$\chi(g) = \sum_{\mu \in \mathrm{IBr}(\mathbf{C}_G(g_p))} d^{g_p}_{\chi\mu}\mu(g_{p'}) \in \mathbb{Q}_m(\rho)$$

because $\mu(g_{p'}) \in \mathbb{Q}_m$ for all $\mu \in \mathrm{IBr}(\mathbf{C}_G(g_p))$.

Now, we claim that $\chi, \psi \in \mathrm{Irr}(B)$ are p-conjugate if and only if there exists a $\sigma \in \mathcal{G}$ such that $\chi^\sigma = \psi$. (By the previous paragraph, we have that $\mathbb{Q}_m(\chi), \mathbb{Q}_m(\psi) \subseteq \mathbb{Q}_n$.) Suppose first that $\chi = \psi^\sigma$ for some $\sigma \in \mathcal{G}$. By elementary Galois theory, we know that restriction defines an onto homomorphism $\mathrm{Gal}(\mathbb{Q}_{|G|}/\mathbb{Q}_m) \rightarrow \mathrm{Gal}(\mathbb{Q}_n/\mathbb{Q}_m)$. Hence, there exists $\tilde{\sigma} \in \mathrm{Gal}(\mathbb{Q}_{|G|}/\mathbb{Q}_m)$ extending σ. It is clear that $\chi = \psi^{\tilde{\sigma}}$. Conversely, if $\chi = \psi^\tau$ for some $\tau \in \mathrm{Gal}(\mathbb{Q}_{|G|}/\mathbb{Q}_m)$ and we write $\sigma = \tau_{\mathbb{Q}_m(\rho)} \in \mathcal{G}$, then $\chi = \psi^\sigma$ because $\mathbb{Q}_m(\psi) \subseteq \mathbb{Q}_m(\rho)$. This proves the claim.

As remarked before the statement of Theorem (11.4), recall that if $\chi \in \mathrm{Irr}(B)$, then the number r_χ of characters which are p-conjugate to χ is $|\mathbb{Q}_m(\chi) : \mathbb{Q}_m|$. Hence, r_χ divides $|\mathbb{Q}_m(\rho) : \mathbb{Q}_m| = p - 1$.

Now, suppose again that $\chi, \psi \in \mathrm{Irr}(B)$ and let $\mathcal{X}$ and $\mathcal{Y}$ be representations of G satisfying conditions (a) and (b) of Theorem (11.11), affording the characters χ and ψ, respectively. If $\chi = \psi$ we take $\mathcal{Y} = \mathcal{X}$. Since $\mathbb{Q}_m(\chi), \mathbb{Q}_m(\psi) \subseteq \mathbb{Q}_n$, we have that the entries of $\mathcal{X}$ and $\mathcal{Y}$ are in S_n.

If $\sigma, \tau \in \mathcal{G}$, we claim that $\mathcal{X}^\sigma$ and $\mathcal{X}^\tau$ are nonsimilar or equal. Suppose that they are similar. Then $\chi^\sigma = \chi^\tau$. Hence, $\chi^\mu = \chi$, where $\mu = \sigma\tau^{-1}$. Thus, μ fixes elementwise the field $\mathbb{Q}_m(\chi)$. This implies that $\mathcal{X}^\mu = \mathcal{X}$ because the entries of $\mathcal{X}$ are in $S_\chi \subseteq \mathbb{Q}_m(\chi)$. The same happens for $\mathcal{Y}$.

From now on, we assume that ψ and χ are not p-conjugate such that ψ^0 and χ^0 have an irreducible Brauer constituent in common or that $\chi = \psi$ and χ^0 has an irreducible Brauer constituent φ with $d_{\chi\varphi} \geq 2$.

Write $\mathcal{X}(g) = (x_{ij}(g))$ and $\mathcal{Y}(g) = (y_{ij}(g))$ for $g \in G$. By assumption and condition (b) of Theorem (11.11), we have that there exist indices i and j such that $x_{ii}(g)^* = y_{jj}(g)^*$ for all $g \in G$. (The elements $x_{ii}(g)^*$ and $y_{jj}(g)^*$ are the upper left corners in the blocks on the diagonal corresponding to the common irreducible Brauer constituent, for instance.) Also, if $\chi = \psi$ then $i \neq j$.

Since the matrices $\mathcal{X}$ and $\mathcal{Y}$ have their entries in S_n, we have that

$$x_{ii}(g) - y_{jj}(g) \in \mathcal{P}_n$$

for all $g \in G$. Also, if $\sigma \in \mathcal{G}$, we have that

$$x_{ii}(g)^\sigma - y_{jj}(g)^\sigma \in \mathcal{P}_n$$

because $\mathcal{P}_n$ is $\mathcal{G}$-invariant (by Theorem (11.7)). Hence, by using Theorem (11.8), we conclude that

$$\sum_{\sigma \in \mathcal{G}} (x_{ii}(g)^\sigma - y_{jj}(g)^\sigma) \in \mathcal{P}_n \cap \mathbb{Q}_m = \mathcal{P}_m$$

for all $g \in G$. Therefore,

$$\sum_{\sigma,\tau \in \mathcal{G}} \left(x_{ii}(g)^\sigma - y_{jj}(g)^\sigma\right)\left(x_{ii}(g^{-1})^\tau - y_{jj}(g^{-1})^\tau\right) \in (\mathcal{P}_m)^2.$$

Now, summing for $g \in G$, we obtain that

$$\sum_{\sigma,\tau \in \mathcal{G}} \left(\sum_{g \in G} x_{ii}(g)^\sigma x_{ii}(g^{-1})^\tau\right) - \sum_{\sigma,\tau \in \mathcal{G}} \left(\sum_{g \in G} y_{jj}(g)^\sigma x_{ii}(g^{-1})^\tau\right)$$

$$+ \sum_{\sigma,\tau \in \mathcal{G}} \left(\sum_{g \in G} y_{jj}(g)^\sigma y_{jj}(g^{-1})^\tau\right) - \sum_{\sigma,\tau \in \mathcal{G}} \left(\sum_{g \in G} x_{ii}(g)^\sigma y_{jj}(g^{-1})^\tau\right) \in (\mathcal{P}_m)^2.$$

Since the characters χ and ψ are not $\mathcal{G}$-conjugate (which is the same as not being p-conjugate) or they are equal (and in this case $\mathcal{X} = \mathcal{Y}$), we deduce that for $\sigma, \tau \in \mathcal{G}$ the representations $\mathcal{Y}^\sigma$ and $\mathcal{X}^\tau$ are nonsimilar or equal (and, if they are equal, then $i \neq j$). We have that

$$\sum_{\sigma,\tau \in \mathcal{G}} \left(\sum_{g \in G} y_{jj}(g)^\sigma x_{ii}(g^{-1})^\tau\right) = \sum_{\sigma,\tau \in \mathcal{G}} \left(\sum_{g \in G} x_{ii}(g)^\sigma y_{jj}(g^{-1})^\tau\right) = 0$$

by applying the Schur relations. We conclude that

$$\sum_{\sigma,\tau \in \mathcal{G}} \left(\sum_{g \in G} x_{ii}(g)^\sigma x_{ii}(g^{-1})^\tau\right) + \sum_{\sigma,\tau \in \mathcal{G}} \left(\sum_{g \in G} y_{jj}(g)^\sigma y_{jj}(g^{-1})^\tau\right) \in (\mathcal{P}_m)^2.$$

Let $\mathcal{G}_\chi = \mathrm{Gal}(\mathbb{Q}_n/\mathbb{Q}_m(\chi))$ and $\mathcal{G}_\psi = \mathrm{Gal}(\mathbb{Q}_n/\mathbb{Q}_m(\psi))$. Now, using that $\mathcal{X}^\sigma$ and $\mathcal{X}^\tau$ are nonsimilar or equal and the Schur relations, we have that

$$\sum_{g \in G} x_{ii}(g)^\sigma x_{ii}(g^{-1})^\tau = 0$$

unless $\sigma \mathcal{G}_\chi = \tau \mathcal{G}_\chi$. In the latter case, $\mathcal{X}^\sigma = \mathcal{X}^\tau$ and then

$$\sum_{g \in G} x_{ii}(g)^\sigma x_{ii}(g^{-1})^\tau = \frac{|G|}{\chi(1)}.$$

(Of course, the same is true with respect to $\mathcal{Y}^\sigma$ and $\mathcal{Y}^\tau$.) Hence, we conclude that

$$|\mathcal{G}||\mathcal{G}_\chi|\frac{|G|}{\chi(1)} + |\mathcal{G}||\mathcal{G}_\psi|\frac{|G|}{\psi(1)} \in (\mathcal{P}_m)^2 \cap \mathbb{Q}.$$

By using Theorem (11.10), we have that

$$\nu\left(|\mathcal{G}||\mathcal{G}_\chi|\frac{|G|}{\chi(1)} + |\mathcal{G}||\mathcal{G}_\psi|\frac{|G|}{\psi(1)}\right) \geq 2.$$

Now, by dividing by the p'-number $|\mathcal{G}|^2 = (p-1)^2$ and using that $|\mathcal{G}| = |\mathcal{G}_\chi|r_\chi = |\mathcal{G}_\psi|r_\psi$, it follows that

$$\nu\left(\frac{|G|}{r_\chi\chi(1)} + \frac{|G|}{r_\psi\psi(1)}\right) \geq 2.$$

Since $\nu(\frac{r_\chi r_\psi \chi(1)\psi(1)}{|G|}) = a - 2$ (because $\nu(\chi(1)) = \nu(\psi(1)) = a - 1$ and r_χ and r_ψ are p'-numbers), it follows that

$$\nu(r_\chi\chi(1) + r_\psi\psi(1)) \geq a. \tag{1}$$

Recall that equation (1) is valid whenever χ and ψ are not p-conjugate characters in $\mathrm{Irr}(B)$ with χ^0 and ψ^0 having an irreducible constituent in common or whenever $\chi = \psi$ and $d_{\chi\varphi} \geq 2$ for some $\varphi \in \mathrm{IBr}(G)$.

If $\chi = \psi$ and $d_{\chi\varphi} \geq 2$ for some $\varphi \in \mathrm{IBr}(G)$, we would deduce that p^a divides $2r_\chi\chi(1)$ by using equation (1). Since p is odd and r_χ is a p'-number (recall that r_χ divides $p - 1$), we get that p^a divides $\chi(1)$. This is a contradiction which proves that $d_{\chi\varphi} \leq 1$ for all $\chi \in \mathrm{Irr}(B)$ and $\varphi \in \mathrm{IBr}(B)$. This proves part (a).

Suppose now that $\varphi \in \mathrm{IBr}(B)$ is a constituent of χ^0, ψ^0 and ξ^0, where χ, ψ and ξ are irreducible characters of B lying in three different $\mathcal{G}$-orbits. By applying the equation (1) three times, we will obtain the congruences

$$r_\chi\chi(1) \equiv -r_\psi\psi(1) \bmod p^a, \quad r_\chi\chi(1) \equiv -r_\xi\xi(1) \bmod p^a$$

and

$$r_\psi\psi(1) \equiv -r_\xi\xi(1) \bmod p^a.$$

This shows that

$$r_\psi\psi(1) \equiv -r_\psi\psi(1) \bmod p^a.$$

Hence, p^a divides $2r_\psi\psi(1)$ and this is impossible. Therefore, each $\varphi \in \mathrm{IBr}(B)$ is an irreducible constituent of at most two families ξ_i and ξ_j of p-conjugate characters. If φ were an irreducible constituent of a single family of p-conjugate characters, say $\mathcal{O}(\xi_i)$, by applying part (a) we would have that

$$\Phi_\varphi = \sum_{\chi \in \mathcal{O}(\xi_i)} \chi.$$

By evaluating at one and applying Dickson's theorem (Corollary (2.14)), we would have that p^a divides $r_{\xi_i}\xi_i(1)$. Since p does not divide r_{ξ_i}, this is impossible. This proves the theorem. $\blacksquare$

To give a complete description of the blocks B of defect one, it remains to analyse the number ω of different families of p-conjugate characters in B. We will do this for groups G with $|G|_p = p$. (A similar argument gives the description in the general case.)

(11.12) NOTATION AND HYPOTHESES. Suppose that $|G|_p = p$, where p is odd. Let $P \in \mathrm{Syl}_p(G)$. Write $N = \mathbf{N}_G(P)$ and $C = \mathbf{C}_G(P)$. Note that N/C is a subgroup of $\mathrm{Aut}(P)$. Hence, N/C is a cyclic group of order dividing $p-1$. By the Schur-Zassenhaus theorem, we may write $C = P \times K$, where $K = \mathbf{O}_{p'}(N) = C^0$. Note that $\mathrm{IBr}(C) = \mathrm{Irr}(K)$, by Theorem (8.21), for instance. Let B be a block of G with defect group P. Let $b \in \mathrm{Bl}(C)$ be a root of B and suppose that $1_P \times \xi \in \mathrm{Irr}(b)$ is the canonical character of B, where $\xi \in \mathrm{Irr}(K)$. (See Theorem (9.12).) Note that $\mathrm{IBr}(b) = \{\xi\}$. Let T be the inertia group of ξ in N and write $e = |T : C|$ and $t = \frac{p-1}{e}$. Since $C \subseteq T \subseteq N$ and $|N : C|$ divides $p - 1$, it follows that t is an integer.

(11.13) THEOREM. *Assume Notation and Hypotheses (11.12).*

(a) The number of ordinary irreducible characters in B is $e + t$. They can be labelled as $\chi_1, \ldots, \chi_e, \theta_1, \ldots, \theta_t$. The characters $\chi_1, \ldots, \chi_e$ (called **nonexceptional***) are all p-rational and have degrees*

$$\chi_i(1) \equiv \epsilon_i \frac{|N : C|}{e}\xi(1) \bmod p$$

for uniquely defined $\epsilon_i = \pm 1$ and $1 \le i \le e$. The characters $\theta_1, \ldots, \theta_t$ (called **exceptional** *if $t \ge 2$) are all p-conjugate. In particular, they have the same values on p-regular elements. Also, there is a unique $\epsilon = \pm 1$ such that*

$$\theta_j(1) \equiv \epsilon|N : C|\xi(1) \bmod p$$

for every $1 \le j \le t$.

(b)

$$\epsilon(\theta_1)^0 = \sum_{i=1}^{e} \epsilon_i(\chi_i)^0.$$

(c) The number of irreducible Brauer characters in B is e. Also, $d_{\chi\varphi} \le 1$ for all $\chi \in \mathrm{Irr}(B)$.

(d) Suppose that $1 \ne x \in P$ and $y \in K$. Then

$$\chi_i(xy) = \epsilon_i \frac{1}{e}(1_P \times \xi)^N(y)$$

for $1 \leq i \leq e$.

(e) *There exist representatives $\lambda_1, \ldots, \lambda_t$ of the action of T on $\mathrm{Irr}(P) - \{1_P\}$ such that*

$$\theta_j(xy) = \epsilon(\lambda_j \times \xi)^N(xy)$$

for $1 \leq j \leq t$, $1 \neq x \in P$ and $y \in K$.

Proof. Let ρ be a primitive pth root of unity and write

$$\mathcal{G} = \mathrm{Gal}(\mathbb{Q}_m(\rho)/\mathbb{Q}_m),$$

where $m = |G|_{p'}$. By elementary Galois theory, observe that restriction defines an isomorphism from $\mathcal{G}$ onto $\mathrm{Gal}(\mathbb{Q}(\rho)/\mathbb{Q})$. Note that $\mathcal{G} = \langle \sigma \rangle$, where the order of σ is $p - 1$.

If H is a subgroup of G and ψ is a generalized character of H, then ψ has its values in $\mathbb{Q}_m(\rho)$. As usual, we will denote by ψ^τ the function $\psi^\tau(h) = \psi(h)^\tau$ for $h \in H$ and $\tau \in \mathcal{G}$.

Now, T/C acts on $\mathrm{Irr}(P) - \{1_P\}$ with orbits of size e because $n \in N$ fixes $1 \neq \lambda \in \mathrm{Irr}(P)$ if and only if $n \in C$. Let $\{\lambda_1, \ldots, \lambda_t\}$ be a complete set of representatives of the orbits of the action of T on $\mathrm{Irr}(P) - \{1_P\}$, so that $t = \frac{p-1}{e}$. Given λ_i, it is clear that there exists $1 \leq n_i \leq p - 1$ such that

$$\lambda_i = (\lambda_1)^{\sigma^{n_i}}$$

because $\mathcal{G}$ acts transitively on $\mathrm{Irr}(P) - \{1_P\}$.

The essence of the proof is to analyse the generalized characters

$$\Gamma_i = (1_P \times \xi - \lambda_i \times \xi)^G$$

for $i = 1, \ldots, t$. Since $\xi^\sigma = \xi$ observe that

$$\Gamma_i = (\Gamma_1)^{\sigma^{n_i}}.$$

Also, if $\lambda \in \mathrm{Irr}(P) - \{1_P\}$, then there exists an i such that

$$(1_P \times \xi - \lambda \times \xi)^G = \Gamma_i$$

because there exist a $z \in T$ and $1 \leq i \leq t$ such that $\lambda^z = \lambda_i$. In this case, $(\lambda \times \xi)^G = ((\lambda \times \xi)^z)^G = (\lambda_i \times \xi)^G$ and $(1_P \times \xi - \lambda \times \xi)^G = \Gamma_i$.

In particular, note that $\mathcal{G}$ acts (transitively) on the set $\{\Gamma_1, \ldots, \Gamma_t\}$

For the reader's convenience, we divide the proof into several steps.

Step 1. $k(B) - l(B) = t$.

Write $l(B) = l$. To prove this step, we will use Theorem (5.12). Since two nontrivial elements of P are G-conjugate if and only if they are N-conjugate and no nontrivial element of N/C fixes any nontrivial element of P, we see that there are exactly $\frac{p-1}{|N:C|}$ conjugacy classes of nontrivial p-elements in G. Given $1 \neq x \in P$, note that there are exactly $|N : T|$ different blocks of $\mathbf{C}_G(x) = C$ inducing B because all of them are N-conjugate to b (by the extended first main theorem (9.7)), b has $|N : T|$ distinct N-conjugates (because T is also the inertia group of b in N) and each N-conjugate of b induces B. Also, if $n \in N$, then $\mathrm{IBr}(b^n) = \{\xi^n\}$ because $\mathrm{IBr}(b) = \{\xi\}$. By Theorem (5.12), we have that

$$k(B) = |N : T|\frac{p-1}{|N:C|} + l = \frac{p-1}{e} + l = t + l,$$

as desired.

Step 2. If $\chi \in \mathrm{Irr}(B)$, $1 \neq x \in P$ and $y \in K$, then

$$\chi(xy) = \sum_{i=1}^{u} d^x_{\chi\xi_i}\xi_i(y)$$

where $\xi_1, \ldots, \xi_u$ are the distinct N-conjugates of ξ. Also, for $n \in N$, we have that

$$d^x_{\chi\xi^{n-1}} = d^{x^n}_{\chi\xi}.$$

By Corollary (5.8), we know that

$$\chi(xy) = \sum_{\mu \in \mathrm{IBr}(C)} d^x_{\chi\mu}\mu(y),$$

where the sum is over $\mu \in \mathrm{IBr}(C)$ lying in blocks of C inducing B. We already discussed in the previous step that there are exactly $|N : T|$ different blocks of $\mathbf{C}_G(x) = C$ inducing B and that these are all the different N-conjugates of b. Since the inertia group of ξ in N is the inertia group of b in N and $\mathrm{IBr}(b^n) = \{\xi^n\}$, we have that

$$\chi(xy) = \sum_{i=1}^{u} d^x_{\chi\xi_i}\xi_i(y).$$

Now let $1 \neq x \in P$ and $n \in N$, and note that $\mathbf{C}_G(x^n) = C$. By using Lemma (5.1), we have that

$$\sum_{\varphi \in \mathrm{IBr}(C)} d^{x^n}_{\chi\varphi}\varphi(y^n) = \chi(x^n y^n) = \chi(xy) = \sum_{\varphi \in \mathrm{IBr}(C)} d^x_{\chi\varphi}\varphi(y).$$

By the uniqueness of the generalized decomposition numbers, we conclude that

$$d^x_{\chi\varphi^{n-1}} = d^{x^n}_{\chi\varphi}$$

for $\varphi \in \mathrm{IBr}(C)$, $n \in N$ and $1 \neq x \in P$.

Step 3. *A character $\chi \in \mathrm{Irr}(G)$ lies in B if and only if $[\Gamma_i, \chi] \neq 0$ for some i. Also, a character $\chi \in \mathrm{Irr}(B)$ is p-rational if and only if $[\Gamma_i, \chi] = [\Gamma_1, \chi]$ for all i. In this case, $[\Gamma_1, \chi]$ is the generalized decomposition number $d^x_{\chi\xi}$ for all $1 \neq x \in P$.*

Let $1 \neq x \in P$. We calculate the decomposition numbers $d^x_{\chi\mu}$ for $\chi \in \mathrm{Irr}(G)$ and $\mu \in \mathrm{IBr}(C)$. Recall that $C^0 = K$ and $\mathrm{IBr}(C) = \mathrm{Irr}(K)$. We may write

$$\chi_C = \sum_{\mu \in \mathrm{Irr}(K)} \psi_\mu \times \mu,$$

where

$$\psi_\mu = \sum_{\lambda \in \mathrm{Irr}(P)} [\chi_C, \lambda \times \mu]\lambda.$$

By the uniqueness of the generalized decomposition numbers, we deduce that

$$d^x_{\chi\mu} = \psi_\mu(x) = \sum_{\lambda \in \mathrm{Irr}(P)} [\chi_C, \lambda \times \mu]\lambda(x). \tag{1}$$

Since $\{\lambda(x) \mid \lambda \in \mathrm{Irr}(P)\} = \{1, \rho, \ldots, \rho^{p-1}\}$ and $1 + x + \ldots + x^{p-1}$ is the minimal polynomial of ρ over $\mathbb{Q}$, it easily follows that

$$d^x_{\chi\mu} = 0 \quad \text{iff} \quad [\chi_C, \lambda \times \mu] = [\chi_C, 1_P \times \mu] \quad \text{for all} \quad \lambda \in \mathrm{Irr}(P).$$

(Note, therefore, that if $d^x_{\chi\mu} = 0$ for some $1 \neq x \in P$, then $d^{x^i}_{\chi\mu} = 0$ for every $1 \leq i \leq p-1$.) In particular, $d^x_{\chi\xi} \neq 0$ if and only if there is a $\lambda \in \mathrm{Irr}(P)$ such that

$$[(1_P \times \xi - \lambda \times \xi)^G, \chi] \neq 0.$$

Given $\lambda \in \mathrm{Irr}(P) - \{1_P\}$, we know that there is an i such that

$$(1_P \times \xi - \lambda \times \xi)^G = \Gamma_i$$

by the comments preceding the statement of Step 1. We deduce that $d^x_{\chi\xi} \neq 0$ if and only if there is an i such that $[\Gamma_i, \chi] \neq 0$. To complete the first part of this step, it suffices to show that $\chi \in \mathrm{Irr}(B)$ if and only if $d^x_{\chi\xi} \neq 0$. If $d^x_{\chi\xi} \neq 0$, then $\chi \in \mathrm{Irr}(B)$ by the second main theorem. Conversely, suppose that $\chi \in \mathrm{Irr}(B)$ and that $d^x_{\chi\xi} = 0$. Then $d^{x^i}_{\chi\xi} = 0$ for $1 \leq i \leq p-1$. Hence, we have that $\chi(xy) = 0$ for every $y \in K$ by Step 2. By the same argument applied to x^i for $1 \leq i \leq p-1$, we deduce that $\chi(x^i y) = 0$ for every $y \in K$. Since every p-singular element is G-conjugate to some $x^i y$ for $1 \leq i \leq p-1$ and $y \in K$, we deduce that χ has defect zero by Theorem (3.18). This contradiction proves the first part of Step 3.

To complete the proof of this step, we show that $\chi \in \mathrm{Irr}(B)$ is p-rational if and only if $[\Gamma_i, \chi] = [\Gamma_1, \chi]$ for all $1 \leq i \leq t$.

Recall that χ is p-rational if and only if χ has its values in $\mathbb{Q}_m$ or, in other words, if $\chi^\sigma = \chi$. It is clear that if χ is p-rational, then $[\Gamma_i, \chi] = [\Gamma_1, \chi]$ for all $1 \leq i \leq t$ because $\Gamma_i = (\Gamma_1)^{\sigma^{n_i}}$. Conversely, assume that $[\Gamma_i, \chi] = [\Gamma_1, \chi]$ for all $1 \leq i \leq t$. We wish to show that χ is p-rational. Since $|G|_p = p$ and $\chi(y) \in \mathbb{Q}_m$ for $y \in G^0$, it suffices to show that $\chi(g) \in \mathbb{Q}_m$ for g in the p-section of $1 \neq x \in P$. By Step 2, it suffices to prove that the generalized decomposition numbers $d^x_{\chi \xi_i} \in \mathbb{Z}$ for all $1 \neq x \in P$ and $1 \leq i \leq u$. By assumption, we have that

$$[\chi, \Gamma_1 - \Gamma_i] = 0$$

for all i. Then

$$[\chi_C, \lambda_i \times \xi] = [\chi_C, \lambda_1 \times \xi]$$

for all i. Hence,

$$[\chi_C, \lambda \times \xi] = [\chi_C, \lambda_1 \times \xi]$$

for all $1 \neq \lambda \in \mathrm{Irr}(P)$ (see the comments preceding the statement of Step 1). Now, by equation (1), we have that

$$d^x_{\chi \xi} = [\chi_C, 1_P \times \xi] + [\chi_C, \lambda_1 \times \xi] \sum_{\lambda \in \mathrm{Irr}(P) - \{1_P\}} \lambda(x)$$

$$= [\chi_C, 1_P \times \xi] - [\chi_C, \lambda_1 \times \xi] = [\chi, \Gamma_1] \in \mathbb{Z}$$

for all $1 \neq x \in P$. If $n \in N$, by Step 2 we have that

$$d^x_{\chi \xi^{n-1}} = d^{x^n}_{\chi \xi} \in \mathbb{Z}.$$

This proves that χ is p-rational and completes the proof of the step.

Step 4. $[\Gamma_i, \Gamma_j] = \delta_{ij} + e$. Also, $(\Gamma_i)_N = (1_P \times \xi - \lambda_i \times \xi)^N$.

First, we check that $(\Gamma_i)^0 = 0$. Note that if $z \in C$ is such that $z_p = 1$, then $(1_P \times \xi - \lambda_i \times \xi)(z) = 0$. Now, if $g \in G^0$, then

$$\Gamma_i(g) = \frac{1}{|C|} \sum_{\substack{u \in G \\ ugu^{-1} \in C}} (1_P \times \xi - \lambda_i \times \xi)(ugu^{-1}) = 0$$

by the preceding remark. By the same argument (applied in N), we have that $(1_P \times \xi - \lambda_i \times \xi)^N (n) = 0$ if $n \in N$ with $n_p = 1$.

To prove the second assertion in Step 4, it suffices to check that the functions $(\Gamma_i)_N$ and $(1_P \times \xi - \lambda_i \times \xi)^N$ coincide on the elements of the form xy, where $1 \neq x \in P$ and $y \in K$. If $g \in G$ is such that $(xy)^g \in N$, then $x^g \in N$ (because x^g is a power of $(xy)^g$). Since $\langle x \rangle$ is the unique Sylow p-subgroup of N, it follows that g normalizes $\langle x \rangle$. Therefore, $g \in N$. By the definition of induced characters, we see that

$$\Gamma_i(xy) = ((1_P \times \xi - \lambda_i \times \xi)^N)^G(xy) = (1_P \times \xi - \lambda_i \times \xi)^N(xy)$$

for $1 \neq x \in P$ and $y \in K$. This proves the second part of this step.

Now, by using Frobenius reciprocity and the second part of this step, we have that

$$[\Gamma_i, \Gamma_j] = [(\Gamma_i)_N, ((1_P - \lambda_j) \times \xi)^N]$$
$$= [((1_P - \lambda_i) \times \xi)^N, ((1_P - \lambda_j) \times \xi)^N].$$

If $\lambda \in \mathrm{Irr}(P) - \{1_P\}$, we claim that $(\lambda \times \xi)^N \in \mathrm{Irr}(N)$. If $n \in N$, we have that $(\lambda \times \xi)^n = \lambda \times \xi$ if and only if $\lambda^n = \lambda$ and $\xi^n = \xi$. However, $\lambda^n = \lambda$ implies that $n \in C$. Hence, the inertia group of $\lambda \times \xi$ in N is C. Therefore, $(\lambda \times \xi)^N$ is irreducible, as claimed.

Next, we claim that for $i \neq j$, we have that

$$[(\lambda_i \times \xi)^N, (\lambda_j \times \xi)^N] = 0.$$

Suppose that $[(\lambda_i \times \xi)^N, (\lambda_j \times \xi)^N] \neq 0$. Since the characters $(\lambda_i \times \xi)^N$ and $(\lambda_j \times \xi)^N$ are irreducible, we will have that $(\lambda_i \times \xi)^N = (\lambda_j \times \xi)^N$. Now, the characters $\lambda_i \times \xi$ and $\lambda_j \times \xi$ would be N-conjugate by Clifford's theorem. In this case, λ_i and λ_j would be T-conjugate. This is a contradiction because $i \neq j$. This proves the claim.

Finally, for every $1 \leq i \leq t$, we have that the character $(1_P \times \xi)^N$ does not contain $(\lambda_i \times \xi)^N$ as an irreducible constituent since P is contained in the kernel of $(1_P \times \xi)^N$ and, therefore, it is contained in the kernel of each of its irreducible constituents. Thus,

$$[\Gamma_i, \Gamma_j] = [(1_P \times \xi)^N, (1_P \times \xi)^N] + \delta_{ij}.$$

By using the fact that T is the inertia group of $1_P \times \xi$ in N and the Clifford correspondence, note that $[(1_P \times \xi)^N, (1_P \times \xi)^N] = e$ because the irreducible character $1_P \times \xi$ extends to T (recall that T/C is cyclic) and has exactly e different extensions. This proves that

$$[\Gamma_i, \Gamma_j] = e + \delta_{ij},$$

as desired.

Step 5. *There are $\theta_1 \in \mathrm{Irr}(B)$ and $\delta = \pm 1$ such that*

$$\Gamma_1 - (\Gamma_1)^\sigma = \delta(\theta_1 - (\theta_1)^\sigma).$$

Recall that $\mathcal{G} = \langle\sigma\rangle$ acts transitively on $\{\Gamma_1, \ldots, \Gamma_t\}$ by the remark preceding the statement of the Step 1.

Suppose first that $t = 1$. In this case, $(\Gamma_1)^\sigma = \Gamma_1$. Let θ_1 be any irreducible character in B and set $\delta = 1$. By Step 3, θ_1 is p-rational and the proof of the step is complete in this case.

For the rest of the proof of this step, suppose that $t \geq 2$.

If we prove that $\Gamma_1 - (\Gamma_1)^\sigma = \delta(\theta_1 - (\theta_1)^\sigma)$ for some $\theta_1 \in \mathrm{Irr}(G)$ with $\theta_1 \neq (\theta_1)^\sigma$, we will have that $[\Gamma_1, \theta_1] \neq 0$ or $[(\Gamma_1)^\sigma, \theta_1] \neq 0$. Since $(\Gamma_1)^\sigma = \Gamma_i$ for some i, we will conclude that $\theta_1 \in \mathrm{Irr}(B)$ by Step 3 and the proof of Step 5 will be complete.

By Step 4, we have that

$$[\Gamma_i - \Gamma_j, \Gamma_i - \Gamma_j] = 2$$

if $i \neq j$. This means that the generalized character $\Gamma_i - \Gamma_j$ is a difference of two irreducible characters (because $(\Gamma_i - \Gamma_j)(1) = 0$). This is the key result for this step, and we will use it without further reference.

Now, $(\Gamma_1)^\sigma \neq \Gamma_1$ (because $t \geq 2$). Therefore,

$$\Gamma_1 - (\Gamma_1)^\sigma = \theta_1 - \theta_2$$

for two different irreducible characters $\theta_1, \theta_2 \in \mathrm{Irr}(G)$. Now,

$$(\Gamma_1)^\sigma - (\Gamma_1)^{\sigma^2} = (\theta_1)^\sigma - (\theta_2)^\sigma.$$

Thus,

$$\Gamma_1 - (\Gamma_1)^{\sigma^2} = \theta_1 - \theta_2 + (\theta_1)^\sigma - (\theta_2)^\sigma.$$

If $\Gamma_1 = (\Gamma_1)^{\sigma^2}$, then

$$0 = \theta_1 - \theta_2 + (\theta_1)^\sigma - (\theta_2)^\sigma$$

and thus $\theta_1 = (\theta_2)^\sigma$ and $\theta_2 = (\theta_1)^\sigma$. The proof of the step is completed in this case. Suppose that $\Gamma_1 \neq (\Gamma_1)^{\sigma^2}$. Then $\Gamma_1 - (\Gamma_1)^{\sigma^2} = \theta_1 - \theta_2 + (\theta_1)^\sigma - (\theta_2)^\sigma$ is a difference of two irreducible characters. Note that $\theta_1 \neq (\theta_1)^\sigma$ because otherwise, $\theta_2 = \theta_1$ or $(\theta_2)^\sigma = \theta_1 = (\theta_1)^\sigma$. In the same way, $\theta_2 \neq (\theta_2)^\sigma$. This easily implies that $(\theta_2)^\sigma = \theta_1$ or that $(\theta_1)^\sigma = \theta_2$. In the latter case, we are done. In the first case, we have that

$$\Gamma_1 - (\Gamma_1)^\sigma = (-1)(\theta_2 - (\theta_2)^\sigma)$$

and we replace θ_1 by θ_2. The proof of Step 5 is now complete.

Step 6. *There are exactly t distinct irreducible characters $\theta_1, \ldots, \theta_t$ which are $\mathcal{G}$-conjugate to θ_1. All of them lie in B. Also, if $\{\chi_1, \ldots, \chi_l\} = \text{Irr}(B) - \{\theta_1, \ldots, \theta_t\}$, then each χ_i is p-rational. Furthermore, there are unique integers ϵ_i for $1 \le i \le e$ and a such that*

$$\Gamma_1 = a(\theta_1 + \ldots + \theta_t) + \sum_{i=1}^{l} \epsilon_i \chi_i + \delta\theta_1 \,.$$

(Recall that δ has been introduced in the previous step.) By Step 5, we have that the character $\Gamma_1 - \delta\theta_1$ is fixed by σ. Recall that each λ_i may be written as $(\lambda_1)^{\sigma^{n_i}}$ for a unique $0 \le n_i \le p-1$. Let $\theta_i = (\theta_1)^{\sigma^{n_i}} \in \text{Irr}(G)$ for $1 \le i \le t$. Since

$$\Gamma_i - \delta\theta_i = (\Gamma_1 - \delta\theta_1)^{\sigma^{n_i}} = \Gamma_1 - \delta\theta_1 = (\Gamma_1 - \delta\theta_1)^{\sigma^{n_j}} = \Gamma_j - \delta\theta_j$$

for $1 \le i, j \le t$ and $\Gamma_i \ne \Gamma_j$ if $i \ne j$, it follows that the characters $\theta_1, \ldots, \theta_t$ are all distinct. Also, all of them lie in B because they are p-conjugate to a character θ_1 lying in B.

Now, if $\tau \in \mathcal{G}$, we have that $(\Gamma_1)^\tau = \Gamma_j$ for some j because $\mathcal{G}$ acts on $\{\Gamma_1, \ldots, \Gamma_t\}$. Since

$$\Gamma_j - \delta\theta_j = \Gamma_1 - \delta\theta_1 = (\Gamma_1)^\tau - \delta(\theta_1)^\tau = \Gamma_j - \delta(\theta_1)^\tau \,,$$

we deduce that $\theta_j = (\theta_1)^\tau$, proving that $\theta_1, \ldots, \theta_t$ are all the distinct $\mathcal{G}$-conjugates of θ_1.

By Step 1, note that we may write $\text{Irr}(B) - \{\theta_1, \ldots, \theta_t\} = \{\chi_1, \ldots, \chi_l\}$.

Next, we analyse the irreducible characters appearing in the decomposition of the generalized character $\Gamma_1 - \delta\theta_1$.

Suppose that $\chi \in \text{Irr}(G) - \{\theta_1, \ldots, \theta_t\}$ is such that $[\Gamma_1 - \delta\theta_1, \chi] \ne 0$. Then, by Step 3, we have that $\chi \in \text{Irr}(B)$ since $[\Gamma_1, \chi] \ne 0$. Furthermore,

$$[\Gamma_1, \chi] = [\Gamma_1 - \delta\theta_1, \chi] = [\Gamma_i - \delta\theta_i, \chi] = [\Gamma_i, \chi]$$

for all $1 \le i \le t$. Therefore, χ is p-rational by Step 3.

Now, suppose that $\chi \in \text{Irr}(B) - \{\theta_1, \ldots, \theta_t\}$. Since $\chi \in \text{Irr}(B)$, we have that $[\Gamma_i, \chi] \ne 0$ for some Γ_i (by Step 3). Hence,

$$0 \ne [\Gamma_i, \chi] = [\Gamma_i - \delta\theta_i, \chi] = [\Gamma_1 - \delta\theta_1, \chi]$$

and χ appears in the decomposition of $\Gamma_1 - \delta\theta_1$.

We have already shown that the characters $\chi_1, \ldots, \chi_l$ are all p-rational and are exactly the irreducible characters different from $\theta_1, \ldots, \theta_t$ appearing in the decomposition of $\Gamma_1 - \delta\theta_1$. Hence, if $a = [\Gamma_1 - \delta\theta_1, \theta_i]$ (observe that a does not depend on i since the characters θ_i are p-conjugate and $\Gamma_1 - \delta\theta_1$ is p-rational), we may write

$$\Gamma_1 = a(\theta_1 + \ldots + \theta_t) + \sum_{i=1}^{l} \epsilon_i \chi_i + \delta\theta_1$$

for some integers ϵ_i.

Step 7. $l = e$ and $\epsilon_i = \pm 1$ for $i = 1, \ldots, e$. If $\epsilon_0 = at + \delta$, then $\epsilon_0 = \pm 1$.

Since $(\Gamma_1)^0 = 0$ by the first paragraph of Step 4, we conclude that

$$(at + \delta)(\theta_1)^0 + \sum_{i=1}^{l} \epsilon_i (\chi_i)^0 = 0$$

because $(\theta_i)^0 = (\theta_1)^0$ for every $1 \le i \le t$. Write $\chi_0 = \theta_1$. Then

$$\sum_{i=0}^{l} \epsilon_i (\chi_i)^0 = 0. \tag{2}$$

Hence, by the linear independence of $\mathrm{IBr}(G)$, we have that

$$\sum_{i=0}^{l} \epsilon_i d_{\chi_i \varphi} = 0$$

for every $\varphi \in \mathrm{IBr}(B)$.

Now, notice that the set $\{\chi_0, \chi_1, \ldots, \chi_l\}$ is a complete set of representatives of the distinct families of p-conjugate elements in $\mathrm{Irr}(B)$ by Step 6.

Now, given $\varphi \in \mathrm{IBr}(B)$, there are exactly two indices $0 \le i, j \le l$ such that $d_{\chi_i \varphi} \ne 0 \ne d_{\chi_j \varphi}$ by Theorem (11.4). In fact, the latter are equal to 1, again by Theorem (11.4). Thus, we conclude that $\epsilon_i + \epsilon_j = 0$. We see that if for some $0 \le u, v \le l$ the characters χ_u and χ_v are linked, it follows then that $\epsilon_u = -\epsilon_v$. Since p-conjugate characters have the same restriction to p-regular elements, note that two irreducible characters χ and ψ of G are linked if and only if any $\mathcal{G}$-conjugate of χ and any $\mathcal{G}$-conjugate of ψ are linked. Thus, every two irreducible characters in B are linked by a chain of characters in $\{\chi_0, \chi_1, \ldots, \chi_l\}$. Since $\mathrm{Irr}(B)$ is a single connected component of the graph defined by linking (Theorem (3.9)), we conclude that

$$(\epsilon_i)^2 = (\epsilon_0)^2$$

for every $0 \leq i \leq l$.

Now, by Step 4 we have

$$[\Gamma_1, \Gamma_1] = e + 1$$

and we deduce that

$$e + 1 = a^2(t - 1) + (a + \delta)^2 + l\epsilon_0^2.$$

Thus,

$$a^2 t + 2a\delta + l\epsilon_0^2 = e$$

and

$$a^2 t^2 + 2at\delta + lt\epsilon_0^2 + 1 = te + 1 = p.$$

Now,

$$a^2 t^2 + 2at\delta + 1 = (at + \delta)^2 = \epsilon_0^2.$$

Therefore,

$$(lt + 1)\epsilon_0^2 = p$$

and we deduce that

$$\epsilon_0^2 = 1 \quad \text{and} \quad lt + 1 = p.$$

Also, $l = \frac{p-1}{t} = e$ and the proof of this step is completed.

(Since $\epsilon_0^2 = 1$ and $l = e$, note that the equation $a^2 t + 2a\delta + l\epsilon_0^2 = e$ implies that $a(at + 2\delta) = 0$.)

Final Step. By Step 6, Step 7 and equation (2), we have that

$$\epsilon_0(\theta_1)^0 + \sum_{i=1}^{e} \epsilon_i(\chi_i)^0 = 0.$$

Now, since χ_i is p-rational, we have that

$$\epsilon_i = [\Gamma_1 - \delta\theta_1, \chi_i] = [\Gamma_1, \chi_i] = d^x_{\chi_i \xi}$$

for every $1 \neq x \in P$ by Step 3. Thus, for $1 \neq x \in P$ and $y \in K$, by Step 2 we have that

$$\chi_i(xy) = \epsilon_i \sum_{j=1}^{u} \xi_j(y), \tag{3}$$

where $\xi_1, \ldots, \xi_u$ are the distinct N-conjugates of ξ. Now, since the character $(1_P \times \xi)$ has exactly e extensions to T (recall that T is the stabilizer in N

of $1_P \times \xi$ and that T/C is cyclic), and their sum is $(1_P \times \xi)^T$, by using the Clifford correspondence it easily follows that

$$\frac{1}{e}(1_P \times \xi)^N(y) = \sum_{j=1}^{u} \xi_j(y).$$

Hence

$$\chi_i(xy) = \epsilon_i \frac{1}{e}(1_P \times \xi)^N(y)$$

for $1 \le i \le e$, $1 \ne x \in P$ and $y \in K$.

Now, $\chi_i(x) \equiv \chi_i(1) \bmod M$ (see the remark after the proof of Lemma (2.4)). By equation (3) applied to $y = 1$, we have that

$$\chi_i(x) = \epsilon_i \frac{|N : C|}{e} \xi(1).$$

We deduce that

$$\chi_i(1) \equiv \epsilon_i \frac{|N : C|}{e} \xi(1) \bmod p.$$

Now,

$$-\epsilon_0 \theta_1(1) = \sum_{i=1}^{e} \epsilon_i \chi_i(1) \equiv \sum_{i=1}^{e} \epsilon_i^2 \frac{|N : C|}{e} \xi(1) \equiv |N : C| \xi(1) \bmod p$$

and we obtain

$$\theta_1(1) \equiv -\epsilon_0 |N : C| \xi(1) \bmod p.$$

(Since the characters θ_j are $\mathcal{G}$-conjugate to θ_1, it follows that $\theta_j(1) = \theta_1(1)$ satisfies the same congruence.)

Now, say $\epsilon = -\epsilon_0$. This completes the proof of parts (a), (b), (c) and (d) of the theorem.

Finally, we prove part (e). We want to show that there exists a set of representatives $\{\mu_1, \ldots, \mu_t\}$ for the action of T on $\mathrm{Irr}(P) - \{1_P\}$ such that

$$\theta_j(xy) = \epsilon(\mu_j \times \xi)^N(xy)$$

for all $1 \ne x \in P$ and $y \in K$.

Now, by Step 4, we know that

$$\Gamma_1(xy) = (1_P \times \xi - \lambda_1 \times \xi)^N(xy)$$

for $x \in P$ and $y \in K$. By the remark after Step 7, we have that $a(at + 2\delta) = 0$. Also, recall that

$$\Gamma_1 = a(\theta_1 + \ldots + \theta_t) + \sum_{i=1}^{e} \epsilon_i \chi_i + \delta \theta_1$$

by Steps 6 and 7. Since

$$\chi_i(xy) = \epsilon_i \frac{1}{e}(1_P \times \xi)^N(y)$$

for $1 \le i \le e$, $1 \ne x \in P$ and $y \in K$, we have that

$$\Gamma_1(xy) - \sum_{i=1}^{e} \epsilon_i \chi_i(xy) = (1_P \times \xi - \lambda_1 \times \xi)^N(xy) - \sum_{i=1}^{e} \epsilon_i^2 \frac{1}{e}(1_P \times \xi)^N(y)$$

$$= -(\lambda_1 \times \xi)^N(xy)$$

for $1 \ne x \in P$ and $y \in K$. (Note that $(1_P \times \xi)^N(xy) = (1_P \times \xi)^N(y)$ because the character $(1_P \times \xi)^N$ has P in its kernel.) Thus,

$$a(\theta_1 + \ldots + \theta_t)(xy) + \delta\theta_1(xy) = -(\lambda_1 \times \xi)^N(xy)$$

for $1 \ne x \in P$ and $y \in K$.

Suppose now that $t = 1$ or $a = 0$. In this case, we clearly have that

$$\theta_1(xy) = \epsilon(\lambda_1 \times \xi)^N(xy)$$

for all $1 \ne x \in P$ and $y \in K$. Next, we prove the same formula for θ_i. (Recall that $\theta_i = (\theta_1)^{\sigma^{n_i}}$ and $\lambda_i = (\lambda_1)^{\sigma^{n_i}}$.) By elementary character theory, if $\tau \in \mathcal{G}$ satisfies $\rho^\tau = \rho^k$ for some $1 \le k \le p-1$ and ψ is a character of G, then

$$\psi(xy)^\tau = \psi(x^k y)$$

for $x \in P$ and $y \in K$. Put $\rho^{\sigma^{n_i}} = \rho^{k_i}$ for some $1 \le k_i \le p - 1$. Then

$$\theta_i(xy) = \theta_1(xy)^{\sigma^{n_i}} = \theta_1(x^{k_i} y) = \epsilon(\lambda_1 \times \xi)^N(x^{k_i} y)$$

$$= \epsilon((\lambda_1 \times \xi)^N)^{\sigma^{n_i}}(xy) = \epsilon(\lambda_i \times \xi)^N(xy).$$

Hence, if $t = 1$ or $a = 0$, the proof of the theorem is complete.

Suppose now that $t \ne 1$ and $a \ne 0$. Then $at + 2\delta = 0$ and we conclude that $t = 2$ and $a = -\delta = \epsilon_0 = -\epsilon$. Now,

$$a(\theta_1 + \ldots + \theta_t)(xy) + \delta\theta_1(xy) = -\delta(\theta_1 + \theta_2)(xy) + \delta\theta_1(xy)$$

$$= -\delta\theta_2(xy) = \epsilon_0\theta_2(xy) = -\epsilon\theta_2(xy).$$

Therefore,

$$\theta_2(xy) = \epsilon(\lambda_1 \times \xi)^N(xy).$$

Observe that in this case, we necessarily have that $(\theta_1)^\sigma = \theta_2$ and $(\Gamma_1)^\sigma = \Gamma_2$ (because $\mathcal{G}$ acts transitively on $\{\Gamma_1, \ldots, \Gamma_t\} = \{\Gamma_1, \Gamma_2\}$ and θ_1, θ_2 are all the $\mathcal{G}$-conjugates of θ_1 by Step 6). Hence, $(\lambda_1)^{\sigma^{-1}}$ cannot be T-conjugate to λ_1 (otherwise, $(\Gamma_1)^\sigma = \Gamma_1$). Thus λ_1 and $(\lambda_1)^{\sigma^{-1}}$ are representatives of the action of T on $\mathrm{Irr}(P) - \{1_P\}$. Suppose that $\rho^{\sigma^{-1}} = \rho^k$ for some $1 \le k \le p-1$. Then

$$\theta_1(xy) = \theta_2(xy)^{\sigma^{-1}} = \theta_2(x^k y) = \epsilon(\lambda_1 \times \xi)^N(x^k y)$$

$$= \epsilon((\lambda_1 \times \xi)^N)^{\sigma^{-1}}(xy) = \epsilon((\lambda_1)^{\sigma^{-1}} \times \xi)^N(xy).$$

The proof of the theorem is now complete. ∎

There is an object which remarkably reflects the properties of the blocks of defect one. We associate $e + 1$ distinct points $P_0, \ldots, P_e$ with the characters $\chi_0, \chi_1, \ldots, \chi_e$, where $\chi_0 = \theta_1$, for instance. We join P_i and P_j when $(\chi_i)^0$ and $(\chi_j)^0$ have an irreducible Brauer character in common. The graph obtained this way is called the **Brauer tree** of the block B. (It is a tree because it is a connected graph with one more vertex than edges.)

Proof of Theorem (11.1). In the case that B is the principal block, we have that the canonical character of B is 1_C by the third main theorem and Theorem (9.12). In the notation of Theorem (11.13), we have that $\xi = 1_K$. The inertia group of ξ in N is N. Thus, $e = |N : C|$. What remains for us to show is part (f). Assume now that $C = \mathbf{C}_G(P) = P$. In this case, the unique block of C is the principal block by Problem (3.1), for instance. By the third main theorem and the extended first main theorem (9.7), we deduce that all the nonprincipal blocks of G have defect zero. ∎

There is a group theoretic consequence for which no purely group theoretic proof is known.

(11.14) COROLLARY. *Suppose that G has a Sylow p-subgroup of order p. Then the number of conjugacy classes of G is greater than or equal to the number of conjugacy classes of $\mathbf{N}_G(P)$.*

Proof. Write $N = \mathbf{N}_G(P)$. If $p = 2$, then G has a normal 2-complement. In this case, $\mathbf{N}_G(P) = \mathbf{C}_G(P)$ and every conjugacy class of N has defect group P. Now, the result easily follows from Lemma (4.16), for instance.

Suppose that p is odd. Now, every block b of N has defect group P because $P \lhd N$. By Theorem (11.13), we have that $\mathrm{k}(b) = \mathrm{k}(b^G)$. The result follows from the first main theorem. ∎

PROBLEMS

(11.1) Calculate the character table of A_5 by using Theorem (11.13).

(11.2) Calculate the character table of $GL(3, 2)$ by using Theorem (11.13).

Notation

RG	the group algebra of the finite group G over the ring R
$\mathrm{Mat}(n, R)$	$n \times n$ matrices with entries in R
$\mathrm{ann}(V)$	if V is an A-module, this is $\{a \in A \mid Va = 0\}$
$\mathbf{Z}(A)$	the center of the algebra A, page 2
$\mathbf{J}(A)$	the Jacobson radical of the algebra A, page 3
a_V	if V is an A-module and $a \in A$, this is the map $a_V(v) = va$ for $v \in V$
$\mathrm{char}(F)$	the characteristic of the field F
$\mathrm{GL}(n, R)$	the group of invertible matrices with entries in the ring R
$\mathbb{Z}$	the integers
$\mathbb{Z}_p$	$\mathbb{Z}/p\mathbb{Z}$, the field of p elements
$\mathbb{Q}$	the rational numbers
$\mathbb{C}$	the complex numbers
$\mathbf{R}$	the ring of algebraic integers in $\mathbb{C}$
M	from Chapter 2, this is a fixed maximal ideal of $\mathbf{R}$ containing $p\mathbf{R}$, page 16
S	the local ring $\{\frac{r}{s} \mid r \in \mathbf{R}, s \in \mathbf{R} - M\}$, page 16
$\mathcal{P}$	the maximal ideal $\{\frac{r}{s} \mid r \in M, s \in \mathbf{R} - M\}$ of S, page 16
F	from Chapter 2, this is the field $\mathbf{R}/M \cong S/\mathcal{P}$
$*$	the natural ring homomorphism $S \to F$ with kernel $\mathcal{P}$ or one of its extension $SG \to FG$, $S[x] \to F[x]$, $\mathrm{Mat}(n, S) \to \mathrm{Mat}(n, F)$
$\mathbf{U}$	$\{\xi \in \mathbb{C} \mid \xi^m = 1$ for some integer m not divisible by $p\}$
G^0	the set of p-regular elements of the finite group G
G_p	the set of p-elements of the finite group G
$\mathrm{Irr}(G)$	the set of irreducible (complex) characters of the finite group G
$\mathrm{IBr}(G)$	the set of irreducible Brauer characters of the finite group G (once M has been chosen), page 18
1_{G^0}	the principal Brauer character, page 18
V^*	the dual of the FG-module V, page 18
$\mathbf{K}$	the algebraic closure of $\mathbb{Q}$ in $\mathbb{C}$

cf(G)	the space of class functions $G \to \mathbb{C}$
cf(G^0)	the space of class functions $G^0 \to \mathbb{C}$
vcf(G)	the space of class functions $G \to \mathbb{C}$ vanishing off G^0
cl(G)	the set of conjugacy classes of G
cl(G^0)	the set of conjugacy classes of G consisting of p-regular elements
χ^0	if $\chi \in$ cf(G), this is the restriction of χ to G^0
o(g)	the order of the element g in a group
$\mathbb{Z}[\text{Irr}(G)]$	the set of $\mathbb{Z}$-linear combinations of Irr(G)
$R[\text{Irr}(G)]$	the set of R-linear combinations of Irr(G), R any ring
$\mathbb{Z}[\text{IBr}(G)]$	the set of $\mathbb{Z}$-linear combinations of IBr(G)
$d_{\chi\varphi}$	the decomposition numbers, page 23
$c_{\varphi\theta}$	the Cartan numbers, page 25
Φ_φ	the projective indecomposable character, page 25
Φ_1	the projective indecomposable character associated to 1_{G^0}
$[\theta, \varphi]^0$	if $\theta, \varphi \in$ cf(G) $\cup$ cf(G^0), this is $\frac{1}{\|G\|} \sum_{x \in G^0} \theta(x)\overline{\varphi(x)}$
cl(x)	the conjugacy class of the element x
$\hat{\theta}$	see Lemma (2.15)
$\tilde{\theta}$	see Lemma (2.15)
χ_H	if $\chi \in$ cf(G), this is the restriction of χ to the subgroup H of G
θ^G	if $\theta \in$ cf(H), this is the induction of θ to G
A^{t}	the transpose of the matrix A
det(A)	the determinant of the matrix A
rank(A)	the rank of the matrix A
$S_{p'}(y)$	the p'-section of y, page 30
$A \otimes B$	the tensor product of the matrices A and B, page 32
$V \otimes W$	the tensor product of the FG-modules V and W, page 32
$\mathcal{X} \otimes \mathcal{Y}$	the tensor product of the representations $\mathcal{X}$ and $\mathcal{Y}$, page 33
$\bar{\varphi}$	usually denotes the complex conjugate of φ; sometimes, it also denotes the character of $\bar{G} = G/N$ corresponding to φ if $N \subseteq$ ker(φ)
ker(φ)	the kernel of the character φ, page 39
$\mathbf{O}_p(G)$	the largest normal p-subgroup of G
$\mathbb{Q}_n$	is $\mathbb{Q}(e^{\frac{2\pi i}{n}})$
Gal($\mathbb{Q}_n/\mathbb{Q}$)	the Galois group of the field extension $\mathbb{Q}_n/\mathbb{Q}$
$\hat{K}$	if $K \subseteq G$, then $\hat{K} = \sum_{x \in K} x$

x_K	a fixed element of the conjugacy class K
ω_χ	the algebra homomorphism $\mathbf{Z}(\mathbb{C}G) \to \mathbb{C}$ defined by $\omega_\chi(\hat{K}) = \frac{\chi(x_K)|K|}{\chi(1)}$
λ_χ	if $\chi \in \mathrm{Irr}(G)$, this is the algebra homomorphism $\mathbf{Z}(FG) \to F$ defined by $\lambda_\chi(\hat{K}) = \omega_\chi(\hat{K})^*$
λ_φ	if $\varphi \in \mathrm{IBr}(G)$, page 48
A_{KLM}	if K, L, M are conjugacy classes of G, this is the set $\{(x,y) \in K \times L \mid xy = x_M\}$
a_{KLM}	this is $|A_{KLM}|$
$\mathrm{Bl}(G)$	the set of p-blocks of G
$\mathrm{Irr}(B)$	the set of ordinary irreducible characters in the block B
$\mathrm{IBr}(B)$	the set of irreducible Brauer characters in the block B
$\mathrm{k}(B)$	the number of ordinary irreducible characters in B
$\mathrm{l}(B)$	the number of irreducible Brauer characters in B
C_B	the Cartan matrix associated to the block B, page 51
e_χ	the primitive central idempotent corresponding to $\chi \in \mathrm{Irr}(G)$, page 51
f_B	$\sum_{\chi \in \mathrm{Irr}(B)} e_\chi$
$f_B(\hat{K})$	the coefficient of $\hat{K}$ in f_B
e_B	is $(f_B)^*$, the block idempotent of B
$a_B(\hat{K})$	the coefficient of $\hat{K}$ in e_B; $f_B(\hat{K})^*$
λ_B	the algebra homomorphism $\mathbf{Z}(FG) \to F$ associated to B
B	a block of G; sometimes, it also denotes $B = e_B FG$
$\mathrm{d}(B)$	the defect of B, page 60
$\mathrm{height}(\chi)$	the height of χ, pages 60, 61
$\nu(\frac{n}{m})$	page 64
θ_B	the B-part of θ, page 72
$\bar{B}$	denotes the dual block of B; sometimes, it also denotes a block of a factor group
$\delta(K)$	the set of defect groups of the conjugacy class K, page 80
D_K	$D_K \in \delta(K)$
$X \subseteq_G Y$	X, Y are subsets of G such that $X \subseteq Y^g$ for some $g \in G$
$X =_G Y$	X, Y are G-conjugate subsets of G
$\mathbf{Z}_D(FG)$	page 80
D_B	a defect group of B; sometimes, also denotes the decomposition matrix associated to B
$\delta(B)$	the set of defect groups of B. Hence, $D_B \in \delta(B)$
Br_P	the Brauer homomorphism, page 85

λ^G	pages 87 and 119
b^G	the block b^G is defined, page 87
$\mathrm{Bl}(G \mid P)$	the set of blocks of G with defect group P
$\mathrm{cl}(G \mid P)$	the set of conjugacy classes of G with defect group P
$\mathrm{cl}(G^0 \mid P)$	the set of conjugacy classes of G of p-regular elements with defect group P
$d^x_{\chi\varphi}$	the generalized decomposition number, pages 100, 135
$\mathrm{supp}(z)$	the support of z, page 91
$S(x)$	the p-section of the p-element x, page 105
$\mathrm{k}(G)$	the number of conjugacy classes of the group G
(x, b)	a subsection, x is a p-element, $b \in \mathrm{Bl}(\mathbf{C}_G(x))$, page 114
$\ker(B)$	the kernel of the block B, page 125
$Z^*(G)$	page 146
W^G	the induced module, page 150
V_H	the restriction of the module V to H, page 150
α^G	if $\alpha \in \mathrm{cf}(H^0)$, page 151
$\mathrm{Irr}(G \mid \alpha)$	the set of irreducible characters χ of G such that α is an irreducible constituent of χ_H
$\mathrm{IBr}(G \mid \alpha)$	the set of irreducible Brauer characters φ of G such that α is an irreducible constituent of φ_H
$\mathbf{I}(U, V)$	page 156
$\mathbf{I}(\tau, \psi)$	page 157
$I_G(\theta)$	is the inertia group of θ in G, page 159
$\mathbf{Z}^2(G, F^\times)$	page 164
$\mathbf{H}^2(G, F^\times)$	page 166
$\mathbf{B}^2(G, F^\times)$	page 166
$F^\alpha G$	the twisted group algebra, page 171
$\alpha \times \beta$	page 176
$\det(\varphi)$	page 178
$\mathrm{o}(\varphi)$	the determinantal order of φ, page 178
(G, N, θ)	a character triple, page 179
$\mathrm{Br}(G \mid \theta)$	page 179
$\mathrm{Ch}(G \mid \theta)$	if (G, N, θ) is a character triple, this is the set of positive integer linear combinations of $\mathrm{Irr}(G \mid \theta)$
$T(b)$	the stabilizer of the block b, page 193
$\mathrm{Bl}(G \mid b)$	the blocks of G covering b, page 193
(Q, b_Q)	a subpair, page 219
$K(\chi)$	smallest subfield containing K and $\{\chi(g) \mid g \in G\}$

Bibliographic Notes

The aim of these notes has been to give an introduction of modular representation theory by using character theoretic methods. The representation theory of finite groups can only be fully understood if these are combined with module theoretic and ring theoretic methods. (The approach followed here does not allow us to develop Green's theory of indecomposable modules, vertices and sources, for instance.)

The books of W. Feit and those of C. Curtis and I. Reiner are the most complete and treat modular theory from the three points of view. (See also [Nagao-Tsushima].)

The book of J. Alperin is a short (but deep) module theoretic introduction to representation theory. The book of P. Landrock shows the ring theoretic methods. (See also [Michler].) The book of D. Goldschmidt is the closest to our point of view.

Other books on the subject are [Dornhoff], [Puttaswamaiah-Dixon] and some parts of [Huppert-Blackburn]. (See also the notes [G-H-K-M-W].)

For an account on R. Brauer's work the reader is refered to the articles of J. A. Green, P. Fong and W. Wong in [Brauer, *Collected Papers*]. Most of the results in these notes are from Brauer. To avoid repetitiveness, I have not given credit to each one of his theorems since they have already become "classical".

My reference for ordinary character theory is the book of M. Isaacs. The introduction to Brauer characters is made as in Chapter 15 of [Isaacs, *Characters*].

Okuyama's theorem (2.33) was comunicated to me by W. Willems and seems not have been published before. Problem (2.6) appears in [Isaacs, 1982].

The Frobenius conjecture (in connection with Theorem (2.28)) is known to be true by using the simple group classification. (I thank J. Olsson and R. Solomon for conversations on the subject.)

Theorem (3.33) appears in [Gow-Willems]. I learned of Passman's result (Problem (3.12)) in [Michler, 1990]. Robinson's theorem (4.20) appears in [Robinson, 1983]. Külshammer's theorem (4.23) appears in [Külshammer, 1984]. The proof of the second main theorem is due to Isaacs and it is based on work by Juhász and Tsushima ([Juhász-Tsushima]). Here, Green's theory and relative projectivity for modules are avoided. Knörr's theorem

(5.16) appears in [Knörr] (with a different proof). Theorem (5.21) is a key step in the proof of [Broué-Puig] on nilpotent blocks. The proof of the third main theorem is based on [Okuyama]. Okuyama's result is a very general version of the third main theorem. Blau's theorem (6.8) appears in [Blau]. The nice formula for the principal block idempotent in Theorem (6.14) appears, as well as Problems (6.1), (6.2) and (6.3), in [Külshammer, 1991].

Dade's theorem (8.13) is from [Dade, 1970]. Gow's proof of Dade's theorem appears in [Gow]. Passman's proof of Dade's theorem is unpublished (although the key lemmas appear in [Isaacs, 1981]). Theorems (8.33) and (8.34), as well as Problem (8.14), appear in [Navarro].

Laradji's theorem (9.13) appears in [Laradji]. Here, Brauer's characterization of characters is the main ingredient of the proof. (There is another similar proof in [Robinson, 96].) Knörr's theorem (9.26) was independently discovered by S. Gagola and here I give Gagola's proof (Knörr's proof uses Green's theory). The Harris-Knörr theorem appears in [Harris-Knörr]. (There is a module theoretic proof due to J. Alperin.) Problem (9.3) appears in [Navarro-Willems].

In Chapter 10, Wolf's proof of the Fong-Swan theorem appears in [Manz-Wolf]. The proofs from (10.9) to (10.19), as well as some of the problems in Chapter 10, are influenced by Isaacs π-partial theory. See [Isaacs, 1986] and [Isaacs-Navarro]. The Kiyota-Okuyama proof of the p-solvable case of the Harada conjecture (see [Harada]) appears in [Kiyota-Okuyama].

The proof of Theorem (11.4) is basically Brauer's proof. (See paper [38] of [Brauer, *Collected Papers*].) The proof of Theorem (11.13) is influenced by [Curtis-Reiner, Methods I, II] and [Goldchsmidt].

References

Most of the books in the list below contain extensive lists of references on modular representation theory. We add a few more items.

[Alperin, Book]
J. L. Alperin, *Local Representation Theory*, Cambridge University Press, 1986.

[Alperin, Weights]
J. L. Alperin, Weights for finite groups, *Proceedings of Symposia in Pure Mathematics* **47** (1987), 369-379.

[*Atlas*]
J. H. Conway, R. T. Curtis, S. P. Norton, R. A. Parker, R. A. Wilson, *Atlas of Finite Groups*, Clarendon Press, Oxford, 1985.

[*Atlas of Brauer*]
C. Jansen, K. Lux, R. Parker, R. Wilson, *An Atlas of Brauer Characters*, London Mathematical Society Monographs, New Series **11**, Clarendon Press, Oxford, 1995.

[Blau]
H. I. Blau, On block induction, *J. Algebra* **104** (1986), 195-202.

[Brauer, *Collected Papers*]
R. Brauer, *Collected papers*, Volumes I, II, III. Edited by P. Fong and W. Wong. The MIT Press, Cambridge, Massachusetts, 1980.

[Broué-Puig]
M. Broué, L. Puig, A Frobenius theorem for blocks, *Invent. Math.* **56** (1980), 117-128.

[Curtis-Reiner]
C. Curtis, I. Reiner, *Representation Theory of Finite Groups and Associative Algebras*, Interscience, New York, 1962.

[Curtis-Reiner, Methods I]
C. Curtis, I. Reiner, *Methods of Representation Theory I* , Wiley, New York, 1981.

[Curtis-Reiner, Methods II]
C. Curtis, I. Reiner, *Methods of Representation Theory II*, Wiley, New York, 1987.

[Dade, 1970]
E. C. Dade, Isomorphisms of Clifford extensions, *Ann. of Math.* **92** (1970), 375-433.

[Dade, 1992]
E. C. Dade, Counting characters in blocks I, *Invent. Math.* **109** (1992), 187-210.

[Dade, 1994]
E. C. Dade, Counting characters in blocks II, *J. reine angew. Math.* **448** (1994), 97-190.

[Dornhoff]
L. Dornhoff, *Group Representation Theory, Part B*, Marcel Dekker, New York, 1972.

[Feit]
W. Feit, *The Representation Theory of Finite Groups*, North-Holland, Amsterdam, 1982.

[G-H-K-M-W]
R. Gow, B. Huppert, R. Knörr, O. Manz, W. Willems, *Representation Theory in Arbitrary Characteristic*, Centro Internazionale per la ricerca matematica, C. Ed. Antonio Milani, Trento, 1993.

[Goldschmidt]
D. Goldschmidt, *Lectures on Character Theory*, Publish or Perish, Berkeley, 1980.

[Gow]
R. Gow, Extensions of modular representations for relatively-prime operator groups, *J. Algebra* **36** (1975), 492-494.

[Gow-Willems]
R. Gow, W. Willems, Quadratic geometries, projective modules and idempotents, *J. Algebra* **160** (1993), 257-272.

[Guralnick-Robinson]
R. M. Guralnick, G. R. Robinson, On extensions of the Baer-Suzuki theorem, *Israel J. Math.* **82** (1993), 281-297.

[Harada]
K. Harada, A conjecture and a theorem on blocks of modular representation, *J. Algebra* **70** (1981), 350-355.

[Harris-Knörr]
M. Harris, R. Knörr, Brauer correspondence for covering blocks of finite groups, *Communications in Algebra* **13** (5) (1985), 1213-1218.

[Huppert-Blackburn]
B. Huppert, N. Blackburn, *Finite Groups II*, Springer-Verlag, Berlin, 1982.

[Isaacs, 1981]
I. M. Isaacs, Extensions of group representations over arbitrary fields, *J. Algebra* **68** (1981), 54-74.

[Isaacs, 1982]
I. M. Isaacs, Fixed points and π-complements in π-separable groups, *Archiv der Mathematik* **39** (1982), 5-8.

[Isaacs, 1986]
I. M. Isaacs, Fong characters in π-separable groups, *J. Algebra* **99** (1986), 89-107

[Isaacs, *Algebra*]
I. M. Isaacs, *Algebra*, Brooks-Cole, New York, 1994.

[Isaacs, *Characters*]
I. M. Isaacs, *Character Theory of Finite Groups*, Dover, New York, 1994.

[Isaacs-Navarro]
I. M. Isaacs, G. Navarro, Weights and vertices for characters of π-separable groups, *J. Algebra* **177** (1995), 339-366.

[Juhász-Tsushima]
A. Juhász, Y. Tsushima, A proof of Brauer's second main theorem and related results, *Hokkaido Math. Journal* **14** (1985), 33-37.

[Kiyota-Okuyama]
M. Kiyota, T. Okuyama, A note on a conjecture of Harada, *Proc. Japan Acad. Ser. A Math. Sci.* **57** (1981), 128-129.

[Knörr]
R. Knörr, Virtually irreducible lattices, *Proc. London Math. Soc.* (3) **59** (1989), 99-132.

[Külshammer, 1984]
B. Külshammer, Bemerkungen über die Gruppenalgebra als symmetrische Algebra, III, *J. Algebra* **88** (1984), 279-291.

[Külshammer, 1991]
B. Külshammer, The principal block idempotent, *Archiv der Mathematik* **56** (1991), 313-319.

[Landrock]
P. Landrock, *Finite Group Algebras and their Modules*, London Mathematical Society Lecture Note Series **84**, Cambridge University Press, 1983.

[Laradji] A. Laradji, On characters with minimal defects, *J. reine angew. Math.* **448** (1994), 27-29.

[Manz-Wolf]
O. Manz, T. Wolf, *Representations of Solvable Groups*, London Mathematical Society Lecture Note Series **185**, Cambridge University Press, 1993.

[Michler]
G. O. Michler, *Blocks and Centers of Group Algebras*, Lecture Notes in Mathematics **246**, Springer-Verlag, (1972), 429-563.

[Michler, 1990]
G. Michler, Trace and defect of a block idempotent, *J. Algebra* **131** (1990), 496-501.

[Nagao-Tsushima]
H. Nagao, Y. Tsushima, *Representations of Finite Groups*, Academic Press, New York, 1987.

[Navarro]
G. Navarro, Two groups with isomorphic group algebras, *Archiv der Mathematik* **55** (1990), 35-37.

[Navarro-Willems]
G. Navarro, W. Willems, When is a p-block a q-block?, *Proc. Amer. Math. Soc.* **125** (1997), 1589-1591.

[Okuyama]
T. Okuyama, On blocks and subgroups, *Hokkaido Math. J.* **10** (1981), 555-563

[Puttaswamaiah-Dixon]
B. M. Puttaswamaiah, J. Dixon, *Modular Representations of Finite Groups*, Academic Press, New York, 1977.

[Robinson, 1983]
G. R. Robinson, The number of blocks with a given defect group, *J. Algebra* **84** (1983), 493-502.

[Robinson, 1996]
G. R. Robinson, Local structure, vertices and Alperin's conjecture, *Proc. London Math. Soc.* (3) **72** (1996), 312-330.

[Robinson-Thompson]
G. R. Robinson, J. G. Thompson, On Brauer's $k(B)$-problem, *J. Algebra* **184** (1996), 1143-1160.

[Wheeler]
W. Wheeler, Extended block induction, *J. London Math. Soc.* (2) **49** (1994), 73-82.

[Willems, 1979]
W. Willems, A note on Brauer's induction theorem, *J. Algebra* **58** (1979), 523-526.

Index

A-homomorphism, 2
algebra, 1
  group algebra, 1
  local, 4
  semisimple, 5
  simple, 5
algebra homomorphism, 2
Alperin, J., 87, 219
  argument, 130
  weight conjecture, 90
Alperin-Broué theorem, 219
Alperin-McKay conjecture, 90
annihilator, 3
augmentation map, 15

B-part, 72, 105
B-subgroup, 219
Baer, R., 39
basic set, 133
Blau, H., 79, 125
block, 49, 57, 63
  admissible, 213
  defined, 87
  dual, 75
  idempotent, 55
  induced, 87
  orthogonality, 106
  principal, 49, 125
  regular, 210
  selfdual, 75

weak orthogonality, 53
weakly regular, 210
Brauer, R., 70, 71, 79, 99, 130, 221,
  248, 277
  character, 17
  correspondence, 86
  first main theorem, 86, 89
  graph, 51
  homomorphism, 85
  irreducible character, 18
  linear character, 46
  second main theorem, 100, 104
  third main theorem, 125
  tree, 271
Brauer-Feit theorem, 70
Brauer-Nesbitt, 36, 151
Brauer-Suzuki theorem, 131, 139
Broué, M., 70, 219, 220
Broué-Puig, 115, 220
Burry, D., 87

Cartan matrix, 25, 29
  of a block, 51, 68, 78
character
  canonical, 204
  conjugate, 157
  exceptional, 243, 259
  faithful, 39
  nonexceptional, 243, 259
Cliff, G. H., 79

Printed in the United States
By Bookmasters